BIOINFORMATICS FOR BIOLOGISTS

The computational education of biologists is changing to prepare students for facing the complex data sets of today's life science research. In this concise textbook, the authors' fresh pedagogical approaches lead biology students from first principles towards computational thinking.

A team of renowned bioinformaticians take innovative routes to introduce computational ideas in the context of real biological problems. Intuitive explanations promote deep understanding, using little mathematical formalism. Self-contained chapters show how computational procedures are developed and applied to central topics in bioinformatics and genomics, such as the genetic basis of disease, genome evolution, or the tree of life concept. Using bioinformatic resources requires a basic understanding of what bioinformatics is and what it can do. Rather than just presenting tools, the authors – each a leading scientist – engage the students' problem-solving skills, preparing them to meet the computational challenges of their life science careers.

PAVEL PEVZNER is Ronald R. Taylor Professor of Computer Science and Director of the Bioinformatics and Systems Biology Program at the University of California, San Diego. He was named a Howard Hughes Medical Institute Professor in 2006.

RON SHAMIR is Raymond and Beverly Sackler Professor of Bioinformatics and head of the Edmond J. Safra Bioinformatics Program at Tel Aviv University. He founded the joint Life Sciences – Computer Science undergraduate degree program in Bioinformatics at Tel Aviv University.

BIOINFORMATICS
FOR BIOLOGISTS

EDITED BY

Pavel Pevzner
University of California, San Diego, USA

AND

Ron Shamir
Tel Aviv University, Israel

CAMBRIDGE
UNIVERSITY PRESS

CAMBRIDGE UNIVERSITY PRESS
Cambridge, New York, Melbourne, Madrid, Cape Town,
Singapore, São Paulo, Delhi, Tokyo, Mexico City

Cambridge University Press
The Edinburgh Building, Cambridge CB2 8RU, UK

Published in the United States of America by Cambridge University Press, New York

www.cambridge.org
Information on this title: www.cambridge.org/9781107011465

First published 2011

Printed in the United Kingdom at the University Press, Cambridge

A catalog record for this publication is available from the British Library

Library of Congress Cataloging in Publication data
Bioinformatics for biologists / edited by Pavel Pevzner, Ron Shamir.
    p.   cm.
Includes index.
ISBN 978-1-107-01146-5 (hardback)
1. Bioinformatics.   I. Pevzner, Pavel.   II. Shamir, Ron.
QH324.2.B5474   2011
572.8 – dc23   2011022989

ISBN 978-1-107-01146-5 Hardback
ISBN 978-1-107-64887-6 Paperback

To Ellina, the love of my life.
(P.P.)

To my parents, Varda and Raphael Shamir.
(R.S.)

CONTENTS

EXTENDED CONTENTS

PREFACE

What is this book?

This book aims to convey the fundamentals of bioinformatics to life science students and researchers. It aims to communicate the computational ideas behind key methods in bioinformatics to readers without formal college-level computational education. It is not a "recipe book": it focuses on the *computational ideas* and avoids technical explanation on running bioinformatics programs or searching databases. Our experience and strong belief are that once the computational ideas are grasped, students will be able to use existing bioinformatics tools more effectively, and can utilize their understanding to advance their research goals by envisioning new computational goals and communicating better with computational scientists.

The book consists of self-contained chapters each introducing a basic computational method in bioinformatics along with the biological problems the method aims to solve. Review questions follow each chapter. An accompanying website (www.cambridge.org/b4b) containing teaching materials, presentations, questions, and updates will be of help to students as well as educators.

Who is the audience for the book?

The book is aimed at life science undergraduates; it does not assume that the reader has a background in mathematics and computer science, but rather introduces mathematical concepts as they are needed. The book is also appropriate for graduate students and researchers in life science and for medical students. Each chapter can be studied individually and used individually in class or for independent reading.

Why this book?

In 1998, Stanford professor Michael Levitt reflected that computing has changed biology forever, even if most biologists did not know it yet. More than a decade later, many biologists have realized that computational biology is as essential for this century's biology as molecular biology was in the last century. Bioinformatics[1] has become an essential part of modern biology: biological research would slow down dramatically if one suddenly withdrew the modern bioinformatics tools such as BLAST from the arsenal of biologists. We cannot imagine forward-looking biological research that does not use any of the vast resources that bioinformatics researchers have made available to the biomedical community.

Bioinformatics resources come in two flavors: databases and algorithms. Thousands of databases contain information about protein sequences and structures, gene annotations, evolution, drugs, expression profiles, whole genomes and many more kinds of biological data. Numerous algorithms have been developed to analyze biological data, and software implementations of many of these algorithms are available to biologists. Using these resources effectively requires a basic understanding of what bioinformatics is and what it can do: what tools are available, how best to use them and to interpret their results, and more importantly, what one can reasonably hope to achieve using bioinformatics even if the relevant tools are not yet available.

Despite this richness of bioinformatics resources and methods, and although sophisticated biomedical researchers draw on these resources extensively, the exposure of undergraduates in biology and biochemistry, as well as of medical students, to bioinformatics is still in its infancy. The computational education of biologists has hardly changed in the last 50 years. Most universities still do not offer bioinformatics courses to life sciences undergraduates, and those that do offer such courses struggle with the question of how and what to teach to students with limited computational culture. In the absence of any preparation in computer science, the generation of biologists that went to universities in the last decade remains poorly prepared for the computational aspects of work in their own discipline in the decades to come. Similarly, medical doctors (who will soon have to analyze personal genomes or blood tests that report thousands of protein levels) are not prepared to meet the computational challenges of future medicine.

Biomedical students typically have a very basic computational background, which leads to a serious risk that bioinformatics courses – when offered – will become technical and uninspired. The software tools are often taught and then used as "black

---

[1] Here and throughout the book, we use the terms bioinformatics and computational biology interchangeably.

boxes," without deeper understanding of the algorithmic ideas behind them. This can lead to under-utilization or over-interpretation of the results that such black-box use produces. Moreover, the students who study bioinformatics at this level will have a much smaller chance of coming up with computational ideas later in their careers when they carry out their own biomedical research. It is therefore essential, in our opinion, that biologists be exposed to deep algorithmic ideas, both in order to make better use of available tools that rely on these ideas, and in order to be able to develop novel computational ideas of their own and communicate effectively with computational biologists later in their careers.

We and others have argued for a revolution in computational education of biologists[2] and noted that the mathematical and computational education of other disciplines have already undergone such revolutions with great success. Physicists went through a computational revolution 150 years ago, and economists have dramatically upgraded their computational curriculum in the last 20 years. As a result, paradoxically, the students in these disciplines are much better prepared for the computational challenges of modern biomedical research than are biology students. Moreover, whatever little mathematical background biologists have, it is mainly limited to classical *continuous* mathematics (such as Calculus) rather than *discrete* mathematics and computer science (e.g. algorithms, machine learning, etc.) that dominate modern bioinformatics. In 2009 we thus came up with a radical prophecy[3] that the education of biologists will soon become as computationally sophisticated as the education of physicists and economists today. As implausible as this scenario looked a few years ago, leading schools in bioinformatics education (such as Harvey Mudd or Berkeley) are well on the way towards this goal.

The time has come for biology education to catch up. Such change may require revising the contents of basic mathematical courses for life science college students, and perhaps updating the topics that are taught. Students' understanding of bioinformatics will benefit greatly from such a change. In parallel, dedicated bioinformatics classes and courses should be established, and textbooks appropriate for them should be developed.

Most undergraduate bioinformatics programs at leading universities involve a grueling mixture of biological and computational courses that prepare students for subsequent bioinformatics courses and research. As a result, some undergraduate bioinformatics courses are too complex even for biology graduate students, let alone

---

[2] W. Byalek and D. Botstein. Introductory science and mathematics education for 21st-Century biologists. *Science*, 303:788–790, 2004.
P. A. Pevzner. Educating biologists in the 21st century: Bioinformatics scientists versus bioinformatics technicians. *Bioinformatics*, 20:2159–2161, 2004.

[3] P. A. Pevzner and R. Shamir. Computing has changed biology – Biology education must catch up. *Science*, 325:541–542, 2009.

undergraduates. This causes a somewhat paradoxical situation on many campuses today: bioinformatics courses are available, but they are aimed at *bioinformatics* undergraduates and are not suitable for *biology* students (undergraduate or graduate). This leads to the following challenge that, to the best of our knowledge, has not yet been resolved:

Pedagogical Challenge. *Design a bioinformatics course that (i) assumes minimal computational prerequisites, (ii) assumes no knowledge of programming, and (iii) instills in the students a meaningful understanding of computational ideas and ensures that they are able to apply them.*

This challenge has yet to be answered, but we claim that many ideas in bioinformatics can be explained at an intuitive level that is often difficult to achieve in other computational fields. For example, it is difficult to explain the mathematics behind the Ising model of ferromagnetism to a student with limited computational culture, but it is quite possible to introduce the same student to the algorithmic ideas (Euler theorem and de Bruijn graphs) behind the genome assembly. Thus, we argue that the recreational mathematics approach (so brilliantly developed by Martin Gardner and others) coupled with biological insights is a viable paradigm for introducing biologists to bioinformatics. This book is an initial step in that direction.

What is in the book?

Each chapter describes the biological motivation for a problem and then outlines a computational approach to addressing the problem. Chapters can be read separately, as each introduces any needed computational background beyond basic college-level knowledge.

The range of biological topics addressed is quite broad: it includes evolution, genomes, regulatory networks, phylogeny, and more. The computational techniques used are also diverse, from probability and graphs, combinatorics and statistics to algorithms and complexity. However, we made an effort to keep the material accessible and avoid complex computational details (those can be filled in by the interested reader using the references). Figure 1 aims to show for each chapter the biological topics it touches upon and the computational areas involved in the analysis. Naturally, many chapters involve multiple biological and computational areas. Not surprisingly, evolution plays a role in almost all the topics covered, following the famous quote from Theodosius Dobzhansky, "Nothing in biology makes sense except in the light of evolution."

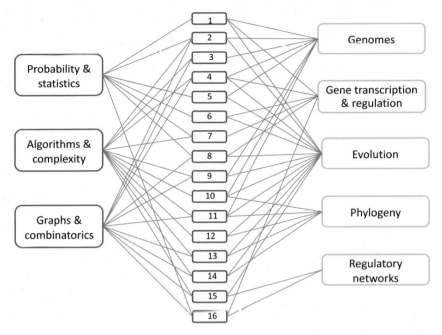

Figure 1 The connections between biological and computational topics for each chapter. The nodes in the middle are chapters, and edges connect each chapter to the biological topics it covers (right) and to the computational topics it introduces (left).

The pedagogical approach, the style, the length, and the depth of the introduced mathematical concepts vary greatly from chapter to chapter. Moreover, even the notation and computational framework describing the same mathematical concepts (e.g. graph theory) across different chapters may vary. As computer scientists say, this is not a bug but a feature: we provided the contributors with complete freedom in selecting the approach that fits their pedagogical goal the best. Indeed, there is no consensus yet on *how* to introduce computer science to biologists, and we feel it is important to see how leading bioinformaticians address the same pedagogical challenge.

 ## How will this book develop?

"Bioinformatics for Biologists" is an evolving book project: we welcome all educators to contribute to future editions of the book. We envision introduction of computational culture to the biological education as an ever-expanding and self-organizing process: starting from the second edition, we will work towards unifying the notation and the pedagogical framework based on the students' and instructors' feedback. Meanwhile, the educators have an option of selecting the specific self-contained chapters they like for the courses they teach.

How to use this book?

Since chapters are self-contained, each chapter can be studied or taught individually and chapters can be followed in any order. One can select to cover, for example, a sample of topics from each of the five biological themes in order to obtain a broader view, or cover completely one of the themes for a deeper concentration. Review questions that follow each chapter are helpful to assimilate the material. Additional resources available at the website will be helpful to teachers in preparing their lectures and to students in deeper and broader learning.

The book's website

The book is accompanied by the website www.cambridge.org/b4b containing teaching materials, presentations, and other updates. These can be of help to students as well as educators.

Contributors

The scientists who contributed to this book are leading computational biologists who have ample experience in both research and education. Some are biologists who have became computational over the years, as their computational research needs developed. Others have formal computational background and have made the transition into biology as their research interests and the field developed. All have experienced the need and the difficulty in conveying computational ideas to biology students, and all view this as an important problem that justifies the effort of contributing to this book. They are all committed to the project.

ACKNOWLEDGMENTS

This book would not be possible without the generous support of the Howard Hughes Medical Institute (provided as HHMI award to Pavel Pevzner).

The editors and contributors also thank the editorial team at Cambridge University Press for their continuous and efficient support at all stages of this project. Special thanks go to Megan Waddington, Hans Zauner, Catherine Flack, Lauren Cowles, Zewdi Tsegai, and Katrina Halliday.

Vineet Bafna would like to acknowledge support from the NSF (grant IIS-0810905) and NIH (grant R01 HG004962).

Kun-Mao Chao would like to thank Phillip Compeau, Yao-Ting Huang, and Tandy Warnow for making several valuable comments that improved the presentation. He is supported in part by NSC grants 97-2221-E-002-097-MY3 and 98-2221-E-002-081-MY3 from the National Science Council, Taiwan.

Phillip Compeau and Pavel Pevzner would like to thank Steffen Heber and Glenn Tesler for very helpful comments, as well as Randall Christopher for his superb illustrations.

Mikhail Gelfand is grateful to Mikhail Roytberg, whose approach to the presentation of the dynamic programming algorithm he has borrowed; to Andrey Mironov and Anatoly Rubinov who do not like this approach and have provided very useful comments and critique; to Phillip Compeau for critique and editing (of course, all remaining errors are the author's); and to Pavel Pevzner for the invitation to participate in this volume and patience over failed deadlines.

He acknowledges support from the Ministry of Education and Science of Russia under state contract 2.740.11.0101.

Andrey Grigoriev would like to thank Joe Martin, Chris Lee, and the editorial team for their careful review of his chapter and many helpful suggestions.

Sridhar Hannenhalli would like to acknowledge the support of NIH grant R01GM085226.

Steffen Heber and Brian E. Howard acknowledge the support of many friends and colleagues, who have contributed to their chapter via extremely helpful discussions and feedback. They would especially like to thank Pavel Pevzner, Glenn Tesler, Jens Stoye, Anne Bergeron, and Max Alekseyev. Their work was supported by Education Enhancement Grant (1419) 2008-0273 of the North Carolina Biotechnology Center.

Eugene V. Koonin, Pere Puigbò, and Yuri I. Wolf wish to thank Jian Ma and Pavel Pevzner for many helpful suggestions. Their research is supported through the intramural funds of the US Department of Health and Human Services (National Library of Medicine).

Christopher Lee wishes to thank Pavel Pevzner, Andrey Grigoriev, and the editorial team for their very helpful comments and corrections.

Ran Libeskind-Hadas recognizes that many people have contributed to the content and exposition of this chapter. However, any omissions or errors are entirely his responsibility. Chris Conow, Daniel Fielder, and Yaniv Ovadia wrote the first version of Jane. The version of Jane used in chapter 12, Jane 2.0, is a significant extension of the original Jane software and was designed, developed, and written by Benjamin Cousins, John Peebles, Tselil Schramm, and Anak Yodpinyanee. Professor Catherine McFadden provided valuable feedback on the exposition of the material in this chapter. The development of Jane 2.0 was funded, in part, by the National Science Foundation under grant 0753306 and from the Howard Hughes Medical Institute under grant 52006301. Finally, Professor Michael Charleston inspired the author to work in this field and has been a patient and generous intellectual mentor.

Jian Ma would like to thank Pavel Pevzner, Eugene Koonin, Ryan Cunningham, and Phillip Compeau for helpful suggestions.

Nataša Pržulj thanks Tijana Milenkovic and Wayne Hayes for comments on the chapter.

Russell Schwartz would like to thank Pavel Pevzner, Sridhar Hannenhalli, and Phillip Compeau for helpful comments and discussion. Dr. Schwartz is supported in part by US National Science Foundation award 0612099 and US National Institutes of Health awards 1R01AI076318 and 1R01CA140214. Any opinions, findings, and conclusions or recommendations expressed in this material are those of the author and do not necessarily reflect the views of the National Science Foundation or National Institutes of Health.

Ron Shamir thanks Hershel Safer for helpful comments, and the support of the Raymond and Beverly Sackler Chair in bioinformatics and of the Israel Science Foundation (grant no. 802/08).

Haixu Tang acknowledges the support of NSF award DBI-0642897.

Tandy Warnow wishes to thank the National Science Foundation for support through grant 0331453; Rahul Suri, Kun-Mao Chao, Phillip Compeau, and Pavel Pevzner for their detailed suggestions that greatly improved the presentation; and Kun-Mao Chao for assistance with making figures for chapter 14.

Tiffani L. Williams and Seung-Jin Sul thank Brian Davis for introducing them to the problem of reconstructing phylogenetic relationships among the big cats. They would also like to thank Danielle Cummings and Suzanne Matthews for their helpful comments on improving this work. Funding for chapter 13 was supported by the National Science Foundation under grants DEB-0629849, IIS-0713618 and IIS-101878.

EDITORS AND CONTRIBUTORS

 ## Editors

 Pavel Pevzner
Department of Computer Science and
 Engineering
University of California at San Diego,
 USA

 Ron Shamir
School of Computer Science
Tel Aviv University, Israel

 ## Contributors

 Vineet Bafna
Department of Computer Science and
 Engineering
University of California at San Diego,
 USA

 Mikhail Gelfand
Department of Bioinformatics
 and Bioengineering
Moscow State University, Russia

 Kun-Mao Chao
Department of Computer Science and
 Information Engineering
National Taiwan University, Taiwan

 Andrey Grigoriev
Department of Biology
Rutgers State University of
 New Jersey, USA

 Phillip Compeau
Department of Mathematics
University of California at San Diego,
 USA

 Sridhar Hannenhalli
Department of Genetics
University of Maryland, USA

Steffen Heber
Department of Computer Science
North Carolina State University, USA

Pere Puigbò
National Center for
    Biotechnology Information
National Library of Medicine
National Institutes of Health,
    USA

Brian Howard
Department of Computer Science
North Carolina State University, USA

Russell Schwartz
Department of Biological
    Sciences
Carnegie Mellon University,
    USA

Eugene Koonin
National Center for Biotechnology
    Information
National Library of Medicine
National Institutes of Health, USA

Seung-Jil Sun
J. Craig Venter Institute
Rockville, USA

Christopher Lee
Department of Chemistry and
    Biochemistry
University of California at
    Los Angeles, USA

Haixu Tang
School of Informatics and
    Computing
Indiana University, USA

Ran Libeskind-Hadas
Department of Computer Science
Harvey Mudd College, USA

Tandy Warnow
Department of Computer
    Sciences
University of Texas at Austin,
    USA

Jian Ma
Department of Bioengineering
University of Illinois at Urbana-
    Champaign, USA

Tiffani Williams
Department of Computer Science
    and Engineering
Texas A&M University, USA

Nataša Pržulj
Department of Computing
Imperial College London, UK

Yuri Wolf
National Center for Biotechnology
    Information
National Library of Medicine
National Institutes of Health, USA

A COMPUTATIONAL MICRO PRIMER

This introduction is a brief primer on some basic computational concepts that are used throughout the book. The goal is to provide some initial intuition rather than formal definitions. The reader is referred to excellent basic books on algorithms which cover these notions in much greater rigor and depth.

Algorithm

An algorithm is a recipe for carrying out a computational task. For example, every child learns in elementary school how to perform long addition of two natural numbers: "add the right-most digits of the two numbers and write down the sum as the right-most digit of the result. But if the sum is 10 or more, write only the right-most digit and add the leading digit to the sum of the next two digits to the left, etc." We have all learned similar simple procedures for long subtraction, multiplication and division of two numbers. These are all actually simple algorithms. Like any algorithm, each is a procedure that works on inputs (two numbers for the problems above) and produces an output (the result). The same procedure will work on any input, no matter how long it is. While we can carry out simple algorithms on small inputs by hand, computers are needed for more complex algorithms or for longer inputs. As with long addition, a complex task is broken down into simple steps that can be repeated many times, as needed. Algorithms are often displayed for human readers in a short form that summarizes their salient features. One aspect of this simplified representation is that a repeated sequence of steps may be listed only once.

Computational complexity

A basic question in studying algorithms is how efficient they are. For a given input, one can time the computation. Since the time depends on the computer being used, a better understanding of the algorithm can be gained by counting the operations (addition, multiplication, comparison, etc.) performed. This number will be different for different inputs. A common way to evaluate the efficiency of a method is by considering *the number of operations required as a function of the input length.* For example, if an algorithm requires $15n^2$ operations on an input of length n, then we know how many operations will be needed for any input. If we know how many operations our computer performs per second, we can translate this to the running time on our machine.

O notation

Suppose our algorithm requires $15n^2 + 20n + 7$ operations on an n-long input. As n grows larger, the contribution of the lower-order terms $20n + 7$ will become tiny compared to the $15n^2$. In fact, as n grows larger, the constant 15 is not very important when it comes to the *rate* of growth of the number of operations (although it affects the run time).[1] Computer scientists prefer to focus only on the main trend and therefore say that an algorithm that takes $15n^2 + 20n + 7$ operations requires "$O(n^2)$" time (pronounced "oh of n squared"), or, equivalently, is "an $O(n^2)$ algorithm." This means that the algorithm's running time increases quadratically with the input length.[2]

Polynomial and exponential complexity

Some problems can be solved using any of several algorithms, and the O notation is used to decide which algorithm is better (i.e. faster). So an $O(n)$ algorithm is better than an $O(n^2)$ algorithm, which in turn is better than an $O(2^n)$ algorithm. This latter complexity, which is called *exponential* (since n appears in the exponent), is

---

[1] Computer scientists do not worry too much about the difference between n^2 and $100n^2$, but they greatly worry about the difference between n^3 and $100n^2$. They will typically prefer $100n^2$ to n^3, since for all inputs of length > 100 the latter will require more time.

[2] To be precise, "$O(n^2)$" means that the algorithm's run time grows *not more* than quadratically. To specify that the run time is exactly quadratic, complexity theory uses the notation "$\Theta(n^2)$." We shall ignore these differences here.

particularly nasty: as the problem size changes from n to $n + 1$, the run time will double! In contrast, for an $O(n)$ algorithm the run time will grow by $O(1)$, and for an $O(n^2)$ algorithm it will grow by $O(2n + 1)$. So no matter how fast our computer is, with an algorithm of exponential complexity we shall very quickly run out of computing time as the problem grows: if the problem size grows from 30 to 40, the run time will grow 1024-fold! The main distinction is therefore between *polynomial* algorithms, i.e. those with complexity $O(n^c)$ for some constant c, and exponential ones.

NP-completeness

Computer scientists often try to develop the most efficient algorithm possible for a particular problem. A primary challenge is to find a polynomial algorithm. Many problems do have such algorithms, and then we worry about making the exponent c in $O(n^c)$ as small as possible. For many other problems, however, we do not know of any polynomial algorithm. What can we do when we tackle such a problem in our research? Computer scientists have identified over the years thousands of problems that are not known to be polynomial, and in spite of decades of research currently have only exponential algorithms. On the other hand, so far we do not know how to prove mathematically that they cannot have a polynomial algorithm. However, we know that if any single problem in this set of thousands of problems has a polynomial algorithm, then all of them will have one. So in a sense all these problems are equivalent. We call such problems *NP-complete*. Hence, showing that your problem is NP-complete is a very strong indication that it is hard, and unlikely to have an algorithm that will solve it exactly in polynomial time for every possible input.[3]

Tackling hard problems

So what can one do if the problem is hard? If a problem is NP-complete this means that (as far as we know) it has no algorithm that will solve every instance of the problem exactly in polynomial time. One possible solution is to develop *approximation algorithms*, i.e. algorithms that are polynomial and can approximately solve the problem, by providing (provably) near-optimal but not necessarily always optimal solutions. Another possibility is *probabilistic algorithms*, which solve the

---

[3] Note that there are problems that were *proven* not to have any polynomial time algorithms, but they are outside the set of established NP-complete problems.

problem in polynomial average time while the worst-case run time can still be exponential. (This would require some assumptions on the probability distribution of the inputs.) Yet another alternative that is often used in bioinformatics is *heuristics* – fast algorithms that aim to provide good solutions in practice, without guaranteeing the optimality or the near-optimality of the solution. Heuristics are typically evaluated on the basis of their performance on the real-life problems they were developed for, without a theoretically proven guarantee for their quality. Finally, *exhaustive algorithms* that essentially try all possible solutions can be developed, and they are often accompanied by a variety of time-saving computational shortcuts. These algorithms typically require exponential time and thus are only practical for modest-sized inputs.

PART I

GENOMES

CHAPTER ONE

Identifying the genetic basis of disease

Vineet Bafna

It is all in the DNA. Our genetic code, or *genotype*, influences much about us. Not only are physical attributes (appearance, height, weight, eye color, hair color, etc.) all fair game for genetics, but also possibly more important things such as our susceptibility to diseases, response to a certain drug, and so on. We refer to these "observable physico-chemical traits" as *phenotypes*. Note that "to influence" is not the same as "to determine" – other factors such as the environment one grows up in can play a role. The exact contribution of the genotype in determining a specific phenotype is a subject of much research. The best we can do today is to measure correlations between the two. Even this simpler problem has many challenges. But we are jumping ahead of ourselves. Let us review some biology.

Background

Why do we focus on DNA? Recall that our bodies have organs, each with a specific set of functions. The organs in turn are made up of tissues. Tissues are clusters of cells of a similar type that perform similar functions. Thus, it is useful to work with cells because they are simpler than organisms, yet encode enough complexity to function autonomously. Thus, we can extract cells into a Petri dish, and they can grow, divide, communicate, and so on. Indeed, the individual starts life as a single cell, and grows up to full complexity, while inheriting many of its parents' phenotypes.

Bioinformatics for Biologists, ed. P. Pevzner and R. Shamir. Published by Cambridge University Press.
© Cambridge University Press 2011.

There must be molecules that contain the instructions for making the body, and these molecules must be inherited from the parents. The cells have smaller subunits (nucleus, cytoplasm, and other organelles) which contain an abundance of three molecules: DNA, RNA, and proteins. Naturally, these molecules were prime candidates for being the inherited material. Of these, proteins and RNA were known to be the machines in the cellular factories, each performing essential functions of the cell, such as metabolism, reproduction, and signal transduction.

This leaves DNA. The discovery of DNA as the inherited material, followed by an understanding of its structure and the mechanism of inheritance, form the major discoveries of the latter half of the twentieth century. DNA consists of long chains of four nucleotides, which we abbreviate as A, C, G, T. Portions of the nucleotides (*genes*) contain the code for manufacturing specific proteins, as well as the regulatory mechanisms that interpret environmental signals, and switch the production on or off. Interestingly, we have two copies of DNA, one from each of our parents. In this way, we produce a similar set of proteins as our parents, and therefore display similar phenotypes, including susceptibility to some diseases. Of course, as we inherit only a randomly sampled half of the DNA from each parent, we are similar but not identical to them, or to our siblings.

On the other hand, if all DNA were identical, it would not matter where we inherited the DNA from. In fact, DNA mutates away from its parent. Often, these mutations are small changes (insertions, substitutions, and deletions of single nucleotides). There are also many additional forms of variation, which are more complex, and include many large-scale changes that are only now being understood. In this chapter, however, we will focus on small mutations as the only source of variation. If we sample DNA from many individuals at a single location (a *locus*) we often find that it is polymorphic (contains multiple nucleotide variants). Clearly, if these mutations occur in a gene, then the protein encoded by the DNA can also change, possibly changing some functional trait in the organism. Therefore, different variants at a locus sometimes present different phenotypes, and are often referred to as *alleles*, after Mendel. Loci with multiple alleles are variously called "segregating sites" (they separate the population), "variants", or "polymorphic markers." If these variants affect single nucleotides, they are also called single nucleotide polymorphisms or SNPs.

We start with a basic instance of a *Mendelian* mutation: individuals present a phenotype if and only if they carry the specific mutation. Our goal is to identify the mutation (or the corresponding genomic locus) from the set. Figure 1.1a shows this with three candidate variants represented by $\diamond$, $\triangle$, and $\circ$. A simple approach to identifying the causal mutation is as follows: (i) determine the genotypes of a collection of individuals that present the phenotype (*cases*), and those that do not (*controls*); (ii) align the genotypes of all individuals, and identify polymorphic locations; (c) for

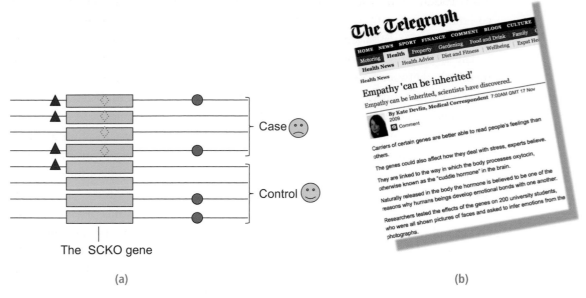

Figure 1.1 Genetic association basics. (a) A Mendelian mutation ◇ that is causal for a phenotype. Other "neutral" variants are nearby. (b) Popular news highlighting the discovery of the gene responsible for a phenotype. In many cases, all that is observed is a correlation between a mutation and the phenotype. The causality is assumed based on some knowledge of the function of the protein encoded by the gene. Figure reprinted by permission. ⓒ Telegraph Media Group Limited 2011.

each polymorphic location, check for a correlation of the variants with case/control status. In Figure 1.1, we see that the occurrence of the ◇ correlates highly with the case status and conclude that the mutation is causal. Given that the mutation lies in the SCKO gene, we conclude that SCKO is responsible. The popular media is peppered with accounts of discoveries of genes responsible for a phenotype.

The intelligent reader will immediately question this premise because these "discoveries" are often not the final confirmation, but simply an observed correlation between the occurrence of the mutation and the phenotype. First, what is the chance that we are even testing with the causal mutation? Typically, genotypes are determined using the technology of DNA chips. The individual DNA is extracted (often from saliva or serum) and washed over the chip. The chip allows us to sample, in parallel, close to 0.5–1 M polymorphic locations, and determine the allelic values at these locations. This fast and inexpensive test allows us to investigate a large population of cases and controls, and makes genetic association possible. However, we do not test *each* location (there are three billion). It is very possible that the causal mutation is not even sampled, and that we may not find correlations even when they exist. Second, even if we do find a correlation, there is no guarantee that we have found the right one. Surely, a simple correlation at one of 1 M markers could have arisen just by chance. How can that be a clue towards the causal gene?

The answer might surprise some. Nature helps us in two ways: first, it establishes a correlation between SNPs that are close to the causal mutation, so any of the SNPs in the region (that contains the relevant gene) are correlated with the mutation. Second, it "destroys" the correlation as the distance from the causal mutation increases. Therefore, a correlation is indeed a strong suggestion that we are in the right location, and any gene in that region is worth a closer look. The next section is devoted to an explanation of the underlying genetic principles, and is followed by a description of the statistical tests used to quantify the extent of the correlation.

Of course, while the basic premise is correct, and simply stated, it is (like everything else in biology) simplistic. In the following sections, we look at issues that can confound the statistical tests for association, and how they are resolved. The resolution of these problems requires a mix of ideas from genetics, statistics, and algorithms.

 ## 2 Genetic variation: mutation, recombination, and coalescence

Dobzhansky famously said that "nothing in biology makes sense except in the light of evolution," and that is where we will start. You might recall from your high-school biology that each of us has two copies of each chromosome, each inherited from one parent.[1] Having two parents makes it tricky to study the ancestral history (the genealogy) of an individual. Therefore, we work with a population of chromosomes, where every individual does have a single parent. In this abstraction, the individual is simply "packaging" for the chromosomes, two at a time. We also make the assumption (absurd, but useful) that all individuals reproduce at the same time. Finally, we assume that the population size does not change from generation to generation. Figure 1.2a shows the basic process. Time is measured in reproductive generations. In each generation, an individual chromosome is created by "choosing" a single parent from the previous generation. To see how this helps, go back in time, starting with the extant population. Every time two chromosomes choose the same parent (coalesce), the number of ancestral chromosomes reduces by 1, and never increases again. Once this ancestry reduces to a single chromosome (the most recent common ancestor, or MRCA), we can stop because the history prior to that event has been lost forever. As each individual has a single parent, the entire history from the MRCA to the extant generation is

---

[1] Not quite, but we will consider recombinations in a bit.

Time   Current (extant) population

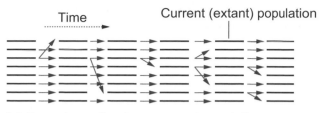

(a) Genealogy of a chromosomal population

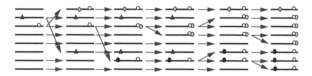

(b) Mutations: drift, fixation, and elimination

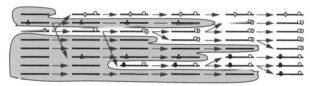

(c) Removing extinct genealogies

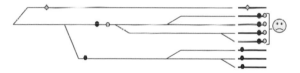

(d) Causal and correlated mutations

Figure 1.2 An evolving population of chromosomes. (a) The Wright Fisher model is an idealized model of an evolving population where the number of individuals stays fixed from generation to generation, and each child chooses a single parent uniformly from the previous generation. (b) Mutations are inherited by all descendants, and drift until they are fixed or eliminated. (c) We only consider the history that connects the existing population to its most recent common ancestor. (d) The underlying data are presented as a SNP matrix (with a hidden genealogy). The genealogy leads to correlations between SNPs.

described by a tree (*the coalescent tree*). Other genealogical events that occurred after MRCA but are not part of the coalescent tree are useless because the lineages died out before reaching the current generation (Figure 1.2c). The only historical events that will concern us are ones in the underlying coalescent tree.

Now, let us consider mutations. Each chromosome is identical to its parent, except when a mutation modifies a specific location. Given the short time frame of evolution of the human population relative to the number of mutating positions, most locations are modified at most once in history. To simplify things, we assume that this is true for all variants (the *infinite sites assumption*): once a location mutates to a new allelic value, it maintains that allele, and all descendants of the chromosome inherit the mutation. As individuals choose their parents and inherit mutations, the frequency of mutations changes (*drifts*) from generation to generation. This principle is illustrated in Figure 1.2b. The mutation denoted by the blue ○ arises before the MRCA, and is therefore *fixed* in the current population. On the other hand, △ arises in a lineage that was *eliminated* and is not observed. Other mutations, such as the ○, arose sometime after the MRCA, and present as polymorphisms when sampled in the existing population. This is illustrated in Figure 1.2d. Here, we have removed the generation information, and represent time simply by the branch-lengths. When we sample a population with DNA microchips, we create a matrix of polymorphisms; rows correspond to individuals, columns represent polymorphic locations, and the entries represent allelic values representing the consequence of historical mutations on the coalescent tree. The tree itself is invisible, although likely trees can be reconstructed using phylogenetic techniques.

What is the point of all this? It is simply that the underlying tree imposes a correlation between mutations. Let the black circle ● in Figure 1.2d represent a causal mutation. Individuals display a phenotype if and only if they carry this mutation. However, every mutation in this matrix is correlated to some extent. For example, the presence of the yellow mutation (which is on the same branch) is equally predictive of the phenotype, and the red ○ (which occurs on a different lineage) implies that the individual does not carry the phenotype. We call this the principle of *linkage*: mutations that are part of an evolutionary lineage are correlated. Thus, it is not necessary to sample all mutations to identify the gene of interest. However, this is not enough. If all SNPs on the chromosome are correlated (albeit to varying degree), they cannot help to narrow the search for the causal locus. We are helped again by the natural phenomenon of *recombination*. In meiosis (production of gametes), a crossing over of the two parental chromosomes might occur. The child therefore gets a mix of the two parental chromosomes, as shown schematically in Figure 1.3a,b. Now consider a population. Recombination events between two locations change the underlying coalescent tree. With increasing distance between loci, the number of historical recombination events increases and destroys the correlations. In Figure 1.3c, the yellow and black ○ are proximal and remain correlated. However, recombination events destroy the correlations (the linkage) between the red ○ and causal (black) ●. This establishes a second principle: *correlation between mutations is destroyed with increasing distance between loci due to the accumulation of recombination events.*

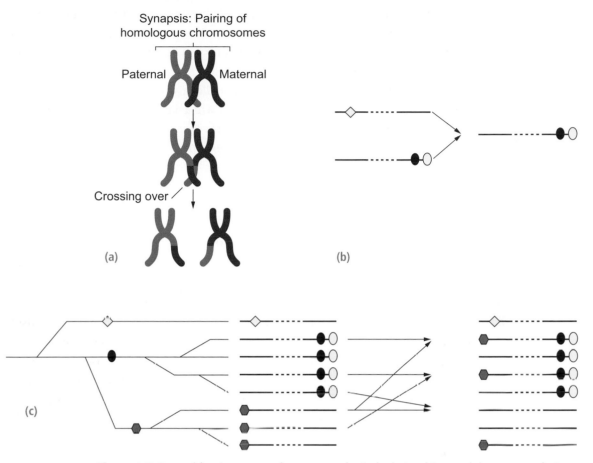

Figure 1.3 Recombination events change genealogical relationships, and destroy correlation between SNPs. (a) Crossover during meiosis. (b) Schematic of a crossover and its effect of linkage between mutations. (c) Multiple recombination events destroy linkage between SNPs.

3 Statistical tests

Let us digress and consider a simple experiment to statistically test for correlation between two events: thunder and lightning. It is intuitively clear that the two are correlated, but we will formalize this. Let $x_i = 1$ indicate the event that we saw lightning on the ith day. Respectively, let $y_i = 1$ indicate the event that we heard thunder on the ith day. Let P_x (respectively, P_y) denote $\Pr(x_i = 1)$ (respectively, $\Pr(y_i = 1)$) for a randomly chosen day. Assume that we see lightning 35 days in a year, so that $P_x = 35/365 \simeq 0.1$. Likewise, let $P_y \simeq 0.1$. What is the chance of seeing both on the same day? Formally, denote the chance of joint occurrence by $P_{xy} = \Pr(x_i = 1 \text{ and } y_i = i)$. If the two were not correlated, we would not observe both very often. In other words,

$P_{xy} = P_x P_y \simeq 0.01$, and so only 3–4 days a year are expected to present both events. If we observe 30 days of thunder and lightning, then we can conclude that they are correlated. What if we observe 10 days of thunder and lightning? This is the question we will consider.

Denote two loci as x, y, and let x_i denote the allelic value for the ith chromosome. If we make the assumption of infinite sites, x_i will take one of two possible allelic values. Without loss of generality, let $x_i \in \{0, 1\}$. The generalization to multi-allelic loci will be considered in Section 4.2. Let P_x denote $\Pr(x_i = 1)$ for a randomly sampled chromosome i at locus x. Correspondingly, $P_{\bar{x}} = 1 - P_x$ represents the probability that $x_i = 0$. Denote the joint probabilities as

$$P_{xy} = \Pr(x_i = 1, y_i = 1) = P_x \Pr(y_i = 1 | x_i = 1)$$

$$P_{\bar{x}y} = \Pr(x_i = 0, y_i = 1) = P_{\bar{x}} \Pr(y_i = 1 | x_i = 0)$$

and so on. If x, y are proximal then $\Pr(y_i = 1 | x_i = 1)$ is very different from P_y. See, for example, the black and yellow ○ in Figure 1.3c. By contrast, if x, y are very far apart so that recombination events have destroyed any correlation, then

$$P_{xy} \simeq P_x P_y$$

$$P_{\bar{x}y} \simeq P_{\bar{x}} P_y.$$

As the recombination events destroy correlation over time, we use the term *Linkage Equilibrium* to denote the lack of correlation. The converse of this, often termed *Linkage Disequilibrium* (LD), or *association*, describes the correlation between the proximal loci. A straightforward statistic to measure LD(x, y) is given by

$$D = P_{xy} - P_x P_y. \tag{1.1}$$

Note that the choice of allele does not matter. The interested reader can verify that

$$|D| = \left| P_{xy} - P_x P_y \right|$$
$$= \left| P_{\bar{x}y} - P_{\bar{x}} P_y \right|$$
$$= \left| P_{x\bar{y}} - P_x P_{\bar{y}} \right|$$
$$= \left| P_{\bar{x}\bar{y}} - P_{\bar{x}} P_{\bar{y}} \right|.$$

The larger the value of $|D|$, the greater the correlation. Apart from its historical significance, the D-statistic is used more as a relative, rather than an absolute measure. Instead, a scaled statistic D' is defined as

$$D' = \frac{D}{D_{\max}} = \begin{cases} \frac{D}{\min\{P_{\bar{x}} P_y, P_x P_{\bar{y}}\}} & D \geq 0 \\ \frac{D}{-\min\{P_x P_y, P_{\bar{x}} P_{\bar{y}}\}} & D < 0 \end{cases}. \tag{1.2}$$

The normalized statistic, D', ranges between 0 and 1, with 0 implying no correlation, and 1 implying perfect correlation. Ultimately, these statistic values are still numbers, however, and it might be hard to say how much better is $D' = 0.7$ (say) than $D = 0.6$. To address these questions, statisticians attempt to compute a p-value for the statistic. The p-value of $D = 0.6$ is the probability that a random experiment would yield a value of $D \geq 0.6$ just by chance if the null hypothesis of $D =$ was true.

To compute the p-value here, we have to use a different normalization for reasons that will become clear. Define $LD(x, y)$ as

$$\rho = \frac{D}{\sqrt{P_x P_{\bar{x}} P_y P_{\bar{y}}}}.$$

(1.3)

The statistic ρ is closely related to the χ^2 test of independence between two variables. Recall that with n chromosomes, the number of chromosomes i with $x_i = 1$ and $y_i = 1$ is given by $P_{xy}n$. The observations of joint occurrences for x, y can be expressed by the 2×2 table:

$x \backslash y$	0	1	Total
0	$P_{\bar{x}\bar{y}}n$	$P_{\bar{x}y}n$	$P_{\bar{x}}n$
1	$P_{x\bar{y}}n$	$P_{xy}n$	$P_x n$
Total	$P_{\bar{y}}n$	$P_y n$	n

If x, y are not correlated (null hypothesis), then the number of individuals in the first cell is expected to be

$$P_{\bar{x}\bar{y}}n = P_{\bar{y}} P_{\bar{x}} n$$

and so on, for all cells. The statistic $(P_{xy}n - P_x P_y n)/\sqrt{P_x P_y n}$ behaves approximately like a normal distribution, and the square $(P_{xy}n - P_x P_y n)^2/P_x P_y n$ behaves like a χ^2 distribution. Under the null hypothesis, the mean value is 0, and the p-value can be obtained simply by looking at pre-computed tables. Finally, we get a p-value for ρ^2 observing that it is the sum of four χ^2 distributed values, as follows:

$$\chi^2_{xy} = \frac{(P_{\bar{x}\bar{y}}n - P_{\bar{x}}P_{\bar{y}}n)^2}{P_{\bar{x}}P_{\bar{y}}n} + \frac{(P_{\bar{x}y}n - P_{\bar{x}}P_y n)^2}{P_{\bar{x}}P_y n} + \frac{(P_{x\bar{y}}n - P_x P_{\bar{y}}n)^2}{P_x P_{\bar{y}}n} + \frac{(P_{xy}n - P_x P_y n)^2}{P_x P_y n}$$

$$= \frac{D^2 n}{P_x P_y P_{\bar{x}} P_{\bar{y}}} = \rho^2 n.$$

(1.4)

A low p-value implies that our assumption is incorrect, implying Linkage Disequilibrium or correlation. The actual inference (correlation, or not) based on probabilities conforms to a "frequentist" interpretation of the data, and is not universally accepted. Nevertheless, the reader will agree that it is a useful tool for interpretation.

3.1   LD and statistical tests of association

Finally, we are ready to put it all together and identify the locus responsible for a specific phenotype. Assume there is a phenotype with a single causal mutation at locus d. For individual i, $d_i = 1$ implies case status; otherwise, the individual is control. Our question can be reformulated as

Find the location of d.
   OR,-
Find known polymorphisms that are located close to d, and are statistically associated.
   OR,-
Find all polymorphisms x s.t. LD(x, d) is high.

However, we have already provided an answer to the last question above. The test described here is but one of a battery of different statistical tests that can be performed. How well a specific test works is calculated by taking a known set (perhaps simulated) and measuring the accuracy of positive and negative results of the test. The test's *power* $(1 -$ false negative rate) after fixing the type I error (false positive) rate can quantify this.

4   Extensions

Let us extend the basic methodology. The actual mutation at d need not be considered, and may not even exist in a Mendelian sense. To generalize, the allelic value $d_i = 1$ simply predisposes an individual towards the case status. Define the relative risk

$$RR = \frac{\Pr(\text{CASE}|d_i = 1)}{\Pr(\text{CASE}|d_i = 0)}.$$

As long as $RR \gg 1$, a similar test of association will work.

4.1   Continuous phenotypes

Recall that phenotype is any trait that can be measured. We assumed categorical values for the phenotype (Case/Control). This is reasonable in some cases (occurrence or non-occurrence of disease), but less applicable to others. For example, obesity (measured by the Body Mass Index), blood pressure (measured by the systolic or diastolic blood pressure measurements), and height all represent phenotypes with continuous values. Testing for association can be somewhat tricky in these circumstances. One simple solution is the categorization of continuous values: for example, all diastolic

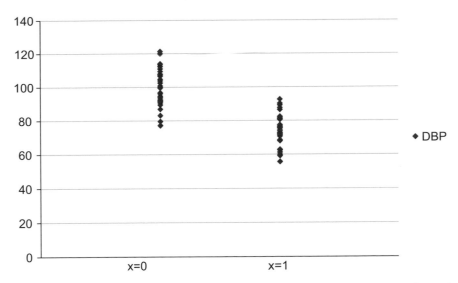

Figure 1.4 Distribution of diastolic blood pressure segregated by the allelic value at locus x. The estimated mean and variances of either class are $(\bar{X}_0, S_0^2) = (103, 109)$, $(\bar{X}_1, S_1^2) = (62, 76)$ for $n = 35$ individuals in each class. The large difference between the means, and the relatively low spread of each distribution, indicates that DBP is correlated with the allelic value at the locus.

blood pressure values over 90 can be considered cases; else, controls. Another way to approach this is through analysis of variance (ANOVA) tests, which we will explain informally with an example. In this case, there are only two segregating classes, so a specific ANOVA test, the Student's t, can be used.

Consider the sketch in Figure 1.4 which plots the diastolic blood pressure (DBP) readings for individuals with different allelic values at locus x. The readings for individuals with $x = 1$ are distinctly higher than the individuals with $x = 0$, providing the intuition that allelic values at locus x are correlated with DBP. Is it better to consider this population as two classes (segregated by the allelic value at x), or as a single class?

We make the assumption that the DBP values are normally distributed. The estimated mean and variances of either class are $(\bar{X}_0, S_0^2) = (103, 109)$, $(\bar{X}_1, S_1^2) = (62, 76)$ for $n = 35$ individuals in each class. We would like to know if the two mean values are significantly different given the underlying variances. Intuitively, an allelic value of 0 implies that the DBP will be at least $103 - 2\sqrt{109} \simeq 82$. On the other hand, the DBP for allelic value 1 is rarely greater than $62 + 2\sqrt{76} \simeq 79$. Given that the allelic values help predict the DBP somewhat tells us that the locus x is associated.

Formally, assuming the null hypothesis of no association between x and DBP, the t-statistic

$$T = \frac{\bar{X}_0 - \bar{X}_1}{\sqrt{\frac{S_0^2}{n} + \frac{S_1^2}{n}}} \tag{1.5}$$

must follow the Student's t distribution, with $2n - 2$ degrees of freedom, and we can use that to compute a p-value. In this case, the t-statistic is $T = 17.8$ (df $= 68$), with a p-value less than 0.0001, and the correlation is very strong.

4.2 Genotypes and extensions

The astute reader has undoubtedly noticed a discrepancy. The phenotype is assigned to an individual containing a pair of chromosomes. However, we are computing associations against a population of chromosomes. To correct this discrepancy, we consider the genotype of an individual. Consider a locus x with two allelic values 0, 1 in a population. Each individual belongs to one of three classes, depending on the allelic pair, 00, 01, and 11. The test for associations can be modified to accommodate this. For case–control tests, we have a 3×2 contingency table, and can measure significance using a χ^2 test with 2 degrees of freedom. For continuous variables, an analog of the t-test for multiple groups (the F-test) is often used.

In fact, these ideas can be extended even further. We had made the assumption that a location is only mutated once in our history. That may not always be. Each locus may have between 2 and 4 alleles, with each individual contributing a pair of alleles. Indeed, there is no reason to restrict ourselves to a single polymorphic locus. We could consider a chain of proximal loci. Having individuals placed in multiple classes (bins) with continuous phenotypes is not technically difficult, but often leads to the problem of under-sampling. The higher the number of bins, the fewer the number of individuals in each bin, and the higher the chance of a false correlation. We explain this principle with a simple example. Consider a fair-coin. If we toss $2n$ coins, and put them appropriately in two bins, HEADS and TAILS, we expect to see a similar number ($\simeq n$) of coins in each bin. If the discrepancy is large, we conclude that the coin is loaded. However, what if we tossed only 1 coin? It must fall in one of the 2 bins, and the discrepancy is 100%. To get around this, we need to increase the number of individuals (increasing the cost of the experiment), or decrease the number of bins. While not possible in this simple example, creative ways to reduce the number of bins are a large part of the design of statistical tests.

4.3 Linkage versus association

Let's revisit the essential ideas from Section 2. One, SNPs are correlated due to a common evolutionary history, starting from the MRCA. Two, this correlation is destroyed among distant loci due to recombination events. In this discussion, we were silent on the actual number of recombination events.

Recombination events can be assumed to be Poisson-distributed, with a rate of r crossovers per generation per base pair (bp). Consider two loci x, y that are ℓ bp apart, and let $D^{(t)}$ denote the LD at time t. If the allele frequencies do not change over generations (the so-called "Hardy–Weinberg equilibrium"), then we can show

$$D^{(t)} = (1 - r\ell)D^{(t-1)} = (1 - r\ell)^t D^{(0)} \simeq e^{-r\ell t} D^{(0)}.$$

Clearly, LD decreases with both time t, and distance ℓ, eventually going to 0 (Linkage Equilibrium). For two randomly chosen individuals, the common ancestor is many generations in the past (indeed, by symmetry arguments, we can see that it is very close to the time of the original MRCA). In practice, this means that two loci only have to be 50–100 Kbp apart to reach linkage equilibrium. Therefore, in order for us not to miss the causal locus, we need to test with a dense collection of markers through the genome. Until recently, this was prohibitively expensive, and researchers looked for ways to reduce the number of recombination events so that distant markers remained in LD.

One approach is to choose individuals who share a recent common ancestor; simply choose case and control individuals from a family. In the family, the time to MRCA is small (a few generations), and LD is maintained even over large ℓ ($\sim$Mbp). For every polymorphic marker (SNP) in the family, researchers test whether an allele cosegregates with the case phenotype. If so, the marker is considered *linked*. Among family-based tests, we have tests for linkage, and for association, but we will not consider these further.

Of course, there is no free lunch here. The long-range LD among family members means that a sparse collection of markers is sufficient for identifying cosegregating or linked markers, implying a cheaper test. On the other hand, the sparsity of markers also implies that after linkage is found, a lot of work needs to be done to zero in on the causal locus. Often, an association test using a dense map of markers in the region from unrelated case–control individuals is necessary for fine mapping. Today, with the ability to use chips to sample multiple locations simultaneously, and to genotype many individuals, genome-wide tests of association are becoming more common. At the same time, family-based tests are still worthwhile, as they are often immune to some of the confounding problems for associations. We will not discuss this in detail, but the interested reader should look to the section on population substructure and rare variants.

5   Confound it

The underlying principles of genetic association are elegant and simple, and indeed can be derived using extensions of Mendel's laws. However, the genetic etiology of complex diseases is, well, complex, and can confound these tests. Understanding confounding factors is central to making the right inferences. We mention a few below.

5.1   Sampling issues: power, etc.

For the test to be successful, it must have a low false-positive (type I) error rate α, and high power, defined as $1 - \beta$, where β is the false-negative rate. Setting a p-value cutoff for association (as discussed in Section 2) is one way to bound α. Typically, one would only consider loci x, whose LD with the case–control status has a p-value no more than α. However, the number of tests (loci) also play into this. For a genome-wide scan, we are testing at many ($m \simeq 500\,\mathrm{K}$) independent loci. A straightforward (Bonferroni) correction is as follows: if the chance of making a false call at a locus is α, the chance of making a false call at some locus is $m\alpha$.

Usually, the strategy is to fix α to some desired value, and to maximize the power of the test. Here is an informal description of estimating power of a case–control test. Let P_ϕ and P denote the minor allele frequencies (MAF) at a locus in controls and cases, respectively. The two should be equal in the absence of association, so one way to restate the association test is to look for loci at which $P \neq P_\phi$. What if there was a small but significant difference? Suppose the number of cases carrying the minor allele is U. Under the null hypothesis (no association, ($P_\phi = P$)), U is normally ($\mathcal{N}(nP_\phi, \sqrt{nP_\phi(1 - P_\phi)})$) distributed. See the blue curve in Figure 1.5. The threshold for significance is chosen based on the type I error α. Suppose the alternative is true, so that $P \neq P_\phi$. The false-negative rate β can be computed as the probability that U is drawn from the red curve but just happens by chance to lie before the threshold, so the null hypothesis cannot be rejected. Formally, the power is the area of the red curve that lies outside the threshold. With increasing sample size, the distance between the mean of the two curves ($n(P - P_\phi)$) increases, while the "spread" of the red curve (described by the s.d. $\sqrt{nP_\phi(1 - P_\phi)}$) does not increase proportionately. Therefore, power is increased by increasing the sample size n.

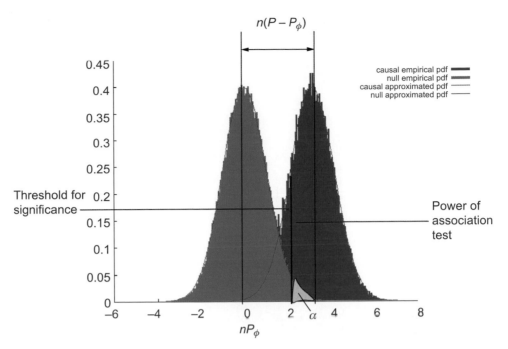

Figure 1.5 Power of an association test. P_ϕ, P denote the minor allele frequencies at a locus for controls and cases, respectively. The distribution of minor allele frequencies for controls and cases is denoted by the blue and red curves. We fail to detect a true association if the sample is drawn from the red curve, but the minor allele frequency is below the threshold of rejecting the null hypothesis.

5.2 Population substructure

Sickle cell anemia is a disease in which the body makes abnormal (sickle-shaped) red blood cells, leading to anemia and many related symptoms. If left untreated, the disease can lead to organ failure and death. It is inherited in a recessive fashion (both alleles need to be mutated in order to present the phenotype), and is common in people of African origin. Consider a typical case–control study as in Figure 1.6. Not surprisingly, a marker in the Duffy locus (which has been implicated previously) shows up with an association to the phenotype. However, we have made a poor design choice in not controlling for structure in populations. Without explicit controls, we find that most case individuals are people of African origin (marked with an A), while most controls are of European origin. Therefore, markers at the locus responsible for skin color also show a strong association with the phenotype, and confound the test.

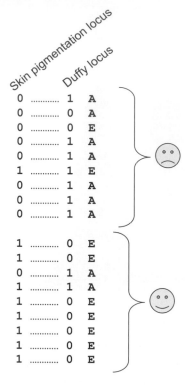

Figure 1.6 Population substructure. As sickle cell anemia is more common in Africans compared to Europeans, the cases and controls can come from different subpopulations. If not corrected, any locus that differentiates between the two subpopulations (such as skin pigmentation) will also correlate with the sickle-cell phenotype, confounding the test.

In general, the problem of population substructure has received much attention. Clearly, care must be taken to choose cases and controls from the same underlying population. As can be imagined, migration and recent admixture of populations can make this difficult, even with self-reported ethnicity. One computational strategy relies on identifying LD between pairs of markers that are too far apart to have significant LD. Long-range LD is indicative of underlying population structure. To deal with population substructure, either we can reduce all observed correlations appropriately, or partition the populations into subpopulations before testing.

5.3 Epistasis

For complex alleles, it could be the case that multiple loci interact to affect the phenotype. Figure 1.7 provides a cartoon illustration of such interactions. Here, compensating mutations in SNPs (T and G, or A and A) allow the encoded proteins to interact, but

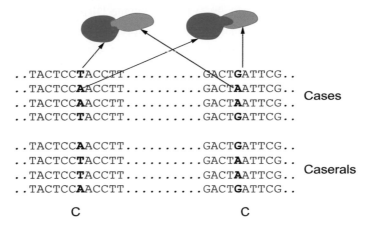

Figure 1.7 Epistatic interactions. Neither *x* nor locus *y* show any marginal association with the phenotype. However, when considered together, the genotype *T . . . G*, and *A . . . A* correlate perfectly with cases. Such interactions pose computational and statistical challenges to identifying genotype phenotype correlations.

individual mutations destroy the lock and key mechanism. Therefore, neither locus *x* nor *y* associates individually with the phenotype. However, if we considered *x*, *y* together, the *T . . . G* and *A . . . A* suggest case status for the individual. Epistasis indeed makes the problem of association much harder. In a genome-wide study with 500 K markers, we would need to test a very large ($2.5 \cdot 10^{11}$) number of possible pairs. More complex *k*-way interactions would be harder. In addition to increasing the computational challenge, the large number of tests would also make it far more likely to create false-positive sets, requiring appropriate statistical corrections.

5.4 Rare variants

It can happen that multiple rare variants (RVs) influence a gene phenotype. For example, the genomic region upstream of a gene acts as a regulatory switch. Transcription factors bind to the upstream DNA, and switch the translation of the gene (production of protein from the gene encoding) on and off. Any mutation in this region could destroy a transcription factor binding site, and therefore the phenotype might be established by a collection of non-specific mutations, each of which has a low frequency but together mediate a large effect (explain the phenotype in a large number of people).

However, several properties of rare variants make their genetic effects difficult to detect with current approaches. As an example, if a causal variant is rare ($10^{-4} \leq$ MAF $\leq 10^{-1}$), and the disease is common, then the allele's Population Attributable Risk (PAR), and consequently the odds ratio (OR), will be low. Additionally, even highly penetrant RVs are unlikely to be in Linkage Disequilibrium (LD) with more common genetic variations that might be genotyped for an association study of a common disease. Therefore, single-marker tests of association, which exploit LD-based associations, are likely to have low power. If the Common Disease Rare Variant (CDRV) hypothesis holds, a combination of multiple RVs must contribute to population risk. In this case, there is a challenge of detecting multi-allelic association between a locus and the disease.

DISCUSSION

The etiology of most (all?) diseases has a genetic basis. In addition, we display a number of phenotypes (eye color) that are inherited. Understanding the genetic basis of phenotypes continues to be a major focus of science today. Until recently, technological limitations made the process arduous. For instance, the identification of the gene for cystic fibrosis in 1989 came after a large multi-year project. Today, with the rapid resequencing of human populations, and an increasing knowledge of gene functions, we are able to focus on complex disorders. In this chapter, we discuss the basics of testing by association, and the problems that can confound these tests.

QUESTIONS

(1) Prove that the LD statistic D for binary alleles does not change depending upon the choice of allele by showing the following:

$$|D| = \left|P_{xy} - P_x P_y\right| = \left|P_{\bar{x}y} - P_{\bar{x}} P_y\right| = \left|P_{x\bar{y}} - P_x P_{\bar{y}}\right| = \left|P_{\bar{x}\bar{y}} - P_{\bar{x}} P_{\bar{y}}\right|.$$

(2) The statistic D' is a scaled measure of linkage disequilibrium. Show that $0 \le D' \le 1$.

(3) The locus X has two alleles, 0 and 1. 100 individuals were genotyped at locus X and also checked for eye color. Their genotypes and eye color segregated as follows: 8 individuals had (00, green), 38 had (01, green), and the remaining 54 individuals had (11, brown). genotype 11 had brown eyes. Does locus X associate with eye color?

FURTHER READING

The treatment here is a simplification of extensive literature from statistical genetics. The basics of the coalesent process can be found in a good review article by Nordborg [1]. The books by Durrett and also Hein, Schierup, and Wiuf cover the topics in greater detail [2, 3]. An excellent overview of statistical association tests is provided by Balding [4].

A classic, although somewhat dated, description of family-based linkage tests is given in the book by Ott [5]. Most algorithms for linkage are derived from Elston and Stewart (large pedigrees, few markers) [6], or Lander and Green (smaller pedigrees, many markers) [7]. The TDT is widely cited as a successful test for family-based association that is immune to population substructure [8].

Population substructure has been addressed in a number of recent papers, and remains an area of active research [9, 10]. Evans and colleagues, and Cordell provide a review of epistasis [11, 12]. Bodmer and Bonilla provide an introduction to analysis with rare variants [13].

REFERENCES

[1] M. Nordborg. Coalescent theory. In: *Handbook of Statistical Genetics*. John Wiley & Sons, 2001.

[2] R. Durrett. *Probability Models for DNA Sequence Evolution*. Springer, New York, 2009.

[3] J. Hein, M. Schierup, and C. Wiuf. *Gene Genealogies, Variation and Evolution: A Primer in Coalescent Theory*. Oxford University Press, Oxford, 2005.

[4] D. J. Balding. A tutorial on statistical methods for population association studies. *Nat. Rev. Genet.*, 7:781–791, 2006.

[5] J. Ott. *Analysis of Human Genetic Linkage*. The Johns Hopkins University Press, Baltimore, 1991.

[6] R. C. Elston and J. Stewart. A general model for the genetic analysis of pedigree data. *Hum. Hered.*, 21:523–542, 1971.

[7] E. S. Lander and P. Green. Construction of multilocus genetic linkage maps in humans. *Proc. Natl Acad. Sci. U S A*, 84(8):2363–2367, 1987.

[8] R. S. Spielman and W. J. Ewens. The TDT and other family-based tests for linkage disequilibrium and association. *Am. J. Hum. Genet.*, 59:983–989, 1996.

[9] A. L. Price, N. J. Patterson, R. M. Plenge, M. E. Weinblatt, N. A. Shadick, and D. Reich. Principal components analysis corrects for stratification in genome-wide association studies. *Nat. Genet.*, 38:904–909, 2006.

[10] J. K. Pritchard, M. Stephens, and P. Donnelly. Inference of population structure using multilocus genotype data. *Genetics*, 155(2):945–959, 2000.

[11] D. M. Evans, J. Marchini, A. P. Morris, and L. R. Cardon. Two-stage two-locus models in genome-wide association. *PLoS Genet.*, 2:e157, 2006.

[12] H. J. Cordell. Genome-wide association studies: Detecting gene–gene interactions that underlie human diseases. *Nat. Rev. Genet.*, May 2009.

[13] W. Bodmer and C. Bonilla. Common and rare variants in multifactorial susceptibility to common diseases. *Nat. Genet.*, 40(6):695–701, 2008.

CHAPTER TWO

Pattern identification in a haplotype block

Kun-Mao Chao

A Single Nucleotide Polymorphism (SNP, pronounced *snip*) is a single nucleotide variation in the genome that recurs in a significant proportion of the population of a species. In recent years, the patterns of Linkage Disequilibrium (LD) observed in the human population reveal a block-like structure. The entire chromosome can be partitioned into high-LD regions, referred to as haplotype blocks, interspersed by low-LD regions, referred to as recombination hotspots. Within a haplotype block, there is little or no recombination and the SNPs are highly correlated. Consequently, a small subset of SNPs, called tag SNPs, is sufficient to distinguish the haplotype patterns of the block. Using tag SNPs for association studies can greatly reduce the genotyping cost since it does not require genotyping all SNPs. We illustrate how to recast the tag SNP selection problem as the set-covering problem and the integer-programming problem – two well-known optimization problems in computer science. Greedy algorithms and LP-relaxation techniques are then employed to tackle such optimization problems. We conclude the chapter by mentioning a few extensions.

 Introduction

A DNA sequence is a string of the four nucleotide "letters" A (adenine), C (cytosine), G (guanine), and T (thymine). The genetic variations in DNA sequences have a major impact on genetic diseases and phenotypic differences. Among various genetic variations, the *Single Nucleotide Polymorphism* (SNP, pronounced *snip*) is one of the

Bioinformatics for Biologists, ed. P. Pevzner and R. Shamir. Published by Cambridge University Press.
© Cambridge University Press 2011.

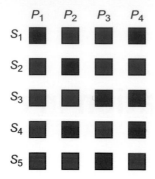

Figure 2.1 A haplotype block containing five SNPs and four haplotype patterns. In this figure, a blue square stands for a major allele and a red square stands for a minor allele.

most frequent forms and has fundamental importance for disease association and drug design. A SNP is a single nucleotide variation in the genome that recurs in a significant proportion of the population of a species. Specifically, a single nucleotide mutation is called a SNP if its minor allele frequency is no less than a given threshold, say 1%. For example, a mutation in the genome in which 85% of the population have a G and the remaining 15% have an A is a SNP. Since tri-allelic and tetra-allelic SNPs are very rare, we often refer to a SNP as a bi-allelic marker: major allele vs. minor allele. Millions of SNPs have been identified and made publicly available.

In recent years, the patterns of *Linkage Disequilibrium* (LD) observed in the human population have revealed a block-like structure. LD refers to the association that particular alleles at nearby sites are more likely to occur together than would be predicted by chance. The entire chromosome can be partitioned into high-LD regions interspersed by low-LD regions. The high-LD regions are usually called "haplotype blocks," and the low-LD ones are referred to as "recombination hotspots." Since there is little or no recombination within a haplotype block, these SNPs are highly correlated. Consequently, a small subset of SNPs, called tag SNPs or haplotype tagging SNPs, is sufficient to categorize the haplotype patterns of the block. It is thus possible to identify genetic variation without genotyping every SNP in a given haplotype block. This can greatly reduce the genotyping cost for genome-wide association studies.

In this study we assume that the haplotype blocks have been delimited in advance, and our objective is to find a minimum set of SNPs which can distinguish all pairs of haplotype patterns in a given block. Figure 2.1 depicts a haplotype block containing five SNPs and four haplotype patterns. To determine which haplotype pattern category a sample belongs to, we may genotype all five SNPs in this block. However, it works just as well if we only genotype SNPs S_1 and S_4, since their combinations can distinguish all pairs of haplotype patterns. For example, if both S_1 and S_4 are major alleles, the sample is categorized as haplotype pattern P_3.

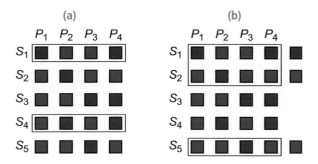

Figure 2.2 Selecting tag SNPs that can distinguish all pairs of haplotype patterns. (a) SNPs S_1 and S_4 form a minimum set of tag SNPs. (b) SNPs S_1, S_2, and S_5 do not form a set of tag SNPs since they cannot distinguish the pair P_1 and P_4.

We show that the tag SNP selection problem is analogous to the minimum test collection problem. We then illustrate how to recast the tag SNP selection problem as the set-covering problem and solve it approximately by a greedy algorithm. Furthermore, it can be formulated as an integer-programming problem, and a simple rounding algorithm can be employed to find its near-optimal solutions. We conclude this chapter by mentioning a few extensions.

 ## The tag SNP selection problem

Assume that we are given a haplotype block containing n SNPs and h haplotype patterns. Let $\mathcal{S} = \{S_1, S_2, ..., S_n\}$ denote the SNP set and let $\mathcal{P} = \{P_1, P_2, ..., P_h\}$ denote the pattern set. A haplotype block is represented by an $n \times h$ binary matrix M whose entries are either a blue square or a red square, representing the major and minor alleles, respectively. Figure 2.1 depicts a 5×4 haplotype block.

We say that SNP S_i can distinguish the pattern pair P_j and P_k if $M[i, j] \neq M[i, k]$, where $1 \leq i \leq n$ and $1 \leq j < k \leq h$. In other words, if one pattern contains a major allele of SNP S_i, and the other contains a minor allele of SNP S_i, then the two patterns can be distinguished by S_i. For instance, in Figure 2.1, SNP S_1 can distinguish patterns P_1 and P_4 from P_2 and P_3 since P_1 and P_4 contain a minor allele of S_1, and P_2 and P_3 contain a major allele of S_1. The goal of the *tag SNP selection problem* is to find a minimum number of SNPs that can distinguish all possible pairwise combinations of patterns. In Figure 2.2, S_1 and S_4 form a set of tag SNPs since they can distinguish all pairs in $\mathcal{P}$, whereas S_1, S_2, and S_5 do not form a set of tag SNPs since they cannot distinguish the pair P_1 and P_4.

In fact, the tag SNP selection problem is analogous to the minimum test collection problem, which arises naturally in fault diagnosis and pattern identification. Given a collection C of subsets of a finite set $\mathcal{A}$ of "possible diagnoses," the minimum test collection problem is to ask for a subcollection $C' \subseteq C$ such that $|C'|$ is minimized and, for each pair $a_j, a_k \in \mathcal{A}$, there exists some set (i.e. a *test*) in C' that contains exactly one of them. In other words, such a test can *distinguish* the pair a_j, a_k. Take Figure 2.1, for example. SNP S_1 can distinguish patterns P_1 and P_4 from others, thus we include $\{P_1, P_4\}$ in C. Similarly, each of SNPs S_2, S_3, S_4, and S_5 can distinguish a particular set of patterns from others. It follows that the instance of the minimum test collection problem for Figure 2.1 is $\mathcal{A} = \{P_1, P_2, P_3, P_4\}$ and $C = \{\{P_1, P_4\}, \{P_2\}, \{P_3, P_4\}, \{P_2, P_4\}, \{P_3\}\}$. Its minimum subcollection C' is $\{\{P_1, P_4\}, \{P_2, P_4\}\}$ since $|C'| = 2$ is minimal and C' can distinguish all pairs in $\mathcal{A}$. The corresponding set of tag SNPs for C' is $\{S_1, S_4\}$.

Unfortunately, the minimum test collection problem has been proved to be NP-hard, which is a technical term that stands for a class of intractable problems for which no efficient algorithms have been found. Nevertheless, we may employ some algorithmic strategies to tackle NP-hard problems by finding near-optimal solutions; in practice, these solutions are often good enough. In the next section, we show that the tag SNP selection problem can be reformulated as the set-covering problem, which is well studied in the field of approximation algorithms. By this reformulation, a simple greedy method for the set-covering problem can be employed for solving the tag SNP selection problem. The algorithm may not always deliver an optimal solution, but we will show that the ratio of its solution to an optimal solution is bounded by a certain factor.

3   A reduction to the set-covering problem

We now recast the tag SNP selection problem as the *set-covering* problem. Given a universal set $\mathcal{U}$ and a collection C of subsets of $\mathcal{U}$, the set-covering problem is to find a minimum-size subcollection of C that covers all elements of $\mathcal{U}$. It is an abstraction of many naturally arising combinatorial problems, such as crew scheduling, committee forming, and service planning. For example, a universal set $\mathcal{U}$ could represent a set of skills required to perform a task. Each person in the candidate pool has certain skills in $\mathcal{U}$. The objective is to form a task force with as few people as possible so that all the required skills are owned by at least one person in the task force. In other words, we wish to recruit a minimum number of persons to *cover* all the requisite skills.

Recall that a haplotype block is represented by an $n \times h$ binary matrix M whose entries are either a blue square (representing a major allele) or a red square (representing a minor allele). To reformulate the tag SNP selection problem as a set-covering problem, let $\mathcal{U} = \{(j, k) \mid 1 \leq j < k \leq h\}$ be the set of all possible pairwise haplotype

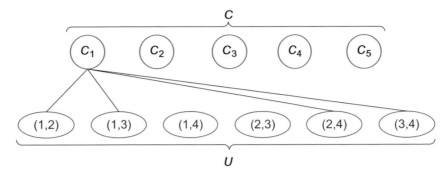

Figure 2.3 The elements covered by C_1, which correspond to the pairs of haplotype patterns distinguished by SNP S_1.

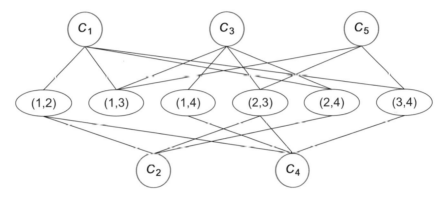

Figure 2.4 The elements covered by each C_i in C.

pattern indexes. Let $C = \{C_1, C_2, ..., C_n\}$, where $C_i = \{(j, k) \mid M[i, j] \neq M[i, k]$ and $1 \leq j < k \leq h\}$ stores the index pairs of haplotype patterns that SNP $S_i \in S$ can distinguish. We show that a subset of S forms a set of tag SNPs if and only if its corresponding subset of C covers all the elements in U. Each element in U represents a pair of haplotype patterns needed to be distinguished. If a subset of C covers all the elements in U, then its corresponding SNP subset of S forms a set of tag SNPs since all pairs of haplotype patterns can be distinguished. Conversely, if a subset of S forms a set of tag SNPs, it can distinguish all pairs of haplotype patterns, which yields that its corresponding subset of C covers all the elements in U.

Now let us consider the example given in Figure 2.1. We have four haplotype patterns, so the universal set U is $\{(1, 2), (1, 3), (1, 4), (2, 3), (2, 4), (3, 4)\}$, which contains all the elements to be covered. Since SNP S_1 can distinguish patterns P_1 and P_4 from P_2 and P_3, we set C_1 to be $\{(1, 2), (1, 3), (2, 4), (3, 4)\}$ (see Figure 2.3). SNP S_2 can distinguish pattern P_2 from P_1, P_3, and P_4, so we set C_2 to be $\{(1, 2), (2, 3), (2, 4)\}$. Figure 2.4 depicts the pairs of haplotype patterns distinguished by each SNP. As a

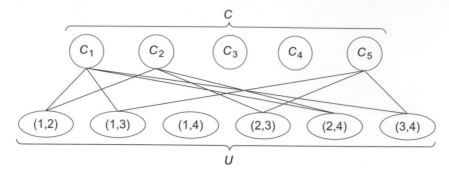

Figure 2.5 An invalid set cover. Element (1, 4) is not covered by C_1, C_2, and C_5.

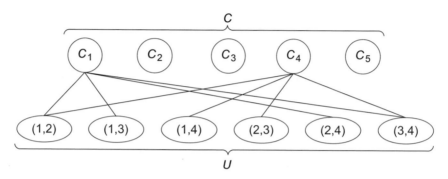

Figure 2.6 A valid set cover. All elements are covered by C_1 and C_4.

consequence, the collection $\mathcal{C}$ of subsets is $\{C_1, C_2, C_3, C_4, C_5\}$, where

$C_1 = \{(1, 2), (1, 3), (2, 4), (3, 4)\}$,

$C_2 = \{(1, 2), (2, 3), (2, 4)\}$,

$C_3 = \{(1, 3), (1, 4), (2, 3), (2, 4)\}$,

$C_4 = \{(1, 2), (1, 4), (2, 3), (3, 4)\}$, and

$C_5 = \{(1, 3), (2, 3), (3, 4)\}$.

As shown in Figure 2.2(b), S_1, S_2, and S_5 do not form a set of tag SNPs since they cannot distinguish the pair P_1 and P_4. In the corresponding set-covering instance, element (1, 4) is not covered by C_1, C_2, and C_5 (see Figure 2.5).

On the contrary, S_1 and S_4 form a set of tag SNPs since they can distinguish all pairs in $\mathcal{P}$. In the corresponding set-covering instance, each element is covered by at least one set in $\mathcal{C}$ (see Figure 2.6).

Now let us consider a greedy method for the set-covering problem. The greedy algorithm iteratively picks the set that covers the most remaining uncovered elements until all elements are covered. In the context of the tag SNP selection problem, the algorithm iteratively chooses the SNP that distinguishes the most remaining undistinguished pairs until all pairs of haplotype patterns are distinguished.

The SET-COVER-GREEDY algorithm takes as an input a universal set $\mathcal{U}$ and a colletion $\mathcal{C}$ of subsets of $\mathcal{U}$. Let $\mathcal{R}$ store the uncovered elements in $\mathcal{U}$, which is initially set to be $\mathcal{U}$ because all elements are uncovered at the beginning of the procedure. $\mathcal{C}'$ stores the selected sets and is initialized as an empty set. While $\mathcal{R}$ is not empty, we choose the set $C_i \in \mathcal{C}$ that can cover the most elements in $\mathcal{R}$. C_i would essentially cover the most uncovered elements in $\mathcal{U}$. Then we include C_i in $\mathcal{C}'$ and remove from $\mathcal{R}$ the elements that are covered by it. Repeat this procedure until all elements are covered.

Algorithm: SET-COVER-GREEDY $(\mathcal{U}, \mathcal{C})$

```
1   R ← U
2   C' ← φ
3   while R ≠ φ do
4         Select a set Cᵢ from C that maximizes |Cᵢ ∩ R|
5         C' ← C' ∪ {Cᵢ}
6         R ← R − Cᵢ
7   endwhile
8   return C'
```

The subcollection of sets, $\mathcal{C}'$, returned by the SET-COVER-GREEDY algorithm is valid as long as each element of $\mathcal{U}$ is covered by at least one set in $\mathcal{C}$. However, the size of $\mathcal{C}'$ may not always be minimal over all possible valid set covers. For example, let $\mathcal{U} = \{1, 2, 3, 4, 5, 6, 7, 8, 9\}$ and $\mathcal{C} = \{C_1, C_2, C_3\}$, where $C_1 = \{2, 3, 4, 5, 6, 7\}$, $C_2 = \{1, 2, 3, 4, 5\}$, and $C_3 = \{5, 6, 7, 8, 9\}$. The greedy algorithm will first pick C_1 since it covers the most elements. After this choice, it will also need to pick C_3 followed by C_2 to form a valid set cover. The resulting $\mathcal{C}'$ is $\{C_1, C_2, C_3\}$. However, for this instance, the minimum set cover is $\{C_2, C_3\}$ since all the elements in $\mathcal{U}$ can be covered by C_2 and C_3 without including C_1.

Although the SET-COVER-GREEDY algorithm may not always deliver the minimum set cover, its solution is in fact not too far away from an optimal one. Assume that C^* is an optimal set cover. Let $|X|$ denote the size (cardinality) of a given set X. We show that $|\mathcal{C}'|$ can be bounded by $|C^*|$ times a reasonable factor. To calculate the bound, we distribute the covering cost of a selected set to the elements it covers. For the example given in the previous paragraph, the covering order of the elements by the greedy algorithm might be $[2, 3, 4, 5, 6, 7, 8, 9, 1]$ because each of the elements in $\{2, 3, 4, 5, 6, 7\}$ is covered for the first time by C_1 in the first iteration, and then $\{8, 9\}$ by C_3 in the second iteration, and $\{1\}$ by C_2 in the last iteration. Since C_1 covers six uncovered elements, each element in $\{2, 3, 4, 5, 6, 7\}$ shares a cost of $1/6$. Similarly, each element in $\{8, 9\}$ shares a cost of $1/2$, and the element in $\{1\}$ shares a cost of 1. The covering cost for each element in order is $[1/6, 1/6, 1/6, 1/6, 1/6, 1/6, 1/2, 1/2, 1]$. Summing these costs would get 3, which is the size of the set cover, $\mathcal{C}'$, delivered by the greedy algorithm.

Let $[u_1, u_2, ..., u_{|\mathcal{U}|}]$ be the elements in the order in which they are covered by the SET-COVER-GREEDY algorithm. A key observation here is that the cost shared by u_k is at most $|\mathcal{C}^*|/(|\mathcal{U}| - k + 1)$ for $1 \leq k \leq |\mathcal{U}|$. In the iteration when u_k is covered, there are at least $|\mathcal{U}| - k + 1$ elements still uncovered, and certainly these uncovered elements can be covered by $\mathcal{C}^*$, which gives an average shared cost of $|\mathcal{C}^*|/(|\mathcal{U}| - k + 1)$. Since the greedy algorithm covers the most uncovered elements, its shared cost for each element in any iteration is the minimum. It follows that the cost shared by u_k is no more than $|\mathcal{C}^*|/(|\mathcal{U}| - k + 1)$. In other words, the covering cost for $[u_1, u_2, ..., u_{|\mathcal{U}|}]$ is no more than $[|\mathcal{C}^*|/|\mathcal{U}|, |\mathcal{C}^*|/(|\mathcal{U}| - 1), \ldots, |\mathcal{C}^*|]$, respectively. Since the size of $\mathcal{C}'$ is the sum of the costs shared by u_k for $1 \leq k \leq |\mathcal{U}|$, we have

$$|\mathcal{C}'| \leq (1 + \frac{1}{2} + \cdots + \frac{1}{|\mathcal{U}|}) \times |\mathcal{C}^*|. \tag{2.1}$$

The series $1 + 1/2 + \cdots + \frac{1}{|\mathcal{U}|}$ is called the harmonic series. It grows very slowly. For instance, it sums approximately to 2.929 when $|\mathcal{U}| = 10$, to 5.187 when $|\mathcal{U}| = 100$, to 7.485 when $|\mathcal{U}| = 1,000$, and to 14.393 when $|\mathcal{U}| = 1,000,000$. As a matter of fact, the harmonic series $1 + 1/2 + \cdots + 1/|\mathcal{U}|$ is bounded by $1 + \int_1^{|\mathcal{U}|} 1/x \, dx$, which yields the bound $\log_e |\mathcal{U}| + 1$. Furthermore, this factor is only a worst-case analysis, and the real approximation ratio could be even better.

 4 ## A reduction to the integer-programming problem

Linear programming is a general formulation of problems involving maximizing or minimizing a linear objective function subject to certain linear constraints. The following is a simple example.

Minimize $x_1 + x_2$

Subject to $x_1 + 2x_2 \geq 2$,

$$3x_1 + x_2 \geq 3,$$

$$x_1 \geq 0,$$

$$x_2 \geq 0.$$

Here the linear objective function is $x_1 + x_2$, and there are four linear constraints $x_1 + 2x_2 \geq 2$, $3x_1 + x_2 \geq 3$, $x_1 \geq 0$, and $x_2 \geq 0$. By graphing the constraints on the plane, we observe that the objective function $x_1 + x_2$ (lines with slope -1, see

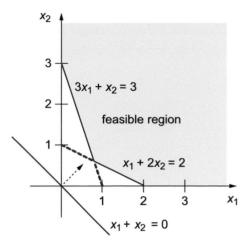

Figure 2.7 A feasible region defined by the four linear constraints $x_1 + 2x_2 \geq 2$, $3x_1 + x_2 \geq 3$, $x_1 \geq 0$, and $x_2 \geq 0$.

Figure 2.7) is minimized when $x_1 = 4/5$ and $x_2 - 3/5$, a corner point where the line $x_1 + 2x_2 = 2$ and the line $3x_1 + x_2 = 3$ intersect.

If we impose the extra constraints that the values of the variables are integers, then the problem is called *integer linear programming* or simply *integer programming*. In the above example, if both x_1 and x_2 are required to be integers, the problem becomes an integer-programming problem.

Now we show how to formulate the tag SNP selection problem as an integer-programming problem. Recall that we are given a haplotype block containing n SNPs and h haplotype patterns. Let us assign a variable x_i for each SNP $S_i \in \mathcal{S}$. Variable x_i is set to be 1 if SNP S_i is selected and set to be 0 otherwise. Define $D(P_j, P_k)$ as the set of SNPs which can distinguish between patterns P_j and P_k, $1 \leq j < k \leq h$. Each pair of patterns must be distinguished by at least one SNP. Therefore, for each set $D(P_j, P_k)$, at least one SNP has to be selected to distinguish between patterns P_j and P_k. The following integer program formulates the tag SNP selection problem whose objective is to minimize the number of selected SNPs.

Minimize $\displaystyle\sum_{i=1}^{n} x_i$

Subject to $\displaystyle\sum_{S_i \in D(P_j, P_k)} x_i \geq 1,$ for all $1 \leq j < k \leq h$,

$\qquad\qquad\quad x_i = 0 \text{ or } 1,$ for all $1 \leq i \leq n$.

In Figure 2.1, the pair P_1 and P_2 can be distinguished by SNPs S_1, S_2, and S_4. Thus, we have $D(P_1, P_2) = \{S_1, S_2, S_4\}$, which yields the constraint $x_1 + x_2 + x_4 \geq 1$. Similarly, $D(P_1, P_3) = \{S_1, S_3, S_5\}$, $D(P_1, P_4) = \{S_3, S_4\}$, $D(P_2, P_3) = \{S_2, S_3, S_4, S_5\}$,

$D(P_2, P_4) = \{S_1, S_2, S_3\}$, and $D(P_3, P_4) = \{S_1, S_4, S_5\}$. By examining all possible pairs of haplotype patterns, we obtain the following integer program for Figure 2.1.

Minimize $x_1 + x_2 + x_3 + x_4 + x_5$

Subject to $x_1 + x_2 + x_4 \geq 1,$

$$x_1 + x_3 + x_5 \geq 1,$$

$$x_3 + x_4 \geq 1,$$

$$x_2 + x_3 + x_4 + x_5 \geq 1,$$

$$x_1 + x_2 + x_3 \geq 1,$$

$$x_1 + x_4 + x_5 \geq 1,$$

$$x_1, x_2, x_3, x_4, x_5 = 0 \text{ or } 1.$$

In the above integer program, if we set x_1 and x_4 to be 1 and the rest of the x_i's to be 0, then all constraints are satisfied. Consequently, the set of SNPs S_1 and S_4 can distinguish all pairs of haplotype patterns and its size is minimized. However, if we set $x_1, x_2,$ and x_5 to be 1 and set x_3 and x_4 to be 0, then the third constraint $x_3 + x_4 \geq 1$ (for distinguishing P_1 and P_4) is not satisfied. This implies that SNPs $S_1, S_2,$ and S_5 do not form a set of tag SNPs since patterns P_1 and P_4 cannot be distinguished.

All variables x_is are required to be 0 or 1. Such an integral constraint makes the problem much harder to solve. In fact, both integer programming and 0–1 integer programming have been shown to be NP-hard as has the set-covering problem. It should be noted, however, that without the integral constraint, this integer program becomes a linear program in which variables can be fractional numbers, and fast algorithms, such as the simplex algorithm by George Dantzig, are available for solving it. A general strategy for solving the 0–1 integer-programming problems is thus to replace the integral constraint that each variable must be 0 or 1 by a weaker constraint that each variable be a number in the interval [0,1]. This process is referred to as a *linear-programming relaxation*. After the relaxation, the solution to the relaxed linear program may assign fractional values to the variables. For the above integer program, if we set $x_1, x_3,$ and x_4 to be 0.5 and set x_2 and x_5 to be 0, all the constraints can be satisfied except the last integral constraint. Several techniques, such as randomized rounding, can cope with the linear-programming relaxation to derive heuristic integral solutions for the original unrelaxed integer program. A widely used idea for rounding a fractional solution is to use their fractions as probabilities for rounding. The heuristic solutions may not be optimal, but often their quality can be assured by a logarithmic approximation ratio.

DISCUSSION

In this chapter, we reformulate the tag SNP selection problem as two well-known optimization problems in computer science – the set-covering problem and the integer-programming problem. Both problems are hard to solve, yet efficient approximation algorithms can be used to find their near-optimal solutions.

In reality, some tag SNPs may be missing, and we may fail to distinguish two haplotype patterns due to the ambiguity caused by missing data. To conquer this, either we genotype a larger set of tag SNPs for tolerating missing data, or re-genotype some auxiliary tag SNPs to resolve the ambiguity on the fly. We can handle these extensions by modifying the formulations.

It should be noted that selecting tag SNPs *within* a haplotype block is only one of the models for selecting tag SNPs. An alternative is to identify a minimum set of bins, each of which contains highly correlated SNPs. Such an approach identifies a minimum set of tag SNPs that can represent all other SNPs which might be far apart, whereas the block-based methods considered in this chapter are mainly focused on representing all other SNPs in a short contiguous region. Furthermore, some methods may assume that the number of tag SNPs is specified as an input parameter and identify tag SNPs which can reconstruct the haplotype of an unknown sample with high accuracy.

QUESTIONS

(1) Let $\mathcal{U} = \{1, 2, 3, 4, 5, 6, 7, 8, 9\}$ and $\mathcal{C} = \{C_1, C_2, C_3, C_4, C_5\}$, where $C_1 = \{2, 3, 4, 5, 6, 7\}$, $C_2 = \{1, 2, 3, 4\}$, $C_3 = \{6, 7, 8, 9\}$, $C_4 = \{1, 3, 5, 7, 9\}$, and $C_5 = \{2, 4, 6, 8\}$. Find a minimum-size subcollection of $\mathcal{C}$ that covers every element of $\mathcal{U}$.

(2) Suppose that a set of skills is needed to accomplish a given task, and we have a list of people, each with their own skills. Our objective is to form a task force with as few people as possible such that for each requisite skill, we can always find someone in the task force having that skill. Formulate this problem as a *set-covering* problem.

(3) Solve the following linear program.

Minimize $x_1 + x_2$

Subject to $x_1 + 2x_2 \geq 4$,

$$3x_1 + x_2 \geq 6,$$

$$x_1 \geq 0,$$

$$x_2 \geq 0.$$

BIBLIOGRAPHIC NOTES AND FURTHER READING

This chapter presents two algorithmic approaches for solving the tag SNP selection problem. Readers can refer to algorithm textbooks for more algorithmic details. For instance, the algorithm book (or "The White Book") by Cormen *et al.* [1] is a comprehensive reference of data structures and algorithms with a solid mathematical and theoretical foundation. The minimum test collection problem was shown to be NP-hard via a reduction from the three-dimensional matching problem by Garey and Johnson [2].

An early review paper by Brookes [3] provides a good orientation for readers who are not familiar with SNPs. Millions of SNPs have been identified, and these data are now publicly available [4–6]. The Phase II HapMap has characterized over 3.1 million human SNPs genotyped in 270 individuals from 4 geographically diverse populations [5]. The dbSNP database is a public-domain archive for a broad collection of SNPs [6].

In a large-scale study of human Chromosome 21, Patil *et al.* [7] developed a greedy algorithm to partition the haplotypes into 4,135 blocks with 4,563 tag SNPs. It was later refined by Zhang *et al.* [8, 9] and Chang *et al.* [10].

REFERENCES

[1] T. H. Cormen, C. E. Leiserson, R. L. Rivest, and C. Stein. *Introduction to Algorithms*, 3rd edn. The MIT Press, Cambridge, MA, 2009.

[2] M. R. Garey and D. S. Johnson. *Computers and Intractability: A Guide to the Theory of NP-completeness*. W. H. Freeman and Co., New York, 1979.

[3] A. J. Brookes. The essence of SNPs. *Gene*, 234:177–186, 1999.

[4] D. A. Hinds, L. L. Stuve, G. B. Nilsen, E. Halperin, E. Eskin, D. G. Ballinger, K. A. Frazer, and D. R. Cox. Whole-genome patterns of common DNA variation in three human populations. *Science*, 307:1072–1079, 2005.

[5] The International HapMap Consortium. A second generation human haplotype map of over 3.1 million SNPs. *Nature*, 449:851–861, 2007.

[6] S. T. Sherry, M. H. Ward, M. Kholodov, J. Baker, L. Phan, E. M. Smigielski, and K. Sirotkin. dbSNP: The NCBI database of genetic variation. *Nucl. Acids Res.*, 29: 308–311, 2001.

[7] N. Patil, A. J. Berno, D. A. Hinds, W. A. Barrett, J. M. Doshi, C. R. Hacker, C. R. Kautzer, D. H. Lee, C. Marjoribanks, D. P. McDonough, B. T. Nguyen, M. C. Norris, J. B. Sheehan, N. Shen, D. Stern, R. P. Stokowski, D. J. Thomas, M. O. Trulson, K. R. Vyas, K. A. Frazer, S. P. Fodor, and D. R. Cox. Blocks of limited haplotype diversity revealed by high-resolution scanning of human chromosome 21. *Science*, 294:1719–1723, 2001.

[8] K. Zhang, F. Sun, M. S. Waterman, and T. Chen. Haplotype block partition with limited resources and applications to human chromosome 21 haplotype data. *Am. J. Hum. Genet.*, 73:63–73, 2003.

[9] K. Zhang, Z. S. Qin, J. S. Liu, T. Chen, M. S. Waterman, and F. Sun. Haplotype block partition and tag SNP selection using genotype data and their applications to association studies. *Genome Res.*, 14:908–916, 2004.

[10] C.-J. Chang, Y.-T. Huang, and K.-M. Chao. A greedier approach for finding tag SNPs. *Bioinformatics*, 22:685–691, 2006.

CHAPTER THREE

Genome reconstruction: a puzzle with a billion pieces

Phillip E. C. Compeau and Pavel A. Pevzner

While we can read a book one letter at a time, biologists still lack the ability to read a DNA sequence one nucleotide at a time. Instead, they can identify short fragments (approximately 100 nucleotides long) called *reads*; however, they do not know where these reads are located within the genome. Thus, assembling a genome from reads is like putting together a giant puzzle with a billion pieces, a formidable mathematical problem. We introduce some of the fascinating history underlying both the mathematical and the biological sides of DNA sequencing.

 Introduction to DNA sequencing

1.1 DNA sequencing and the overlap puzzle

Imagine that every copy of a newspaper has been stacked inside a wooden chest. Now imagine that chest being detonated. We will ask you to further suspend your disbelief and assume that the newspapers are not all incinerated, as would assuredly happen in real life, but rather that they explode cartoonishly into tiny pieces of confetti (Figure 3.1). We will concern ourselves only with the immediate journalistic problem at hand: what did the newspaper say?

This "newspaper problem" becomes intellectually stimulating when we realize that it does not simply reduce to gluing the remnants of newspaper as we would fit together the disjoint pieces of a jigsaw puzzle. One reason why this is the case is that we

Bioinformatics for Biologists, ed. P. Pevzner and R. Shamir. Published by Cambridge University Press.
© Cambridge University Press 2011.

stack of *NY Times*,
June 27, 2000

stack of *NY Times*, June 27,
2000 on a pile of dynamite

this is just hypothetical

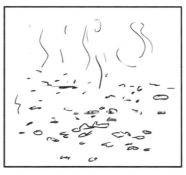

so, what did the June 27, 2000
NY Times say?

Figure 3.1 The exploding newspapers.

have probably lost some information from each copy (the content that was blown to smithereens). However, we can also see that because the chest contained many identical copies of the same newspaper, different shreds of paper may overlap and therefore contain some of the same information. The newspaper problem therefore induces what we will call an *overlap puzzle*.

We reiterate that our analogy of exploding newspapers is far-fetched, but the newspaper problem nevertheless captures the essence of fragment assembly in DNA sequencing. The technology for "reading" an entire genome nucleotide by nucleotide, like reading a newspaper one letter at a time, remains unknown. At the same time, researchers can indirectly interpret short sequences of DNA, which are referred to as *reads*; the most popular modern technology produces reads that are only 100 nucleotides long (Figure 3.2). The idea behind DNA sequencing, then, is to generate many reads from multiple copies of the same genome, which results in a giant overlap puzzle. For instance, a three billion-nucleotide mammalian genome requires an overlap puzzle with a billion (overlapping) pieces, the largest such puzzle ever assembled.

The problem of genome sequencing therefore reduces to *read generation* (a biological problem) and *fragment assembly* (an algorithmic problem). Read generation

Multiple Genome Copies

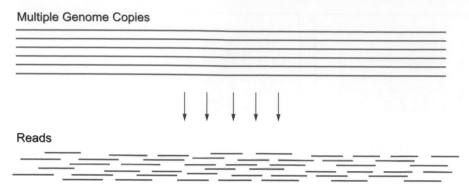

Reads

Figure 3.2 In DNA sequencing, multiple (typically more than a billion) copies of a genome are broken in random locations to generate much shorter reads.

has its own long and tangled history that dates to the 1970s, when Walter Gilbert and Fred Sanger won the Nobel Prize for inventing the first read generation technology. In the early 1990s, modern DNA sequencing machines hit the market and the era of high-throughput DNA sequencing began. In 2000, a few hundred such machines working around the clock for over a year eventually generated enough reads to enable the fragment assembly of the human genome, which was completed within a few months by some of the world's most powerful supercomputers.

1.2 Complications of fragment assembly

Although we shall discuss read generation in some detail at the end of the chapter, our primary target is the computational problem of fragment assembly, or using the generated reads to infer the original genome.

We begin by noting that although we have seen that both the newspaper problem and fragment assembly reduce to solving an overlap puzzle, fragment assembly is substantially more difficult for several reasons, and not simply because of the sheer scale of reconstructing a genome from a billion reads. First, keep in mind that a newspaper is written in some understood language, whose rules will provide us with context clues as to how different shreds of paper may or may not be connected, regardless of whether these shreds overlap (see Figure 3.3a). Yet the rules for the "language" of DNA still mostly elude biologists, and so it is practically impossible to determine how two non-overlapping reads might be connected.

A second complication of fragment assembly is that the underlying nucleotide "alphabet" for DNA contains only four letters: A, T, G, and C. Working with a small

(a)

e murder occurred at approximately 5:2

oodie , appr   1

ce have not yet named any suspects, alt

nation is welc   e ca

(b)

mentalists ha ve ked low levels of oz

e world's mos

zone as a contributing facto

t th   as a c

(c)

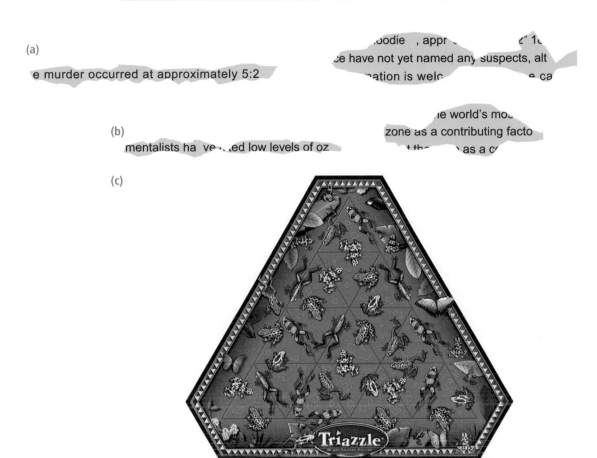

(d) TAGGCCATGTCAGATG
CATGTCAGATGCGTAG

Figure 3.3 Complications of fragment assembly. (a) In the newspaper assembly problem, we can see that even though these two shreds do not overlap they are nevertheless probably connected, because we know that "murder" and "suspect" are highly correlated words. (b) In the newspaper problem, "oz" and "zone" are likely the remnants of "ozone," and we can connect these two shreds even though they overlap in just one letter. In the DNA assembly problem, with only four letters in the underlying alphabet, such clues are not available. (c) Repeated regions complicate assembly, as demonstrated by the Triazzle®. Note that every frog in the Triazzle appears at least three times. (d) DNA sequencing machines are not perfect. Here, the red 'T' was incorrectly sequenced and should be a 'C'; this mistake of only one nucleotide may cause these two reads to be interpreted as overlapping when they are not.

alphabet actually complicates the reconstruction of the original sequence, because we will observe a greater amount of fragment overlap that is purely attributable to randomness. See Figure 3.3b.

Third, any DNA sequence contains a significant number of "conserved regions," or information that is repeated many times with minor changes. For example, the approximately 300-nucleotide long *Alu* sequence occurs over a million times in the human genome, with only a few nucleotides changed each time due to insertions, deletions, or substitutions. Therefore, for any one particular fragment, it can become difficult to identify the specific conserved region to which it belongs within the genome. An appropriate illustration of this difficulty is the once-popular Triazzle® puzzle. Even though a Triazzle is a jigsaw puzzle with only 16 pieces, it contains identical figures shared by multiple pieces, making a Triazzle much more difficult than an ordinary puzzle. See Figure 3.3c.

Last but not least, modern sequencing machines are not perfect, and the reads they generate often contain errors; thus, reads which do not overlap in the genome may be incorrectly interpreted as overlapping (see Figure 3.3d).

With the pitfalls of DNA sequencing established, we next must introduce a rigorous mathematical framework in order to attack fragment assembly.

2   The mathematics of DNA sequencing

2.1   Historical motivation

Before we jump headlong into mathematics, let us take two historical detours in order to provide our mathematical discussion with some necessary context. We begin in the eighteenth century and the Prussian city of Königsberg.[1] Königsberg was formed of opposing banks of the Pregel River, as well as two river islands; joining these four parts of the city were seven bridges (see Figure 3.4a). Now, Königsberg's residents enjoyed taking walks, and they were curious if they could stroll through the city, cross each of the seven bridges exactly once, and return back to their starting point. Their quandary became known as the "Königsberg Bridge Problem," and it was solved once and for all in 1735 by the great Swiss mathematician Leonhard Euler[2] (Figure 3.14a). Euler's result, which we discuss below, is profound because it applies not only to the bridges of Königsberg, but in fact to *any possible* network of bridges.

---

[1] Present-day Kaliningrad, Russia.
[2] Pronounced "oiler."

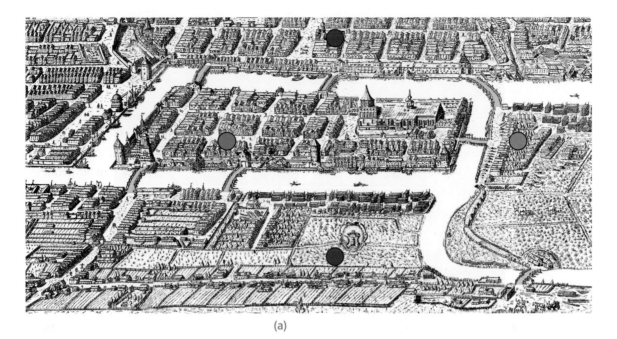

(a)

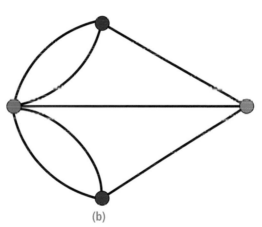

(b)

Figure 3.4 (a) Map of old Königsberg, adapted from Joachim Bering's 1613 illustration. The seven bridges have been highlighted to make them easier to see. (b) The "Königsberg Bridge Graph," formed by compressing each of four land areas to a vertex and representing each of the seven bridges as an edge.

Our second historical detour takes place in Dublin, with the creation in 1857 of the Icosian Game by the Irish mathematician William Hamilton (Figure 3.14b). This "game," which even by contemporary standards could not possibly have been very enjoyable, consisted of a wooden board with 20 pegholes and some lines connecting the holes, as well as 20 numbered pegs (see Figure 3.5a). The game's objective was to

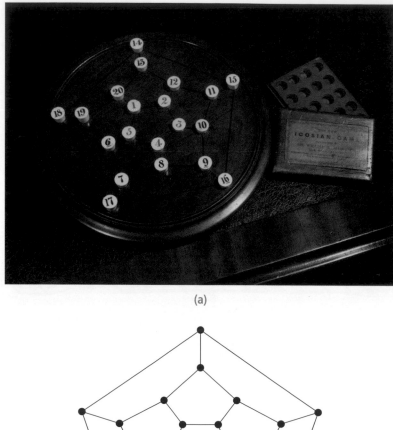

(a)

(b)

Figure 3.5 (a) The Icosian Game, along with (b) the corresponding graph.

place the numbered pegs in the holes in such a way that Peg 1 would be connected by a line on the board to Peg 2, which would in turn be connected by a line to Peg 3, and so on, until finally Peg 20 would be connected by a line back to Peg 1. In other words, if we follow the lines on the board from peg to peg in ascending order, we reach every peg exactly once and then arrive back at our starting peg.

2.2 Graphs

With these two historical asides complete, we are ready to define a "graph" simply as a collection of "vertices" and a collection of "edges," for which each edge pairs two vertices. The abstractness of this definition may be initially offputting, so we quickly clarify that we can always think about a graph as a network or even a map, in which the vertices are cities and the edges are roads connecting the vertices.

The benefit of providing ourselves with such a general definition is that "graph theory," or the branch of mathematics concerned with the study of graphs, can be applied to many different types of problems. Applications of graph theory certainly include road and communications networks; however, graph theory also extends to less obvious examples, such as understanding the spread of disease or modeling the webpage connectivity of the internet.

In particular, graph theory applies to both our historical examples. In the Königsberg Bridge Problem, we obtain a graph K by assigning each of the four sectors of the city to a vertex and then connecting two given vertices (sectors) with one edge for every bridge that connects the two sectors (see Figure 3.4b). As for the Icosian Game, we obtain a graph I by representing each peghole by a vertex and then turning the lines that connect pegholes into edges that connect the corresponding vertices (see Figure 3.5b).

2.3 Eulerian and Hamiltonian cycles

Now we will generalize our two historical problems to *all* graphs. So assume that we are given any graph, which we call G, and consider an ant standing on a vertex of G. Just as the residents of Königsberg walk between the different parts of the city via bridges, the ant may walk along edges from vertex to vertex. If the ant returns to where it started, the result of its walk is a "cycle" of G. We will ask two questions about the cycles of G:

1　Does there exist a cycle of G in which the ant walks along each *edge* exactly once?
2　Does there exist a cycle of G in which the ant travels to every *vertex* exactly once?

Fittingly, Question 1 is called the *Eulerian Cycle Problem* (ECP): note that solving the ECP when our graph is K corresponds to solving the Königsberg Bridge Problem.[3] We therefore define an "Eulerian cycle" in a graph G as a cycle of G which traverses every edge in G once and only once.

The second question is called the *Hamiltonian Cycle Problem* (HCP), because when the underlying graph is I, we can solve the HCP by "winning" Hamilton's Icosian

---

[3]　We call your attention to what we mean by "solving" an ECP: because a solution corresponds to a "Yes" or "No" answer to Question 1, the ECP is considered solved when we have provided either an Eulerian cycle in the graph, or definitive proof that no such cycle exists.

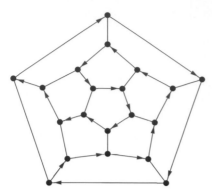

Figure 3.6 A Hamiltonian cycle in the graph I, which provides a solution to Hamilton's Icosian Game.

game (see Figure 3.6). Naturally then, a "Hamiltonian cycle" in a graph G is a cycle of G which travels to each vertex once and only once.

Finally, we define a "connected" graph as one in which an ant standing on any vertex can reach any other vertex by walking through the graph. For our purposes, it only makes sense to study the ECP and HCP for connected graphs. This is because a graph that is not connected automatically contains neither an Eulerian nor a Hamiltonian cycle, in which case the ECP and HCP are both trivial questions. Therefore, every graph in this chapter will be assumed to be connected.

2.4   Euler's Theorem

The decision to extend our historical problems to questions about graphs in general may be confusing, but this decision turns out to be key. While the ECP and HCP are superficially very similar, computer scientists have discovered that the two problems have a fundamentally different algorithmic fate: the ECP can be solved quickly even for huge graphs, while an efficient algorithm for solving the HCP for large graphs remains unknown and may not even exist.

First, we will discuss the ECP. Recall that when we introduced the Königsberg Bridge Problem, we mentioned that Euler's solution could be extended to any possible collection of bridges. What we meant by this was that Euler's solution actually provided a simple condition to solve the ECP for any graph.

Before stating Euler's result, we first need a definition. For a vertex v in a graph G, define the *degree* of v to be the number of edges connecting v to other vertices. For example, for the Königsberg graph K in Figure 3.4b, the top, bottom, and right vertices all have degree 3, while the left vertex (representing the main island of Königsberg) has degree 5. In particular, observe that since a vertex v in K represents a sector of the city, the degree of v is equal to the number of bridges connecting that sector to other parts of the city.

Theorem (Euler's Theorem I). *An equivalent condition to a graph G having an Eulerian cycle is that the degree of every vertex of G is even.*

We call your attention to what two conditions being "equivalent" really means. In a sense, it means that if one is true, then the other is necessarily true as well (and vice versa). In the case of Euler's Theorem, the equivalence of the degree condition and the cycle condition is profound because it implies that for a given graph G, we can determine if G has an Eulerian cycle *without ever having to draw any cycles*. Instead, we simply need to check the degree of every vertex, a relatively simple computational task (even for a large graph).

Let us notice that Euler's Theorem immediately solves the Königsberg Bridge Problem. We have seen above that it is not the case that every vertex of K has even degree. Therefore, K does not contain an Eulerian cycle, and so we conclude that the walk for which the citizens of Königsberg had yearned does not exist.

Since the eighteenth century, much has changed in the layout of Königsberg, and it just so happens that the same graph drawn today for the present-day city of Kaliningrad still does not contain an Eulerian cycle (see Figure 3.7); however, this graph does contain an *Eulerian path*, which means that a denizen of Kaliningrad can cross every bridge exactly once, but cannot do so and return to where he started. Thus, the citizens of Kaliningrad finally achieved at least a small part of the goal set by the citizens of Königsberg. Yet it is also worth noting that strolling around Kaliningrad is not as pleasant as it would have been in 1735, since the beautiful old Königsberg was ravaged by the combination of Allied bombing in 1944 and dreadful Soviet architecture in the years following World War II.

2.5 Euler's Theorem for directed graphs

We need a slightly reworked statement of Euler's Theorem in order to handle the impending application of graph theory to fragment assembly. So first assume that we instead have a "directed graph," which is simply a graph in which all edges are provided with an orientation, so that an edge connecting v to w is not the same as an edge connecting w to v. We might like to think of a directed graph as a network in which all the edges are "one-way streets," in which case our original undirected graph is a network in which all the edges are "two-way streets." Accordingly, an Eulerian cycle in a directed graph G is simply an Eulerian cycle which always travels down the streets in the correct direction. A Hamiltonian cycle in G is defined analogously. See Figure 3.8.

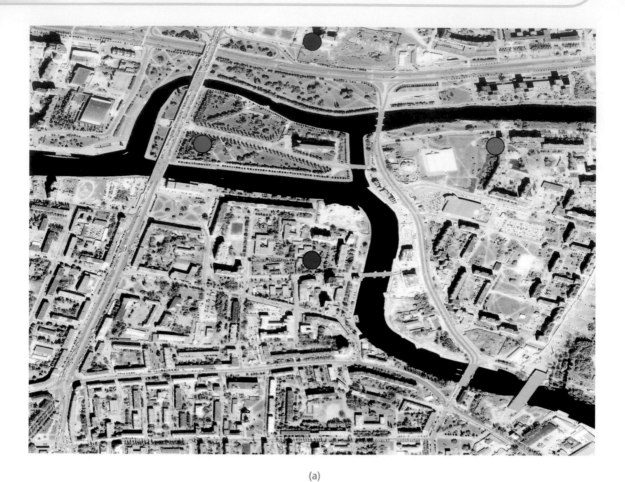

(a)

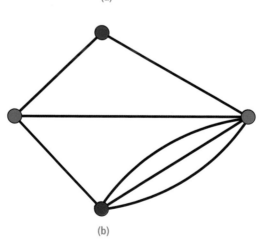

(b)

Figure 3.7 (a) Satellite map of present-day Kaliningrad, with its bridges highlighted. (b) The graph for "Kaliningrad Bridge Problem." Here is a challenge question: where could the city council of Kaliningrad construct new bridges so that the resulting graph will contain an Eulerian cycle?

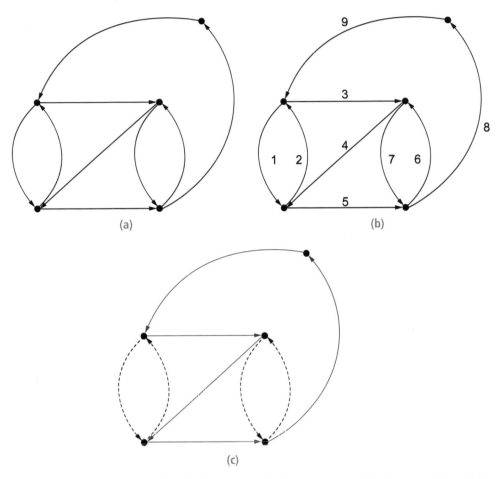

Figure 3.8 (a) A basic example of a directed graph. The arrows provide the orientations of the edges, so that we can see the directions of the "one-way streets." (b) An illustration of an Eulerian cycle in the directed graph. The edges of the graph are numbered to indicate their order in the cycle. (c) An illustration of a Hamiltonian cycle (red edges) in the directed graph.

For any vertex v in a directed graph G, we define the "indegree" of v as the number of edges leading into v and the "outdegree" of v as the number of edges leading out from v. We are now ready to state the application of Euler's result to directed graphs.

Theorem (Euler's Theorem II). *An equivalent condition to a directed graph G having an Eulerian cycle is that for every vertex v in G, the indegree and outdegree of v are equal.*

A proof of Euler's Theorem is provided at the end of the chapter, as well as a discussion of how we can find an Eulerian cycle "quickly" in the parlance of computers. The key point is that we do not have to test every possible cycle in a directed graph

G in order to determine whether G contains an Eulerian cycle. We need only find the indegree and outdegree of each vertex. If for each vertex, the indegree and outdegree match, then finding an Eulerian cycle will be easy; on the other hand, if there is *any* vertex for which the indegree and outdegree do not match, then we know that finding an Eulerian cycle is impossible.

2.6   Tractable vs. intractable problems

Inspired by Euler's Theorem, we should wonder whether there exists such a simple result governing a quick solution of the HCP. Yet although it is easy to win the Icosian Game, a solution to the HCP for an *arbitrary* graph has remained hidden.

The key challenge is that while we are guided by Euler's Theorem in solving the ECP, an analogous simple condition for the HCP remains unknown. Of course, you could always employ the method of "brute force" to solve the HCP, in which you have a computer explore all walks through the graph and report back if it finds a Hamiltonian cycle. This method is simple enough to understand, yet think about a huge graph that does not contain a Hamiltonian cycle. For this graph, the computer would have to test *every* walk through the graph before reporting back that no Hamiltonian cycle exists. The cataclysmic problem with this method is that for the average graph on just a thousand vertices, there are more walks through the graph than there are atoms in the universe!

The HCP was one of the first algorithmic problems that eluded all attempts to solve it by some of the world's most brilliant researchers. After years of fruitless effort, computer scientists began to wonder whether the HCP is *intractable*, or in other words that their failure to find a quick algorithm was not attributable to a lack of cleverness, but rather because an efficient algorithm for solving the HCP simply does not exist. Moreover, in the 1970s, computer scientists discovered thousands more algorithmic problems with the same fate as the HCP: while they are superficially simple, no one has been able to find efficient algorithms for solving them. A large subset of these problems, along with the HCP, are now collectively known as "*NP-complete*."

What has only exacerbated the frustration caused by the failure to find a simplifying condition for the HCP is that while all the NP-complete problems are different, they turn out to be *equivalent* to each other: if you find a fast algorithm for one of them, you will be able to automatically find a fast algorithm for all of them! The problem of efficiently solving NP-complete problems (or finally proving that they are intractable) is so fundamental to both computer science and mathematics that it was named on the list of "Millennium Problems" by the Clay Mathematics Institute in the year 2000: find an efficient algorithm for *any* NP-complete problem, or show that *any* NP-complete problem is in fact intractable, and this institute will award you a prize of one million dollars.

Henceforth, we will simply think of the ECP as "easy" and the HCP as "difficult." Keep this distinction between the two problems in mind, as it will shortly become critical.

 3 ## From Euler and Hamilton to genome assembly

3.1 Genome assembly as a Hamiltonian cycle problem

Equipped with all the mathematics that we need, we return to fragment assembly. Having generated all our reads, we will henceforth make three simplifying assumptions about the problem at hand in order to streamline our work:

1 The genome we are reconstructing is cyclic.
2 Every read has the same length l (a string of l nucleotides is called an "l-mer").
3 All possible substrings of length l occurring in our genome have been generated as reads.
4 The reads have been generated without any errors.

It turns out that we can relax each of these assumptions, but the resulting solution to fragment assembly winds up being far more technical than what is suitable for this text.

In the early days of DNA sequencing, the following idea for fragment assembly was proposed. Construct a graph H by forming a vertex for every read (l-mer); we connect l-mer R_1 to l-mer R_2 by a directed edge if the string formed by the final $l-1$ characters of R_1 (called the *suffix* of R_1) matches the string formed by the first $l-1$ characters of R_2 (called the *prefix* of R_2). For instance, in the case $l=5$, we would connect GGCAT to GCATC by a directed edge, but not vice versa. An example of such a graph H is provided in Figure 3.9a.

Now, consider a cycle in H. It will begin with an l-mer R_1, and then proceed along a directed edge to a different l-mer R_2; let us think of walking along this edge as beginning with R_1 and tacking on the lone non-overlapping character from R_2 in order to form a "superstring" S of length $l+1$. To continue our above example, if we walk from GGCAT to GCATC, then our superstring S will be GGCATC. Observe that the first l characters of S will be R_1, and the final l characters of S will be R_2. At each new vertex that we reach, we append one new character to S and notice that the final l characters of our superstring will represent the read at the present vertex. At the end of the cycle, our (cyclic) superstring S will therefore contain every l-mer that we reached

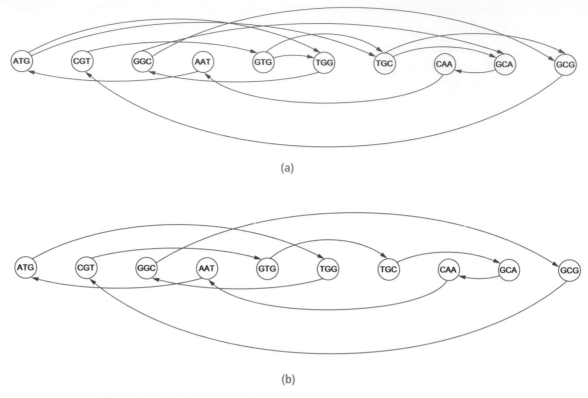

Figure 3.9 (a) The graph H for the set of 3-mers ATG, CGT, GGC, AAT, GTG, TGG, TGC, CAA, GCA, and GCG. (b) A Hamiltonian cycle in H. What is the cyclic "superstring" DNA sequence corresponding to this Hamiltonian cycle?

along the way. Extending this reasoning, a *Hamiltonian* cycle in H, which travels to every vertex in H, must correspond to a superstring of nucleotides which contains every one of our l-mers. Furthermore, *every* substring of length l in S will correspond to an l-mer, so S is as short as possible and therefore provides us with a candidate DNA sequence! See Figure 3.9b.

The problem with this method is that although it is elegant, it nevertheless rests upon solving the HCP, so that it is impractical unless our graph H is small. Therefore, this method is unsuitable for the graph obtained from a genome, which may have billions of vertices.

3.2 Fragment assembly as an Eulerian cycle problem

Yet all is not lost. Instead of assigning each read to a vertex, let us make the admittedly counterintuitive decision to assign each read to an *edge*. To this end, consider all prefixes and suffixes of all reads. Note that different reads may share suffixes and

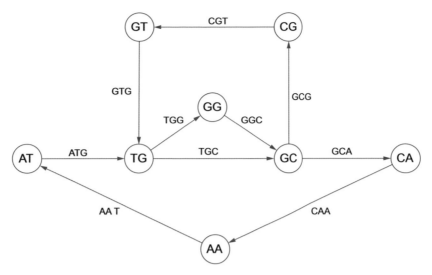

Figure 3.10 The graph E for the same set of 3-mers as in Figure 3.9. Can you find an Eulerian cycle in E? What is the "superstring" DNA sequence corresponding to your Eulerian cycle?

prefixes; for example, reads CAGC and CAGT of length 4 share the prefix CAG. We construct a graph E with each *distinct* prefix or suffix represented by a vertex; connect an $(l-1)$-mer A to an $(l-1)$-mer B via a directed edge if there exists a read whose prefix is A and whose suffix is B. See Figure 3.10 for an example using the same set of reads from Figure 3.9.

Here, then, is the critical question: what does a cycle in E represent? Once again, imagine that you are an ant starting at some vertex of E and that you walk along a directed edge to another vertex. As with H, the result is the creation of a superstring S by tacking on the non-overlapping characters from the second vertex to those of the first. However, in this case S is just the read representing the edge connecting the two vertices. Note that in Figure 3.10, we have labeled each edge with the appropriate 3-mer.

This process repeats itself as the ant walks through E; with each new edge, we append one additional nucleotide to the superstring S, but we also gain one additional read. Therefore, an Eulerian cycle in E will induce a (cyclic) superstring S that contains all our reads with maximum overlap, and so S is also a candidate DNA sequence. Yet in contrast to our above graph H, we have no computational troubles: by Euler's Theorem, the ECP is easy to solve. Hence we have reduced fragment assembly to an easily solved computational problem!

Nevertheless, the reduction of fragment assembly to solving the ECP on our graph E carries one vital concern: how do we know from the start that E even *contains* an Eulerian cycle? After all, E was constructed with no thought as to whether it might

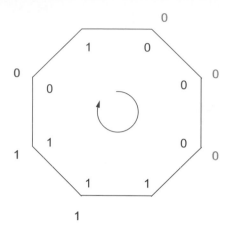

Figure 3.11 The minimal superstring problem. Here we show the circular superstring 00011101 along with illustrations of the location of the 3-digit binary numbers 000 and 110. Note that we can locate all 3-digit binary numbers in the superstring with no repeats, so 00011101 is as short as possible.

have an Eulerian cycle; if it does not, then the construction of E was simply nonsense, and the process of creating a superstring by concatenating nucleotides as we progress through E will not result in a candidate DNA sequence. In order to resolve this potential quagmire, we will tell a third and final mathematical tale.

3.3 De Bruijn graphs

In 1946, the Dutch mathematician Nicolaas de Bruijn[4] (see Figure 3.14c) was interested in the problem of designing a circular superstring of minimal length that contains all possible l-digit binary numbers as substrings. For example, the circular string 00011101 contains all 3-digit binary numbers: 000, 001, 010, 011, 100, 101, 110, and 111. It is easy to see that 00011101 is the shortest such superstring, because it does not contain any "extra" digits, meaning that each 3-digit substring of 00011101 is the unique occurrence of one of the 3-digit binary numbers listed above. See Figure 3.11.

De Bruijn analyzed a specific class of graphs, defined as follows. Consider an alphabet of n characters, as well as some fixed number l. Form all n^{l-1} possible "words" of length $l-1$, where a word is just a string of $l-1$ letters from our alphabet.[5] De Bruijn constructed a graph $B(n, l)$ (now known as the *de Bruijn graph*[6]) whose vertices

---

[4] In contrast to Euler, the anglophone will find the pronunciation of "de Bruijn" very difficult: it is similar to "brine," except with a slight 'r' sound between the 'i' and the 'n.'

[5] There are n^{l-1} such words because there are n choices for the first letter, n choices for the second letter, and so on. Since there are $l-1$ letters to choose, we wind up with n^{l-1} total possibilities.

[6] This nomenclature is a bit cruel to the British mathematician I. J. Good, who independently discovered de Bruijn graphs.

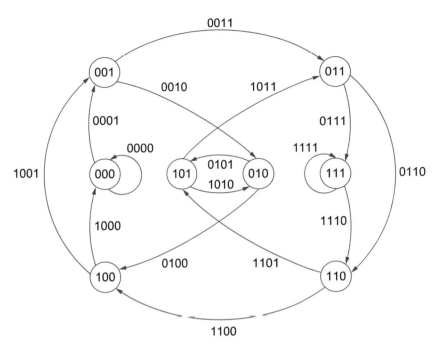

Figure 3.12 The de Bruijn graph $B(2, 4)$, where our 2-character "alphabet" is composed of just the digits 0 and 1. Observe that by Euler's Theorem, this graph must have an Eulerian cycle; we will find such a cycle for this graph in Figure 3.19.

are all n^{l-1} words of length $l - 1$; a directed edge connects word w_1 to word w_2 if there exists an l-letter word W whose prefix is w_1 and whose suffix is w_2. See Figure 3.12.

The crucial property shared by all de Bruijn graphs is that every one of them will always contain an Eulerian cycle. For example, in Figure 3.12 we can see that there are two edges entering every vertex and two edges leaving every vertex of $B(2, 4)$, implying that it has an Eulerian cycle. To see why the same is true for *any* de Bruijn graph $B(n, l)$, consider a vertex w corresponding to a word of length $l - 1$. There exist n words of length l whose prefix is w (each such word is obtained by adding one of n letters to the end of w) and thus the outdegree of each vertex in $B(n, l)$ is n. Similarly, there exist n words of length l whose suffix is w (each such word is obtained by adding one of n letters to the beginning of w) and thus the indegree of each vertex in $B(n, l)$ is also n. Hence every vertex of $B(n, l)$ has indegree and outdegree both equal to n, and so Euler's Theorem implies that $B(n, l)$ must have an Eulerian cycle.

The biological connection arises when we realize that our graph E above will be contained in the de Bruijn graph $B(4, l)$, because whereas the vertices of E are all $(l - 1)$-mers occurring as prefixes or suffixes of our reads, the vertices of $B(4, l)$ are

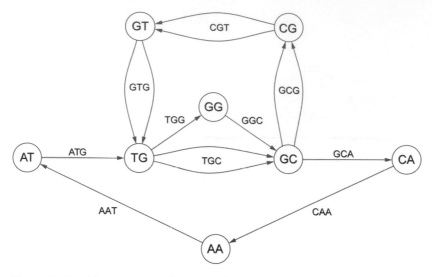

Figure 3.13 This more general version of the graph from Figure 3.10 allows for the case that the same read occurs in more than one location in the genome. The good news is that this generalization does not make the problem any more difficult to solve: an Eulerian cycle in this graph will still correspond to a candidate DNA sequence.

all *possible* $(l-1)$-mers. Furthermore, it can be demonstrated that E itself has an Eulerian cycle!

3.4 Read multiplicities and further complications

Imagine for a moment that our genome is ATGCATGC. Then we will obtain four reads of length 3: ATG, TGC, GCA, and CAT; however, this might lead us to reconstruct the genome as ATGC. The problem is that each of these reads actually occurs *twice* in the original genome. Therefore, we will need to adjust genome reconstruction so that we not only find all l-mers occurring as reads, but we also find how many times each such l-mer occurs in the genome, called its "l-mer multiplicity." The good news is that we can still handle fragment assembly in the case l-mer multiplicities are known.

We simply use the same graph E, except that if the multiplicity of an l-mer is k, we will connect its prefix to its suffix via k edges (instead of just one). Continuing our ongoing example from Figure 3.10, if during read generation we discover that each of the four 3-mers TGC, GCG, CGT, and GTG has multiplicity 2, and that each of the six 3-mers ATG, TGG, GGC, GCA, CAA, and AAT has multiplicity 1, we create the graph shown in Figure 3.13. In general, it is easy to see that the graph resulting from adding multiplicity edges is Eulerian, as both the indegree and outdegree of a vertex

(a)   (b)   (c)

Figure 3.14 The three mathematicians. (a) Leonhard Euler. (b) William Hamilton. (c) Nicolaas de Bruijn.

(represented by an $(l-1)$-mer) equals the number of times this $(l-1)$-mer appears in the genome.

In practice, information about the exact multiplicities of $(l-1)$-mers in the genome may be difficult to obtain, even with modern sequencing technologies. However, computer scientists have recently found a way to reconstruct the genome even when this information is unavailable. Furthermore, DNA sequencing machines are prone to errors, our reads will have varying lengths, and so on. However, with every variation to fragment assembly, it has proven fruitful to apply some cousin of de Bruijn graphs in order to transform a question involving Hamiltonian cycles into a different question about Eulerian cycles.

 4   A short history of read generation

4.1   The tale of three biologists: DNA chips

While Euler, Hamilton, and de Bruijn could not possibly meet each other, their mathematical fates got intricately criss-crossed. In 1988, three other Europeans would find their fates intertwined (Figure 3.15). Radoje Drmanac (Serbia), Andrey Mirzabekov (Russia), and Edwin Southern (UK) simultaneously and independently developed the futuristic and at the time completely implausible method of *DNA chips* as a proposal for read generation. None of these three biologists knew of the work of Euler, Hamilton, and de Bruijn; none could have possibly imagined that the implications of his own

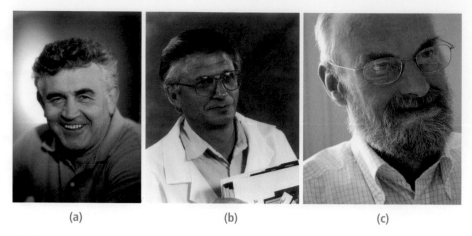

Figure 3.15 The three biologists. (a) Radoje Drmanac. (b) Andrey Mirzabekov.
(c) Edwin Southern.

experimental research would eventually bring him face to face with these giants of mathematics.

In 1977 Fred Sanger and colleagues sequenced the first virus, the tiny 5,375 nucleotide long bacteriophage ϕX174. However, while biologists in the late 1980s were routinely sequencing viruses containing hundreds of thousands of nucleotides, the idea of sequencing bacterial (let alone human) genomes seemed preposterous, both experimentally and computationally. Drmanac, Mirzabekov, and Southern realized that one main problem with the original DNA sequencing technology developed in the 1970s is the fact that it is not cost-effective for larger genomes. Indeed, generating a single read in the late 1980s cost more than a dollar, and thus sequencing a mammalian genome would have been a billion-dollar enterprise.[7] Due to such a high cost, it was infeasible to generate all l-mers from a genome, one of our conditions for the successful application of the Eulerian approach. DNA chips were therefore invented with the goal of cheaply generating *all* l-mers from a genome, albeit with a smaller read length l than the original DNA sequencing technology. For example, whereas traditional sequencing techniques generated reads containing approximately 500 nucleotides, the inventors of DNA arrays aimed at producing reads with around 15 nucleotides.

DNA chips work as follows. One first synthesizes all 4^l possible l-mers (i.e. all DNA fragments of length l) and attaches them to a *DNA array*, which is a grid on which each l-mer is assigned a unique location. We next take an (unknown) DNA fragment,

---

[7] Even in 2000, when the cost of read generation reduced substantially, sequencing the human genome still cost a few hundred million dollars.

AAA	AGA	CAA	CGA	GAA	GGA	TAA	TGA
AAC	AGC	CAC	CGC	GAC	GGC	TAC	TGC
AAG	AGG	CAG	CGG	GAG	GGG	TAG	TGG
AAT	AGT	CAT	CGT	GAT	GGT	TAT	TGT
ACA	ATA	CCA	CTA	GCA	GTA	TCA	TTA
ACC	ATC	CCC	CTC	GCC	GTC	TCC	TTC
ACG	ATG	CCG	CTG	GCG	GTG	TCG	TTG
ACT	ATT	CCT	CTT	GCT	GTT	TCT	TTT

Figure 3.16 A schematic of the DNA array containing all possible 3-mers. Ten fluorescently labeled 3-mers represent complements of the 10 3-mers from Figures 3.9 and 3.10. In order to obtain our reads from this array, we simply take the complements of the highlighted 3-mers. For example, CAC is highlighted, which means that GTG (the complement of CAC) is one of our reads. Note that this DNA array provides no information regarding *l*-mer multiplicities.

fluorescently label it, and apply a solution containing this fluorescently labeled DNA to the DNA array. The upshot is that the nucleotides in the DNA fragment will hybridize (bond) to their complements on the array (A will bond to T, and C to G). All we need to do is use spectroscopy to analyze which sites on the array emit the greatest fluorescence; the complement of the *l*-mer corresponding to such a site on the array must therefore be one of our reads. See Figure 3.16 for an illustration of the DNA array for our recurring set of reads.

At first, almost no one believed that the idea of DNA arrays would work, because both the biochemical problem of synthesizing millions of short DNA fragments and the mathematical problem of sequence reconstruction appeared too complicated. In 1988, *Science* magazine wrote that given the amount of work required to synthesize a DNA array, "using DNA arrays for sequencing would simply be substituting one horrendous task for another." It turned out that *Science* was wrong: in the mid 1990s, a number of startup companies perfected technologies for designing large DNA arrays. However,

DNA arrays ultimately failed to realize the dream that motivated their inventors. Arrays are incapable of sequencing DNA, because the fidelity of DNA hybridization with the array is too low and because the value of l is too small.

Yet the failure of DNA arrays was a spectacular one: while the original goal (DNA sequencing) was out of reach for the moment, two new unexpected applications of DNA arrays emerged. Today, arrays are used to measure gene expression, as well as to analyze genetic variations. These new applications transformed DNA arrays into a multi-billion dollar industry that included Hyseq (founded by Radoje Drmanac) and Oxford Gene Technology (founded by Sir Edwin Southern).

4.2   Recent revolution in DNA sequencing

After founding Hyseq, Radoje Drmanac did not abandon his dream of inventing an alternative DNA sequencing technology. In 2005 he founded *Complete Genomics*, which recently developed the technology to generate (nearly) all l-mers from a genome, thus at last enabling the method of Eulerian assembly. While his *nanoball arrays* technology is quite different from the DNA chip technology he proposed in 1988, one can still recognize the intellectual legacy of DNA chips in nanoball arrays, a testament that good ideas do not die even if they fail. Moreover, a number of other companies, including *Illumina* and *Life Technologies*, are competing with Complete Genomics by using their own technologies to generate (nearly) all l-mers from a genome. While DNA arrays failed to generate accurate reads even 15 nucleotides long, the next generation sequencing technologies generate reads of length 25 nucleotides and longer (and producing hundreds of millions such reads in a single experiment). These developments in *next-generation sequencing technologies* in the last five years have revolutionized genomics, and biologists are presently preparing to assemble the genomes of *all* the mammals on Earth (Figure 3.17) ... while still relying on the grand idea that Leonhard Euler developed in 1735.

 5   Proof of Euler's Theorem

We now will prove Euler's Theorem. First, let us restate his result for the case of undirected graphs, which we may recall are graphs for which the edges are "two-way streets."

Theorem (Euler's Theorem I).   *An equivalent condition to a graph G having an Eulerian cycle is that the degree of every vertex of G is even.*

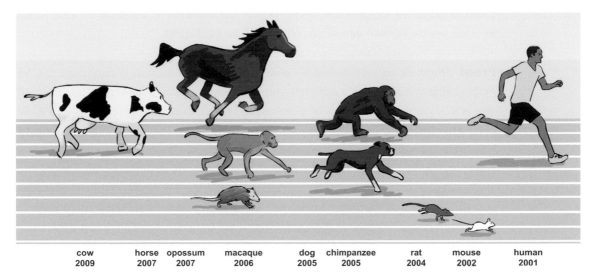

cow	horse	opossum	macaque	dog	chimpanzee	rat	mouse	human
2009	2007	2007	2006	2005	2005	2004	2002	2001

Figure 3.17 At the moment, only nine mammals have had their genomes sequenced: human, mouse, rat, dog, chimpanzee, macaque, opossum, horse, and cow. This is all about to change.

We shall only prove the second version of Euler's Theorem for directed graphs (in which the edges are "one-way streets"), which is ultimately more relevant to the themes of this chapter. We urge you to read through the proof we provide carefully, and then see if you can prove Euler's Theorem I for yourself. Do not be terrified. The overall structure of the two proofs is identical, except for a few details. Simply follow the proof of Euler's Theorem II and fit in the appropriate details for undirected graphs.

Here, then, is the restatement of Euler's Theorem for directed graphs.

Theorem (Euler's Theorem II). *An equivalent condition to a directed graph G having an Eulerian cycle is that for every vertex v in G, the indegree and outdegree of v are equal.*

Recall that two conditions being "equivalent" means that if one is true, then the other must be true. In this specific instance, our equivalent conditions are as follows for a given directed graph G:

1 G has an Eulerian cycle.
2 Each vertex of G has equal indegree and outdegree.

So in order to prove that these two conditions are equivalent, we simply need to demonstrate two statements. First, we need to show that if (1) is true for a directed graph G, then so is (2). Second, we must show that if (2) is true for a directed graph G, then so is (1). If these two statements hold, then there is no way that we can have a directed graph for which condition (1) is true and condition (2) is false, or vice versa. In other words, our two conditions above will be equivalent.

Proof   First we will show that if condition (1) is true, then so is condition (2). So assume that we are given a directed graph G which contains an Eulerian cycle; our aim is to show that each vertex of G has equal indegree and outdegree. Every time we enter a vertex in the Eulerian cycle of G, we leave it via a different edge. If a vertex v is used k times throughout the course of the cycle, then we enter v via a total of k edges and leave v via a total of k edges. All $2k$ edges are distinct, because since our cycle is Eulerian, no edge can be used more than once. Furthermore, these $2k$ edges constitute all edges touching this vertex, since an Eulerian cycle uses every edge in G. Therefore the indegree and outdegree of v are both equal to k. We can iterate this argument on every vertex in G to obtain that every vertex in G has equal indegree and outdegree, as needed.

Conversely, we need to show that if condition (2) is true, then so is condition (1). So assume that we are given a directed graph G for which each vertex has indegree equal to its outdegree. We will actually *form* an Eulerian cycle in G by the following procedure. Choose any vertex v in G, and choose any edge leaving v. Travel down this edge to the next vertex. Continue this process of choosing any unused edge to walk down, creating what is called a "random walk," while making sure only that we never use the same edge twice. Eventually, we will reach our original vertex v, creating a cycle which we call C_1. We should be suspicious of why a random walk in G is guaranteed to produce a cycle; this fact is ensured by the assumed condition that every vertex has equal indegree and outdegree, so that every time we arrive at a vertex, we must be able to find an unused edge leaving it (i.e. we cannot get "stuck" along our walk).

Now, once we have formed our cycle C_1, there are two possibilities for it. Either C_1 is an Eulerian cycle, in which case we are finished, or C_1 is not Eulerian. In the latter case, remove C_1 from G to form a new graph H. Because every vertex of C_1 (a cycle) must have indegree equal to its outdegree, condition (2) must also hold for every vertex in H. Since G is connected, we are guaranteed to have some vertex w in H that contains edges in both H and C_1. So since condition (2) holds for H, we can start at w and form an arbitrary cycle C_2 in H via a random walk in H.

We now have two cycles, C_1 and C_2, which do not share any edges but which both pass through w. We can therefore consolidate C_1 and C_2 to form a single "supercycle," which we call C. See Figure 3.18 for a brief illustration of how we form C.

In turn, we test if C is Eulerian, and if not we can iterate the above procedure indefinitely. If at any step our supercycle C becomes an Eulerian cycle, then we are

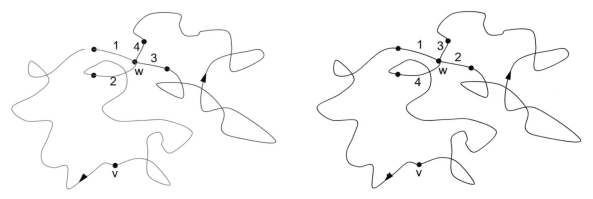

Figure 3.18 Cycle consolidation. If we have two cycles passing through the same vertex *w*, then we can combine them into a single cycle simply by changing the order in which we choose edges leaving *w*.

finished. The only concern is that C might never become Eulerian. However, this is impossible: there are only finitely many edges in the original graph G, so that since we remove some edges at each step, eventually we must reach a step at which we run out of edges. When we consolidate cycles at this step, our supercycle will use every edge in G without using any edges more than once, which is precisely the definition of an Eulerian cycle in G. Therefore G has an Eulerian cycle, which is what we set out to show.

The brilliant facet of this proof (as well as the proof of Euler's Theorem 1) is that it serves as an example of what mathematicians call a "constructive proof," or a proof that not only proves the desired result, but also delivers us with a very precise method for actually *constructing* what we need, which in this case is an Eulerian cycle. Therefore, if we are given a graph and asked to find an Eulerian cycle in it, we can easily test to see if each vertex has indegree equal to its outdegree (or if the degree of each vertex is even, as in the case of undirected graphs). If this condition fails, then the graph contains no Eulerian cycle; if it holds, we simply follow the idea outlined in the proof and form an arbitrary sequence of cycles that do not share any edges, combining the cycles into a single "supercycle" at each step, and iterating this process until an Eulerian cycle is inevitably obtained.

Let us conclude by illustrating the power of our constructive proof. In Figure 3.19, we apply Euler's Theorem to find an Eulerian cycle in the de Bruijn graph from Figure 3.12. Keep in mind that the same method will work for genome graphs containing billions of edges. At last, we have definitively solved our giant puzzle!

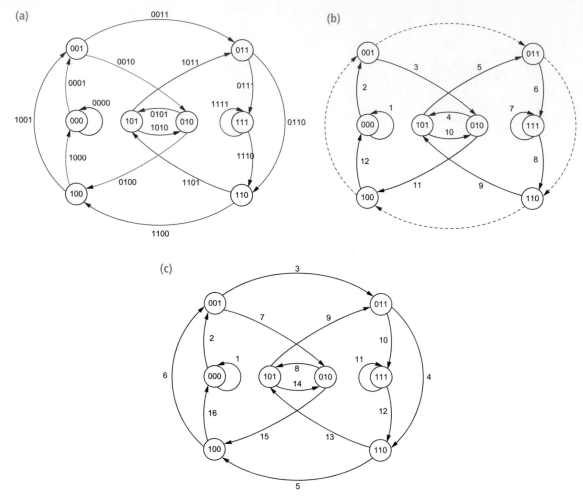

Figure 3.19 Obtaining an Eulerian cycle from a graph in which all vertices have the appropriate degrees. Here, we find an Eulerian cycle in the directed graph $B(2, 3)$ from Figure 3.12. (a) We first find three arbitrary cycles in the graph at hand (here shaded with three different colors). Once we have chosen the green cycle, we remove it from the graph and choose the blue cycle, which we then remove from the graph and choose the red cycle. (b) We next consolidate the green and blue cycles into a single cycle (black). The edge numberings give the order of the edges if we start at vertex 000. Note that the red cycle is dashed to indicate that it is not yet part of our supercycle. (c) Finally, we add the red cycle into our supercycle, which is Eulerian. The edges are renumbered as needed. The resulting Eulerian cycle spells the cyclic superstring 0000110010111101.

DISCUSSION

We have met three mathematicians of three different centuries, Euler, Hamilton, and de Bruijn, spread out across the European continent, each with his own queries. We might be inclined to feel a sense of adventure at their work and how it converged to this singular point in modern biology. Yet the first biologists who worked on DNA sequencing had no idea of how graph theory could be applied to this subject; what's more, the first paper combining the trio's mathematical ideas into fragment assembly was published lifetimes after the deaths of Euler and Hamilton, when de Bruijn was in his seventies. So perhaps we might think of these three men not as adventurers, but instead as lonely wanderers. As is so often the mathematician's curse, each man passionately pursued questions in the abstract mathematical world while having no idea where the answers might one day lead without him in the real world.

NOTES

Euler's solution of the Königsberg Bridge Problem was presented to the Imperial Russian Academy of Sciences in St. Petersburg on August 26, 1735. Euler was the most prolific writer of mathematics of all time: besides graph theory, he first introduced the notation $f(x)$ to represent a function, i for the square root of -1, and π for the circular constant. Working very hard throughout his entire life, he became blind. In 1735, he lost the use of his right eye. He kept working. In 1766, he lost the use of his left eye and commented: "Now I will have fewer distractions." He kept working. Even after becoming completely blind, he published *hundreds* of papers.

After Euler's work on the Königsberg Bridge Problem, graph theory was forgotten for over a hundred years, but was revived in the second half of the nineteenth century by prominent mathematicians, among them William Hamilton. Graph theory flourished in the twentieth century, when it became an area of mainstream mathematical research.

DNA sequencing methods were invented independently and simultaneously in 1977 by Frederick Sanger and colleagues [1] as well as Walter Gilbert and colleagues [2]. The Hamiltonian cycle approach to DNA sequencing was first outlined in 1984 [3] and further developed by John Kececioglu and Eugene Myers in 1995 [4]. Advances in DNA sequencing led to the sequencing of the entire 1800 kb *H. influenzae* bacterial genome in the mid 1990s. The human genome was sequenced using the Hamiltonian approach in 2001.

DNA arrays were proposed simultaneously and independently in 1988 by Radoje Drmanac and colleagues in Yugoslavia [5], Andrey Mirzabekov and colleagues in Russia [6], and Ed Southern in the UK [7]. The Eulerian approach to DNA arrays was described in [8]. The Eulerian approach to DNA sequencing was described in [9] and further developed in 2001 [10], when hardly anybody believed it could be made practical.

At roughly the same time, Sydney Brenner and colleagues introduced the *Massively Parallel Signature Sequencing* (MPSS) method [11], which brought in the era of next generation sequencing with short reads. Throughout the last decade, MPSS in addition to technologies developed by such companies as Complete Genomics, Illumina, and Life Technologies revolutionized genomics. Next-generation techniques produce rather short reads, which vary in length from 30 to 100 nucleotides and result in a challenging fragment assembly problem. To address this challenge, a number of assembly tools have been developed [12–15], all of which follow the Eulerian approach.

QUESTIONS

(1) Does the graph I representing the Icosian Game contain an Eulerian cycle? Why or why not?

(2) Construct the de Bruijn Graph $B(3, 3)$ and find an Eulerian cycle in it.

(3) Give three Eulerian cycles in the graph of Figure 3.13 along with their corresponding cyclic superstrings.

(4) From the following set of reads of length 4, use the ideas of this chapter to provide a (cyclic) candidate DNA sequence: AACG, TCGT, GATC (multiplicity 2), TATC, ATCG, CCCG, ATCC (multiplicity 2), CGGA, CCCT, GTAT, CCGA, CTAA, TCCC (multiplicity 2), GGAT, CCTA, TAAC, CGAT, CGTA, ACGG.

(5) Prove Euler's Theorem I.

REFERENCES

[1] F. Sanger, S. Nicklen, and A. R. Coulson. DNA sequencing with chain-terminating inhibitors. *Proc. Natl Acad. Sci. U S A*, 74:5463–5467, 1977.

[2] A. M. Maxam and W. Gilbert. A new method for sequencing DNA. *Proc. Natl Acad. Sci. U S A*, 74:560–564, 1977.

[3] H. Peltola, H. Soderlund, and E. Ukkonen. SEQAID: A DNA sequence assembling program based on a mathematical model. *Nucl. Acids Res.*, 12:307–321, 1984.

[4] J. Kececioglu and E. W. Myers. Combinatorial algorithms for DNA sequence assembly. *Algorithmica*, 13:7–51, 1995.

[5] R. Drmanac, I. Labat, I. Brukner, and R. Crkvenjakov. Sequencing of megabase plus DNA by hybridization: Theory of the method. *Genomics*, 4:114–128, 1989.

[6] Y. Lysov, V. Florent'ev, A. Khorlin, K. Khrapko, V. Shik, and A. Mirzabekov. DNA sequencing by hybridization with oligonucleotides. *Dok. Acad. Nauk USSR*, 303:1508–1511, 1988.

[7] E. Southern. United Kingdom patent application gb8810400. 1988.

[8] P. A. Pevzner. *l*-tuple DNA sequencing: Computer analysis. *J. Biomol. Struct. Dyn.*, 7:63–73, 1989.

[9] R. Idury and M. Waterman. A new algorithm for DNA sequence assembly. *J. Comput. Biol.*, 2:291–306, 1995.

[10] P. A. Pevzner, H. Tang, and M. Waterman. An Eulerian path approach to DNA fragment assembly. *Proc. Natl Acad. Sci. U S A*, 98:9748–9753, 2001.

[11] S. Brenner, M. Jonson, J. Bridgham, *et al.* Gene expression analysis by massively parallel signature sequencing (MPSS) on microbead arrays. *Nat. Biotech.*, 18:630–634, 2000.

[12] M. J. Chaisson and P. A. Pevzner. Short read fragment assembly of bacterial genomes. *Genome Res.*, 18:324–330, 2008.

[13] D. R. Zerbino and E. Birney. Velvet: Algorithms for de novo short read assembly using de Bruijn graphs. *Genome Res.*, 18:821–829, 2008.

[14] J. Butler, I. MacCullum, M. Kieber, *et al.* ALLPATHS: De novo assembly of whole-genome shotgun microreads. *Genome Res.*, 18:810–820, 2008.

[15] J. T. Simpson, K. Wang, S. D. Jackman, *et al.* ABySS: A parallel assembler for short read sequence data. *Genome Res.*, 19:1117–1123, 2009.

CHAPTER FOUR

Dynamic programming: one algorithmic key for many biological locks

Mikhail Gelfand

Dynamic programming is an algorithm that allows one to find an optimal solution to many important bioinformatics problems without explicit consideration of all possible solutions. This chapter provides a description of the algorithm in the graph-theoretical language, and shows how it is applied to such diverse areas as DNA and protein alignment, gene recognition, and polymer physics.

Introduction

A major part of computational biology deals with the similarity of sequences, be they DNA fragments or proteins. There are four aspects to this problem: defining the measure of similarity, calculating this measure for given sequences, assessing its statistical significance, and interpreting the results from the biological viewpoint. Biologists are interested in the latter: similar sequences may have a common origin, as well as similar structure and function. However, here we shall deal with a formal problem: how to discover similarity.

Consider two sequences from a finite alphabet (e.g. 4 nucleotides or 20 amino acids) written one under the other, possibly with gaps. This is called an *alignment* (Figure 4.1).

Bioinformatics for Biologists, ed. P. Pevzner and R. Shamir. Published by Cambridge University Press.
© Cambridge University Press 2011.

(a) `gelfand`

`  +····+··`

`  gandalf`

(b) `g---elfand`

`  +----·++---`

`  gandalf---`

(c) `gelfand---`

`  +---+++---`

`  g---andalf`

Figure 4.1 Three (of many) alignments of two sequences. Plus denotes a match; dot, a mismatch, minus, a gap. (a) Two matches, five mismatches, (b) three matches, one mismatch, two gaps of size three (six indels, that is, one-nucleotide insertions/deletions), (c) four matches, two gaps of size three (six indels).

We can calculate the number of matching symbols (nucleotides or amino acids), the number of mismatches, and the number and size of gaps. If we assign a positive weight (premium) to a match, and negative weights (penalties) to a mismatch and a gap of a given size, we can calculate the total score as the sum of all weights. Depending on the weights, different alignments will have the highest score. For instance, in Figure 4.1, alignment (c) is clearly better than alignment (b), as it has the same number of gaps, but no mismatches and more matches, whereas the choice between (c) and (a) depends on the gap penalty: if gaps are assumed to be much worse than mismatches, (a) is better than (c).

So, for a pair of sequences, we want to find the best alignment in terms of the scoring function; that is, to introduce gaps so that the similarity between the sequences is maximized. One way to do so is to consider all possible alignments, score each one, and find the one with the maximal score. However, the number of possible alignments is enormous: for two sequences of length N it is approximately proportional to $(1+\sqrt{2})^{2N+1}/\sqrt{N}$, in mathematical notation, $\mathcal{O}((1 + \sqrt{2})^{2N+1}/\sqrt{N})$. This is a very large number. For $N = 1{,}000$ it is about 10^{767} (for comparison, the number of the elementary particles in the Universe is estimated as 10^{80}). For a smaller N, say, $N = 100$, this number is about 10^{76}. This may look better, but assuming one operation per alignment and a supercomputer doing 10^{12} operations per second, we shall need 10^{57} years to complete the construction. That does not look promising.

Another well-known problem is segmentation of a sequence into functionally or statistically homogeneous regions. The most important variant of this problem is gene recognition: given a DNA sequence, map its protein-coding and non-coding regions. It was observed about 30 years ago that the statistical properties of coding and non-coding regions are different. Indeed, amino acid frequencies in proteins are not uniform, and codons corresponding to frequent amino acids such as alanine and lysine are encountered more frequently than codons for tryptophan and histidine. Moreover, synonymous codons encoding the same amino acid also are not used evenly (this is related to the cellular concentration of corresponding tRNAs and other reasons). Hence, the frequency of codons in protein-coding regions is not the same as the frequency of nucleotide triplets in non-coding regions. We can introduce a measure for the "coding potential": how similar the frequencies of nucleotide triplets in a DNA fragment are to those expected in a coding region compared to a non-coding one. To do that, we can assign a weight to each triplet, dependent on how frequently the triplet serves as a codon compared to its background (non-coding) frequency.

In prokaryotes, gene recognition is relatively straightforward, at least from the computational point of view. We simply calculate the coding potential of all open-reading frames, and whenever two open-reading frames happen to overlap, select the higher-scoring one. However, in eukaryotes the problem is complicated by the *exon–intron structure*. *Introns* do not code for proteins and are spliced out from the transcript. *Splicing* creates a mature mRNA consisting of ligated *exons*. Individual exons are too short for reliable estimation of their coding potential. We can try to predict *splice sites*, that is, boundaries between 5′-exons and 3′-introns (called *donor sites*) or 3′-introns and 5′-exons (*acceptor sites*), but this cannot be done reliably: in order not to lose any true sites, we have to use a weak rule that produces numerous false-positives. A combined procedure works as follows: we start with site prediction and then consider all possible exon–intron structures, calculating the statistical score for each. This score is the sum of the total coding potential of exons and the non-coding potential of introns. The latter term measures the similarity to statistical properties of non-coding regions.

Again, we run into a computational problem, since the number of possible exon–intron structures is very large. Indeed, the number of candidate sites is roughly proportional to the sequence length. Assuming that each site might be included into an exon–intron structure, we find that the number of possible structures is exponential in the sequence length. In fact, not all sets of sites yield legitimate structures (e.g. all odd sites must be donor sites and all even sites must be acceptor sites), but this and other corrections still retain the exponential dependence.

We see that in both cases direct scoring of all possible configurations (alignments or exon–intron structures) is not feasible. But do we need to score all of them?

Consider the following toy example. Suppose we have two sets of positive integers $x_1, ..., x_m$ and $y_1, ..., y_n$, and we need to calculate the sum of all pair products

$$x_1 \cdot y_1 + x_1 \cdot y_2 + \ldots + x_1 \cdot y_n + x_2 \cdot y_1 + x_2 \cdot y_2 + \ldots + x_2 \cdot y_n + \ldots + x_m \cdot y_1$$
$$+ x_m \cdot y_2 + \ldots + x_m \cdot y_n.$$

How many operations do we need? Easy: mn multiplications and $mn - 1$ additions. But maybe we can do better? We simply rewrite our sum as

$$x_1 \cdot (y_1 + y_2 + \ldots + y_n) + x_2 \cdot (y_1 + y_2 + \ldots + y_n) + \ldots + x_m \cdot (y_1 + y_2 + \ldots + y_n)$$
$$= (x_1 + x_2 + \ldots + x_m) \cdot (y_1 + y_2 + \ldots + y_n). \tag{4.1}$$

Now we need $m + n - 2$ additions and just one multiplication. I shall rewrite this calculation using the standard mathematical notation:

$$\sum_{i=1...m, j=1...n} x_i \cdot y_j = \sum_{i=1...m} x_i \cdot \sum_{j=1...n} y_j. \tag{4.2}$$

Quiz 1

How many multiplications do we need to calculate

$$x_1^{y_1} \cdot x_1^{y_2} \cdot \ldots \cdot x_1^{y_n} \cdot x_2^{y_1} \cdot x_2^{y_2} \cdot \ldots \cdot x_2^{y_n} \cdot \ldots \cdot x_m^{y_1} \cdot x_m^{y_2} \cdot \ldots \cdot x_m^{y_n} = \prod_{i=1...m, j=1...n} x_i^{y_j}$$

$$\tag{4.3}$$

if we are (a) naïve?, (b) sophisticated? (c) What if in addition to multiplication, we have an operation "taking to the power"? (d) If we may perform not only multiplication, but also addition?

Lesson Restructuring the order of calculations using properties of the data may sharply decrease the number of operations.

So, why not try something similar with our problems? In order to do so we need a mathematical object called a *graph*. We will develop an efficient algorithm for a rather abstract problem on graphs, and then we will apply it to the biological problems of alignment and gene recognition.

2 Graphs

A *graph* consists of two sets, a set of *vertices* (primary objects) and a set of *arcs*, which are pairs of vertices (Figure 4.2). We will consider *oriented graphs*, so that each arc $a_n = (b_n, e_n)$ has a start vertex b_n and an end vertex e_n. We will require that the graph contains neither multiple arcs with the same starts and ends (Figure 4.2d), nor *loops*, that is, arcs whose start and end vertices coincide (Figure 4.2e).

A *walk p* of length N is an ordered set of N arcs $p = (a_1, ..., a_N)$ such that the end vertex of arc $a_n = (b_n, e_n)$ coincides with the start vertex of arc a_{n+1}, $e_n = b_{n+1}$, for all $n = 1, ..., N - 1$. In a graph without loops and multiple arcs, each walk may also be defined as an ordered set of vertices $p = (v_1, ..., v_{N+1})$ such that for each pair of adjacent vertices v_n, v_{n+1} there is an arc $a_n = (v_n, v_{n+1})$, $n = 1, ..., N$. A walk is a *path* if no arc is passed twice. We will also use *non-oriented* paths obtained by disregarding the direction of arcs.

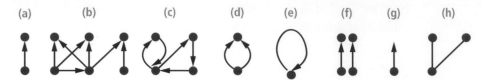

Figure 4.2 (a, b) Graphs. (c) Graph with cycles. (d) Graph with double arcs. (e) Graph with a loop. (f) Graph with two components. (g) Not a graph (hanging arc). (h) Non-oriented graph.

A graph is *connected* (or *consists of one component*) if there is a non-oriented path between any two vertices, and we will consider only such graphs. A non-connected graph is shown in Figure 4.2f . A path is called a *cycle* if the end vertex of the last arc a_N coincides with the start vertex of the first arc a_1, $e_N = b_1$, and we will consider only *acyclic* graphs that contain no cycles (compare an acyclic graph in Figure 4.2b and a graph with cycles in Figure 4.2c).

Quiz 2

(a) Draw all acyclic connected oriented graphs with three vertices (up to vertex labels).
(b) How many oriented graphs will there be if we label vertices with symbols A, B, and C?

A vertex is called a *source* if it is not an end vertex for any arc, and a *sink* if it is not a start vertex for any arc. Unless specified otherwise, we shall assume that a graph has a single source and a single sink and consider only paths starting at the source and ending at the sink, but the algorithms presented below do not depend on this assumption, and in any case we can always perform a technical trick of creating a new source (or sink) and linking it with all initial sources (respectively, sinks), see Figure 4.3. Finally, we shall assign each arc with a number called a *weight*. For a given path, its *path score* is defined as the sum of the weights of its arcs.

Quiz 3

(a) Prove that in an acyclic graph there is at least one source and at least one sink.
(b) Draw sinks and sources in the graphs of Quiz 2.

 ## Dynamic programming

Now we are ready to formulate our problem.

Problem 1　Given a weighted acyclic graph, find the highest scoring path.

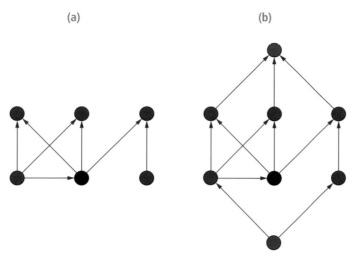

Figure 4.3 (a) Graph with two sources and three sinks (red). (b) Graph with artificially added single source and single sink (blue).

We do not want to enumerate all paths, since their number is very high even for relatively simple graphs; in general, it is exponential in the number of arcs. However, if we have two paths that have several common arcs at the beginning, we do not need to calculate the score of this common subpath twice. Even more importantly, if two subpaths P and Q end at the same vertex v, and the score of P is larger than the score of Q, then for all pairs of paths P^* and Q^* that start with P and Q, respectively, and coincide after v, the score of P^* is higher than the score of Q^*. Hence, we do not need to consider all paths, as it is sufficient to construct the highest-scoring subpath from the source to each vertex, finishing at the sink.

For example, let's do this for the graph shown in Figure 4.4. The entire procedure is shown in Figure 4.5. We start at the source and process all arcs originating at it: these are our initial subpaths. At each end vertex we collect the score of the best (highest-scoring) already considered subpath ending at the vertex and mark the last arc of this subpath. Then we select a vertex with all incoming arcs already processed (at step 2 there is only one such vertex, marked by a star). Again, we process all outgoing arcs. The process is repeated until we come to the sink. Note that we may come to a situation in which there are several vertices with all incoming arcs processed (e.g. at step 5): we select an arbitrary one.

Quiz 4

At what steps in Figure 4.5 do we have more than one vertex with all incoming arcs processed?

(a)   (b)

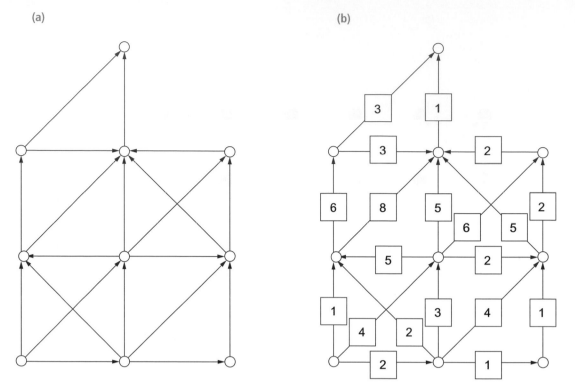

Figure 4.4 Sample graph for construction of the highest scoring path. (a) The structure of the graph, (b) the arc weights.

When all vertices have been processed and we arrive at the sink, we *backtrack*, moving in the opposite direction, each time using the marked arc. Recall that the marked arc is the last arc of the highest-scoring subpath. Hence, when we return to the source, we shall have constructed the highest-scoring path from the source to the sink. A formal algorithm is given in Figure 4.6.

How many operations do we need for this process? The limiting procedure is processing vertices and adding arcs to paths, and we consider each arc only once, hence the number of operations is linear in the number of arcs A: the run time of the algorithm is $O(A)$, meaning approximately proportional to A if A is large.

Do we really need to check every arc? What if we simply start at the source and select the highest-weighted arc at each step? This strategy is called the *greedy algorithm*. Unfortunately, as shown in Figure 4.7, where it is applied to the same graph, we cannot guarantee that we shall construct the highest scoring path by this algorithm.

(a)

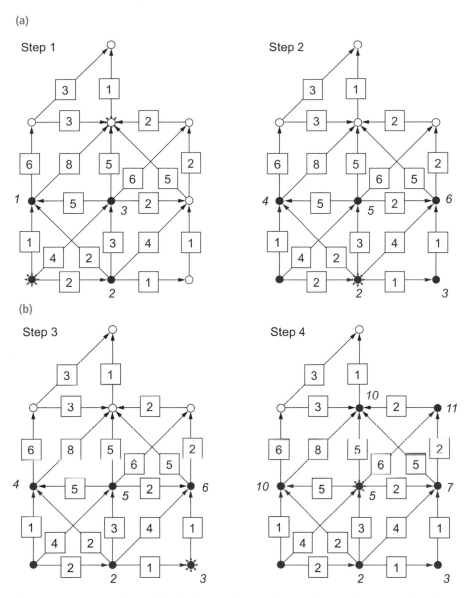

Figure 4.5 Construction of the highest-scoring path. Star denotes the currently active vertex; red vertices represent those for which construction of the highest-scoring subpath has been completed; blue vertices are the ones for which construction of the subpath has started but not yet completed. Blue arrows denote processed arcs. Red arrows, one for each vertex, denote the last arc of the highest-scoring subpath coming to this vertex. Large green arrows denote the highest-scoring path constructed at the last (backtracking) step. A number at a vertex denotes the highest score of already considered subpaths ending at this vertex.

(c)

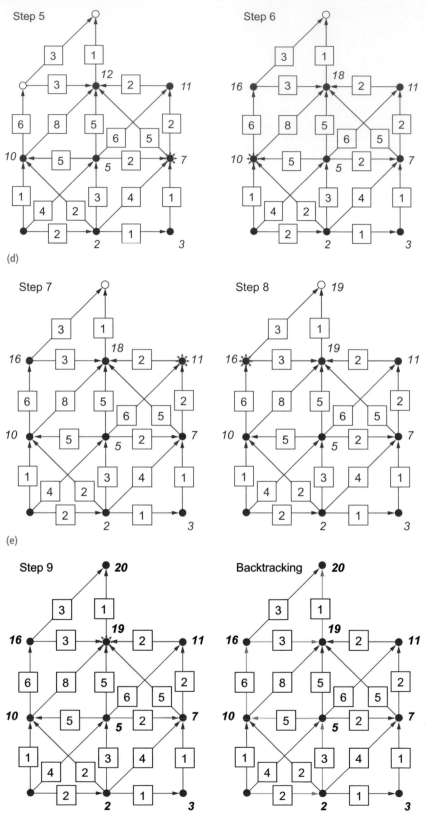

(d)

(e)

Figure 4.5 (Cont.)

Data types and definitions:

vertices: *v*, *u*, *Source*, *Sink*;

arcs: (*v*, *u*), *a*;

start vertex of arc *a*: *B*(*a*);

weight of arc (*v*, *u*): *W*(*v*, *u*);

path: *BestPath*; // defined as a set of arcs

the highest score of subpath ending at *v*: *S*(*v*);

the highest score of subpath ending at *u* and coming through (*v*, *u*): *T*(*v*, *u*);

the last arc of the highest scoring subpath ending at *u*: *L*(*u*);

Initialize: **for** each vertex *v*: *S*(*v*) := minus_infinity.

Forward process: **while** There are unprocessed vertices:

  v := arbitrary unprocessed vertex with all incoming arcs processed;

  for each arc (*v*, *u*): // consider all arcs starting at *v*

    T(*v*, *u*) := *S*(*v*)+*W*(*v*, *u*);

    if *T*(*v*, *u*)>*S*(*u*) // subpath coming through *v* is better
                      than the current best subpath ending at *u*

    then: // update the data for *u*

      S(*u*) := *T*(*v*, *u*);

      L(*u*) := (*v*, *u*);

    endif;

    (*v*, *u*) := processed_arc;

  endfor;

  v := processed_vertex;

endwhile.

Backtracking:

  BestPath = empty_set; // initialize

  v := *Sink*; // go from the sink backwards by marked arcs

  until *v*=*Source*

    Add *L*(*v*) to *BestPath*; // add the last arc of the best path
                      ending at the current vertex

    v := *B*(*L*(*v*)); // go to the start vertex of this arc

  enduntil.

Output *BestPath*.

Figure 4.6 Dynamic programming algorithm for construction of the highest-scoring path.

Quiz 5

(a) Construct the simplest possible graph in which the greedy algorithm yields the highest-scoring path. (b) Construct a graph with three vertices in which the greedy algorithm **does not** yield the highest-scoring path. (c) Construct a graph with three vertices in which the greedy algorithm **does** yield the highest-scoring path. (d) Assign new weights to the arcs of the graph from Figure 4.4a so that the greedy algorithm will yield the highest-scoring path.

Quiz 6

Write an algorithm for construction of the path with the maximum number of arcs and apply it to the graph from Figure 4.4. Hint: do not change the algorithm, set proper arc weights.

Quiz 7

(a) Modify the maximum score algorithm so as to construct the path with the minimal score and find this path for the graph from Figure 4.4. (b) Provide a greedy algorithm for finding the path of minimal score in a graph, and apply it to the graph from Figure 4.4. (c) For the graph in Figure 4.4, find the path with the minimal number of arcs.

Note   One may think that the dynamic programming algorithm is applicable to all path optimization problems. Unfortunately, this is not so. For example, it does not work for the famous *traveling salesman problem*. Given a non-oriented graph with weighted arcs, we need to construct the lowest-scoring path passing through all the vertices (the salesman needs to visit all cities with travel time between the cities given by the arc weights, while spending the least amount of time traveling). The condition that all cities need to be visited in a single trip makes it an example of a so-called *NP-complete* problem, for which no efficient algorithms are known. While it has not been formally proven, most computer scientists believe that for all NP-complete problems the number of operations required to provide an optimal solution is exponential in the problem size.

Lesson   The generic dynamic programming algorithm may be applied to different problems. The common feature of these problems is that each one can be decomposed into an ordered set of smaller subproblems, and to solve a more complex subproblem one needs to know only the solutions of the simpler ones, but not the entire set of possibilities.

Alignment

Return now to the alignment problem.

Problem 2 We are given two symbol sequences (in biological applications, the symbols usually being nucleotides or amino acids) of lengths M and N, and we want to set a correspondence between these sequences so that some symbols are set in pairs, matching or mismatching, whereas other symbols are ignored (deleted). The order of corresponding symbols in the subsequences should coincide (we cannot align TG to GT so that T corresponds to T and G corresponds to G simultaneously). The alignment score is the sum of match premiums r per matching pair minus the sum of mismatch penalties p per mismatching pair and deletion penalties q per ignored symbol. The goal is to construct the highest-scoring alignment.

Note The underlying assumption making this formal problem biologically relevant is that an alignment reflects the process of evolution: aligned symbols have a common ancestor, whereas mismatches, insertions, and deletions reflect evolutionary events, mutations that change nucleotides (and as a consequence, for protein-coding genes, amino acids of the encoded protein), and insertion or removal of gene fragments.

Quiz 8
What are the scores of the alignments in Figure 4.1?

It turns out that the alignment problem elegantly reduces to the highest-scoring path problem, for which, as we have already seen, there exists an efficient dynamic programming algorithm. Indeed, consider a graph whose vertices correspond to pairs of positions (Figure 4.7). Each pair may be of three types: match or mismatch ($M \cdot N$ arcs), deletion in the first sequence ($M \cdot (N + 1)$ arcs), and deletion in the second sequence (($M + 1) \cdot N$) arcs). These arcs are assigned weights of r or ($-p$) for matches and mismatches, respectively, and ($-q$) for deletions (Figure 4.8). There is a one-to-one correspondence between paths from source to sink in the graph and possible alignments (Figure 4.9). By construction, the path score equals the alignment score. Hence, finding the highest-scoring alignment is equivalent to finding the highest-scoring path. Application of the dynamic programming algorithm to the alignment graph produces the highest-scoring alignment in $O(MN)$ time.

We have just solved the so-called *global alignment* problem. There exist other types of alignments. For example, if there are reasons to expect that the aligned sequences may not be complete, we should not penalize hanging ends in any one sequence at both sides. This is achieved by setting all penalties on the "sides" of the rectangular

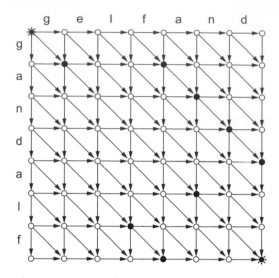

Figure 4.7 Graph for the alignment construction. Diagonal arcs correspond to symbol pairings, with matches shown by red arrows; horizontal and vertical arcs correspond to deletions in the horizontal and vertical sequence, respectively. Source and sink vertices are shown by stars.

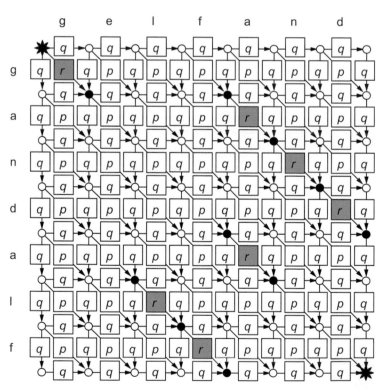

Figure 4.8 Alignment graph of Figure 4.6, with arc weights. Matches (weight of match premium is r) are pink.

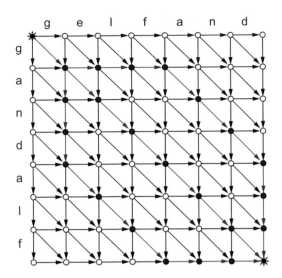

Figure 4.9 Alignment graph of Figure 4.6 with three paths corresponding to the alignments from Figure 4.1 shown by colored arrows. Red arrows: matches; blue arrows: mismatches (diagonal) and deletions (horizontal and vertical).

alignment graph to 0 or, equivalently, removing these side arcs and introducing zero-weight arcs from the source to all vertices at the left and upper sides and from all vertices at the bottom and right side to the sink.

Quiz 9

Construct the hanging ends alignment graphs for the pairs of sequences (a) "gelfand" and "elf" and (b) "gelfand" and "angel", and construct the optimal alignments.

The most important variant of the alignment is the *local alignment*, when both sequences may have hanging ends at both sides, and the goal is to find a region with maximal similarity. This is what one should look for, e.g. in distant proteins retaining similarity only at a fraction of domains. Again, a simple tweak of the alignment graph produces the desired result: we need to add zero-weight arcs from the source to all vertices (not only side ones, as in the "hanging-ends" case) and from all vertices to the sink.

Another direction of modification is playing with the weights. For example, it is well known that some amino acids are similar by their physico-chemical properties (e.g. aspartate and glutamate or leucine and valine), whereas others are rather different (e.g. glycine and tryptophan or alanine and proline). This is also seen in evolutionary analyses: when aligning *homologous* (having common origin) proteins, one often sees aspartate glutamate pairs, but rarely glycine–tryptophan pairs. Hence we should set different penalties to different mismatching pairs. This is done in a general way: we use the *matrix of amino acid match weights*, and assign weights to the alignment graph arcs equal to the weight of the corresponding pair. At that, our old premium-penalty system has the matrix with premiums r on the main diagonal and penalties $(-p)$ in all off-diagonal cells.

One more modification is the use of so-called *affine gap penalties*. A gap of length g is penalized not by qg, as above, but by $c + dg$, where the *gap opening* penalty c is relatively large, whereas the *gap extension* penalty d is small. Again, this may be done by a proper restructuring of the alignment graph. The underlying biological reason is that from the analysis of natural sequences we know that a deletion or insertion of size g is more likely than several independent deletions (respectively, insertions) of total size g.

Quiz 10

For the alignments of Figure 4.1, assuming match premium $r = 10$, what combinations of mismatch and deletion penalties would yield optimal alignments (a), (b), and (c)?

Note The problem of selecting proper gap penalties is important. For random sequences, dependent on the gap penalties, the length of the optimal local alignment of two sequences of the same length may be linear in the sequence length (for gap penalties that are small compared to match premiums) or logarithmic in the sequence length (for prohibitively large gap penalties). In the limit of zero gap penalty, the former case reduces to the *maximum common subsequence* problem, whereas in the limit of infinitely large gap penalty, the latter case is the *maximum common subword* problem. To select reasonable gap penalties for protein alignment, we should study homologous proteins with known 3D structures: a good alignment is one that sets in correspondence structurally equivalent amino acids. After training our parameters on a set of "gold standard" structural alignments, we can apply them to proteins with unknown structures.

Finally, we can apply the algorithm to the alignment of several sequences. For example, if three sequences are aligned, instead of a graph with a square (2D) lattice, we construct a graph with a cube (3D) lattice. The number of arcs, and hence the run time, is now $O(N^3)$, N being the length of all three sequences. Similarly, the run time for K sequences of length N is $O(N^K)$, becoming prohibitively large even for the alignment of a few short sequences. Many heuristics have been suggested to construct *multiple alignments* in reasonable time by reducing the problem to a series of pairwise alignments. They do not guarantee that the constructed alignment will have the highest score, but aim at producing biologically plausible alignments.

Lesson Weights matter. The same graph with differently assigned arc weights will yield different types of alignment.

5 Gene recognition

Another important problem is gene recognition, that is, decomposition of a sequence into *exons* (protein-coding regions) and *introns* (non-coding regions). The definitions in parentheses are somewhat inexact "bioinformatics" ones; for a biologically proper definition, consult a molecular biology textbook.

Problem 3   Define a gene as a sequence fragment consisting of exons and introns. The boundaries between them are *donor sites* (between exons and introns) and *acceptor sites* (between introns and exons). Each exon and intron is assigned a weight, measuring coding affinity (respectively, non-coding affinity) of its sequence. A gene's score is the sum of weights of constituent exons and introns. Our goal is, given a sequence and a set of candidate donor and acceptor sites, to construct the highest-scoring *exon–intron structure* for a gene.

There exist many programs for the identification of splice sites, but unfortunately, all of them are very unreliable and produce numerous false candidates. Hence we need to select the best exon–intron structure among a huge number of possibilities.

Again, we construct a graph. Its vertices correspond to candidate sites, and arcs correspond to possible exons and introns (Figure 4.10a); we shall call it the *exon–intron graph*. The exon arcs go from acceptor site vertices to donor site ones. The intron arcs go from donor site vertices to acceptor site vertices.

There is a one-to-one correspondence between exon–intron structures and paths of the exon–intron graph (Figure 4.10b). Hence, assigning each arc a weight equal to the weight of the corresponding exon or intron, we reduce the problem of finding the highest-scoring exon–intron structure to the problem of finding the highest-scoring path, which we know we can find by dynamic programming.

As we already know, the number of operations is proportional to the number of arcs in the graph. Assuming that candidate sites occur more or less uniformly along the sequence, their number is $O(L)$, where L is the sequence length. Since each pair of donor and acceptor sites generates a candidate exon or intron, the number of arcs is $O(L^2)$.

Note   In this description we leave out cumbersome technical details such as keeping the proper reading frame, the fact that protein-coding regions start and end at specific codons, taking into account restrictions on the minimal exon and intron lengths, the possibility that a sequence fragment may contain several genes, etc.

For long sequence fragments the quadratic run time may become prohibitively large. However, do we need all these arcs? An exon may be a part of a larger exon, and it is

(a)

(b)

Figure 4.10 (a) Exon–intron graph. Donor sites are shown by marked **gt** in the sequence and blue vertices (bottom row) in the graph. Acceptor sites are shown by marked **ag** in the sequence and black vertices (top row) in the graph. Exon arcs go from vertices at the top row to the ones in the bottom row, intron arcs go from the bottom row to the top row. The source and sink, corresponding to the beginning and end of the sequence, respectively, are represented by yellow stars. (b) One possible decomposition of the sequence into exons and introns and the corresponding path. Exons are shown by capitals.

reasonable to assume that the weight of the larger exon is a sum of the weight of the smaller one and the weight of the remaining segment. It would look unnatural to define the gene score by the sum of exon weights, while at the same time making exon weight different from the sum of weights of constituent segments. Indeed, in most cases exon weights are defined by additive measures of coding affinity. The same holds for introns.

If we restrict ourselves to additive weighing functions, we can construct a more efficient representation. We shall call it the *segment graph* (Figure 4.11). Again, vertices correspond to sites, but now each site corresponds to two vertices. Arcs are of two types: arcs between vertices corresponding to the same site represent exon–intron boundaries and are not assigned any weight, whereas arcs between vertices corresponding to adjacent sites of the same type represent exon or intron segments. The key is that we have only arcs between adjacent sites, hence, their number is linear to the number of sites, and we have $O(L)$ arcs. Using the same trick of avoiding multiple calculation of the same value, we have sharply decreased the computational complexity of the algorithm.

(a)

actg**ag**actgc**ag**acggac**gt**acggcactgac**gt**ata**ag**ccccac**ag**tccttac**gt**ctga

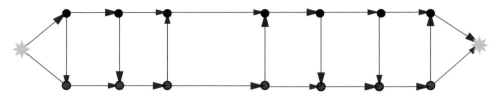

(b)

actgagactgc**ag**ACGGACGTACGGCACTGAC**gt**ata**ag**CCCCACAGTCCTTAC**gt**ctga

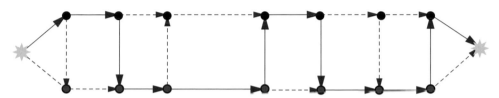

Figure 4.11 (a) Segment graph. Notation as in Figure 4.9. Exon fragments are in the bottom row, while intron fragments are in the top row. Vertical arcs at sites are possible exon–intron and intron–exon boundaries; note that the direction depends on the site type, see the text. (b) The same decomposition of the sequence into exons and introns and the corresponding path.

Quiz 11

There are two paths in the segment graph that describe exon–intron structures not represented in the exon–intron graph. What are they? What arcs need to be added to the exon–intron graph to represent these structures?

Lesson Structure matters. The same problem may be represented by different graphs, and the conceptually simplest representation is not necessarily the most efficient one.

6 Dynamic programming in a general situation. Physics of polymers

Let's return to our toy problem. Again, we have two sets of positive integers $x_1, ..., x_m$ and $y_1, ..., y_n$, but this time we want to calculate the product of all pair sums, $\prod_{i=1...m, j=1...n} (x_i + y_j)$. Can we use the same trick that we did before? Unfortunately, no. The reason for this is the properties of addition and multiplication: we have relied on the identity $x \cdot z + y \cdot z = (x + y) \cdot z$, but now we need $(x + z) \cdot (y + z) = x \cdot y + z$, and this generally is not true.

Quiz 12

When is $(x + z) \cdot (y + z) = x \cdot y + z$?

In our graph problems we were using two operations: calculating the path score (as the sum of the arc weights) and selecting the best path ending at a vertex (as the path of the maximum weight). We used the fact that if the score of a path P is larger than the score of a path Q, then for any arc a, the score of the path P with appended arc a, denoted (P, a), is larger than the score of the path (Q, a). Hence, at each vertex it was sufficient to retain the highest-scoring path ending at this vertex.

To write this condition more formally, let $\otimes$ be the operation of calculating the path score S given arc weights W. We require that this operation is associative, so that $(x \otimes y) \otimes z = x \otimes (y \otimes z)$; this obviously holds in all considered cases. Hence we may write simply $a \otimes b \otimes c$, without bothering about the order of operations, and thus $S(P) = \otimes_{a \in P} W(a)$ (this corresponds to $\sum_{a \in P} W(a)$ when the path score is defined as the sum of arc weights as above).

Let Ψ be the set of all paths from the source to the sink. We now slightly change the focus, and instead of constructing the best path, simply calculate its score, assuming this to be the total graph score $\Omega = \max_{P \in \Psi} S(P)$. Denote the operation of combining paths, which in all above paragraphs has been selecting the path of a higher score, by $\oplus$. We require that this operation is associative, $(x \oplus y) \oplus z = x \oplus (y \oplus z) = x \oplus y \oplus z$, and commutative, $x \oplus y = y \oplus x$.

In our new notation, $\Omega = \oplus_{P \in \Psi} S(P) = \oplus_{P \in \Psi} \otimes_{a \in P} W(a)$. The crucial property of path scores that has allowed for efficient computations, $\max(x + z, y + z) = \max(x, y) + z$, is rewritten as the *distribution law*

$$(x \otimes z) \oplus (y \otimes z) = (x \oplus y) \otimes z \tag{4.4}$$

(technically speaking, since we have not required $\otimes$ to be commutative, we also need $(x \otimes y) \oplus (x \otimes z) = x \otimes (y \oplus z)$).

Why is this new notation useful? Because now we can consider an even more general class of problems. To apply the standard dynamic programming algorithm for finding the maximum path score in a graph, it is sufficient to check that operations are commutative, associative, and satisfy the distribution law. The dynamic programming algorithm in this new notation is given in Figure 4.12. A trivial observation is that if $\oplus$ is the operation of taking the minimum, we immediately obtain the minimal score of a path from the source to the sink. A more interesting case is the following.

Data types:

```
vertices: v, u, Source, Sink;
arcs: (v,u);
weight of arc (v,u): W(v,u);
the current score of vertex v: S(v);
```

Initialize: for each vertex v: $S(v)$:= undefined;

Forward process: while There are unprocessed vertices:

```
v := arbitrary unprocessed vertex with all incoming arcs processed;
for each arc (v,u): // consider all arcs starting at v
    S(u) := S(u) ⊕ (S(v) ⊗ W(v,u)); // update the score of v
    (v,u) := processed_arc;
endfor;
v := processed_vertex;
```

endwhile.

Output $S(Sink)$.

Figure 4.12 General dynamic programming algorithm.

Problem 4 For a linear polymer chain of $L + 1$ monomers $k = 0, ..., L$, let each monomer assume N states $\sigma(k) \in \{\sigma_i | i = 1, ..., N\}$, and let the energy of interactions between adjacent monomers be defined by an $N \times N$ matrix $\xi(\sigma_i, \sigma_j)$ (measured in the KT units). For a particular conformation of the chain P, defined by the states of the monomers $\{\sigma(0), \sigma(1), ..., \sigma(L)\}$, let the exponent of its energy, $E(P)$, be the product of the exponents of its local interaction energies: $S(P) = e^{-E(P)} = \prod_{k=1...L} e^{-\xi(\sigma(k-1),\sigma(k))}$. Let Ψ be the set of all conformations. We need to calculate the *partition function of the set of all conformations* $\Omega = \sum_{P \in \Psi} S(P)$.

We construct a graph whose vertices correspond to monomer states, so that their number is $(L + 1) \cdot N + 2$ (two additional vertices are the source and the sink, corresponding to the virtual start and end of the chain), the arcs link vertices corresponding to adjacent monomers, and arc weights are the interaction energies. Paths through this graph exactly correspond to the chain conformations. If we set $\otimes$ to be ordinary multiplication, and $\oplus$ to be addition, the path score becomes the product of arc weights, and the total graph score is the sum of these products: this is exactly what we need, and we may immediately apply dynamic programming.

Quiz 13

(a) How many operations shall we need? (b) How many operations shall we need if we calculate the partition function directly?

 Quiz 14

Provide an algorithm for calculating the number of paths in a graph. Hint: recall Quiz 6.

 Quiz 15

What will Ω be if both $\otimes$ and $\oplus$ are the operation of taking the maximum?

We shall end with describing, without detail, one last problem of the polymer physics.

Problem 5   In the conditions of Problem 4, calculate the minimum energy and the number of conformations with the minimum energy.

This is solved as follows: arc weights are pairs $[1, \xi]$, with ξ as defined above, and path scores are pars $[n, \varepsilon]$, where ε is the energy, and n is the number of conformations having this energy. When two physical systems are combined, the resulting energy is the sum of the systems' energies, whereas the number of states is the product of the numbers of states. Hence, dynamic programming with $[n_1, \varepsilon_1] \otimes [n_2, \varepsilon_2] = [n_1 \cdot n_2, \varepsilon_1 + \varepsilon_2]$, and

$$[n_1, \varepsilon_1] \oplus [n_2, \varepsilon_2] = \begin{cases} [n_1, \varepsilon_1] & \text{if } \varepsilon_1 < \varepsilon_2, \\ [n_1 + n_2, \varepsilon], & \text{if } \varepsilon_1 = \varepsilon_2 = \varepsilon, \\ [n_2, \varepsilon_2], & \text{if } \varepsilon_1 > \varepsilon_2, \end{cases} \qquad (4.5)$$

solves the problem.

Lesson   Generalizations are useful.

Note   Not all problems that can be solved by dynamic programming have a simple graph representation. For example, reconstruction of the secondary structure of an RNA molecule given its sequence can be decomposed into simpler, embedded problems and can be solved by a variant of the dynamic programming algorithm, but in the language of this paragraph it requires slightly more complicated objects called hypergraphs.

Answers to Quiz

1   (a) $(y_1 + ... + y_n) \cdot m - 1$; (b) $(y_1 + ... + y_n) + m - 2$; (c) mn taking to the power and $mn - 1$ multiplications, or, better, n taking to the power and $m + n - 2$ multiplications; (d) one taking to the power, $m - 1$ multiplications, $n - 1$ additions.

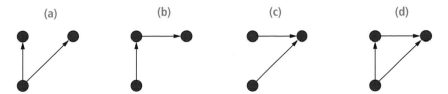

Figure 4.13 All connected acyclic graphs with three vertices.

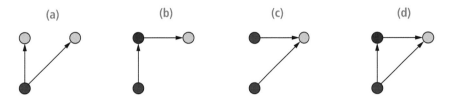

Figure 4.14 Sources are shown by blue circles; sinks, by yellow circles.

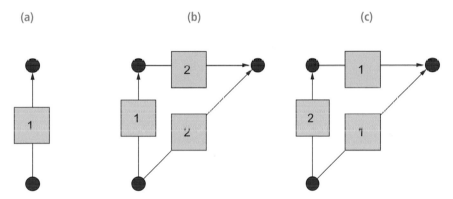

Figure 4.15 In (a) and (c) the greedy algorithm constructs the highest-scoring path; in (b) it does not.

2   (a) See Figure 4.13. (b) 18 graphs: 3 of type (a), 6 of type (b), 3 of type (c), 6 of type (d). The types are defined in Figure 4.13.

3   (a) Consider an arbitrary vertex. If it is an end of an arc, move to the start vertex of this arc. Continue in this manner. If you arrive at a vertex which is not the end for any arc, it is a source. Otherwise you will arrive at one of the already considered vertices and hence construct a cycle, in contradiction to the graph being acyclic. A similar construction works for the sinks. (b) See Figure 4.14.

4   Steps 5, 6, 7.

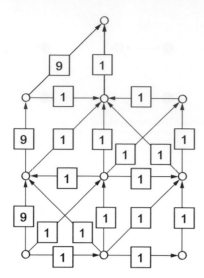

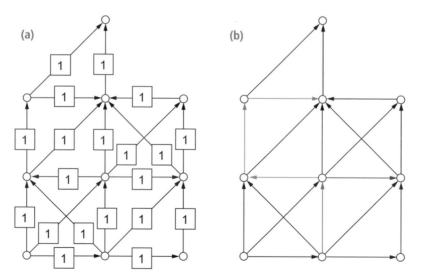

Figure 4.16 For this graph the greedy algorithm and the dynamic programming algorithm construct the same highest-scoring path.

Figure 4.17 (a) Arc weights for constructing the longest path. (b) Three different longest paths, shown by different types of colored arrows with mixed colors corresponding to common parts (green = yellow + blue; violet = blue + red; brown = yellow + blue + red).

5   (a–c) See Figure 4.15. (d) See Figure 4.16.
6   See Figure 4.17.
7   See Figure 4.18.

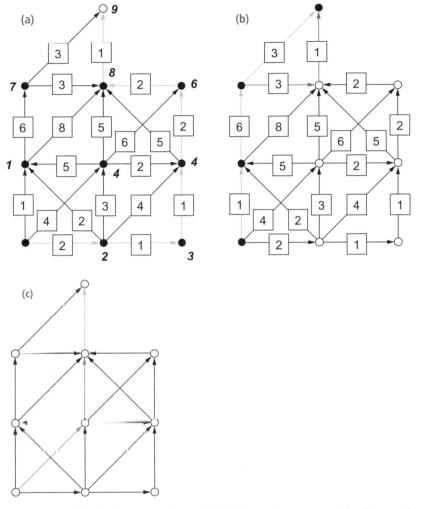

Figure 4.18 (a) The lowest-scoring path. (b) The path constructed by the greedy algorithm (note that there is a variant shown by dark green arcs). (c) Three different shortest paths (shown by different types of colored arrows). Notation in (a) and (b) as in Figure 4.5; color code in (c) as in Figure 4.17.

8 (a) $2r - 5p$; (b) $3r - p - 6q$; (c) $4r - 6q$.

9 See Figure 4.19.

10 (a) is optimal if $6q - 5p > 20$, (c) is optimal if $6q - 5p < 20$, (a) and (c) are tied if $6q - 5p = 20$. (b) is never optimal, since for a positive mismatch penalty of p it is always inferior to (c).

11 The path going through all top vertices (the entire sequence fragment is an intron) and the path going through all bottom vertices (the entire fragment is an exon). We need

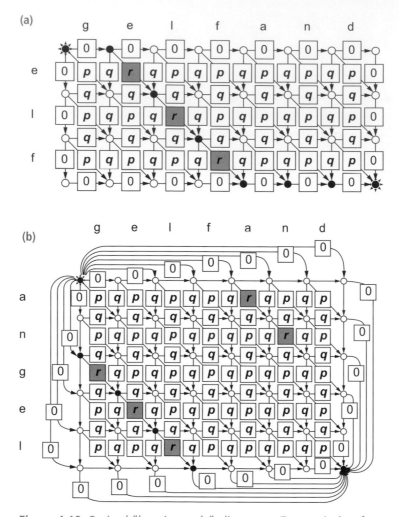

Figure 4.19 Optimal "hanging-ends" alignments. Two equivalent forms are given with (a) weights of side arcs set to 0; and (b) zero-weight arcs from source to side vertices and from side vertices to sink. Highest-scoring paths are shown by black vertices.

two arcs going from the source to the sink, one assigned an intron weight, and the other assigned an exon weight.

12 When $z = 0$ or $x + y + z = 1$.

13 (a) There are K^2 arcs between each layer of vertices corresponding to pairs of adjacent, interacting monomers, and there are L pairs, hence, $O(LK^2)$. (b) $O(L^K)$.

14 Set all arc weights to 1, $\otimes$ to be ordinary multiplication, and $\oplus$ to be addition. Each path weight is now exactly 1, and the sum of all path weights is the sum of 1s, whose number is the number of paths.

15 Maximal arc weight.

HISTORY, SOURCES, AND FURTHER READING

There exists a huge body of literature on the application of dynamic programming to biological problems, and this paragraph mentions only the first or best-known papers, or those that explicitly influenced the text above.

The dynamic programming algorithm was suggested by Bellman [1]. The matrix technique was introduced by Kramers and Wannier [2] and has been used in biophysics, in particular, for the analysis of helix–coil transitions in proteins by Zimm and Bragg [3] and in DNA by Vedenov et al. [4].

One of the first applications to molecular biology is due to Tumanyan, who used it to predict the RNA secondary structure given sequence [5]. The global alignment algorithm was developed by Needleman and Wunsch [6], and the local alignment was developed by Smith and Waterman [7]. Amino acid substitution matrices were first constructed by Dayhoff [8].

The idea of gene recognition using statistics of protein-coding and non-coding regions was introduced by Fickett [9] and Staden [10], and the dynamic programming was applied to this problem by Snyder and Stormo [11] as well as Roytberg and Gelfand [12].

The exposition here follows Finkelstein and Roytberg [13], and that paper contains several additional examples. The general algorithmic treatment in the formal language of semirings can be found in a textbook by Aho et al. [14]. A modern, closely related area using many similar approaches, Hidden Markov Models, is covered in a book by Durbin et al. [15].

REFERENCES

[1] R. E. Bellman. *Dynamic Programming*. Princeton University Press, Princeton, NJ, 1957.

[2] H. A. Kramers and G. H. Wannier. Statistics of the one-dimensional ferromagnet. *Zeitschr. Phys.*, 31:253–258, 1941.

[3] B. H. Zimm and J. R. Bragg. Theory of the phase transitions between helix and random coil in polypeptide chains. *J. Chem. Phys.*, 31:526–535, 1959.

[4] A. A. Vedenov, A. M. Dykhne, A. D. Frank-Kamenetsky, and M. D. Frank-Kamenetsky. To the theory of the transitions helix–coil in DNA. *Mol. Biol. (USSR)*, 1:313–318, 1967.

[5] V. G. Tumanyan, L. E. Sotnikova, and A. V. Kholopov. On identification of secondary RNA structure from the nucleotide sequence. *Doklady Biochemistry*, 166:63–66, 1966.

[6] S. B. Needleman and C. D. Wunsch. A general method applicable to the search for similarities in amino acid sequence of two proteins. *J. Mol. Biol.*, 148:443–453, 1970.

[7] T. F. Smith and M. S. Waterman. Identification of common molecular subsequences. *J. Mol. Biol.*, 147:195–197, 1981.

[8] M. O. Dayhoff, R. Schwartz, and B. C. Orcutt. A model of evolutionary change in proteins. In: *Atlas of Protein Sequence and Structure*, Vol. 5, Suppl. 3. National Biomedical Research Foundation, Washington, DC, 1978, 345–358.

[9] J. W. Fickett. Recognition of protein coding regions in DNA sequences. *Nucl. Acids Res.*, 10:5303–5318, 1982.

[10] R. Staden and A. D. McLachlan. Codon preference and its use in identifying protein coding regions in long DNA sequences. *Nucl. Acids Res.*, 10:141–156, 1982.

[11] E. E. Snyder and G. D. Stormo. Identification of coding regions in genomic DNA sequences: An application of dynamic programming and neural networks. *Nucl. Acids Res.*, 21:607–613, 1993.

[12] M. S. Gelfand and M. A. Roytberg. Prediction of the exon–intron structure by a dynamic programming approach. *BioSystems*, 30:173–182, 1993.

[13] A. V. Finkelstein and M. A. Roytberg. Computation of biopolymers: A general approach to different problems. *BioSystems*, 30:1–19, 1993.

[14] A. Aho, J. Hopcroft, and J. Ullman. Design and analysis of computer algorithms. Addison-Wesley, Reading, MA, 1976.

[15] R. Durbin, S. R. Eddy, A. Krogh, and G. J. Mitchison. *Biological Sequence Analysis: Probabilistic Models of Proteins and Nucleic Acids.* Cambridge University Press, Cambridge, 1998.

CHAPTER FIVE

Measuring evidence: who's your daddy?

Christopher Lee

Single nucleotide polymorphisms (SNPs) are widely used as a genetic "fingerprint" for forensic tests and other genetic screening. For example, they can be used to measure evidence for paternity. To understand how scientists measure the strength of such evidence, we introduce basic principles of statistical inference using Bayes' Law, and apply them to simple genetics examples and the more challenging case of paternity testing. But first, just to make it personal, Maury and I have a little revelation for you ...

 ## Welcome to the Maury Povich Show!

On camera, your mom just told you that your dad, Bob, isn't your real dad! And Maury has just introduced you to the two men who both claim to be your father: Rocco, an aging biker dude with lots of tatoos; and Jacques, a chef in whose restaurant your mom waitressed 18 years ago. But is either of them actually your father? Once again it's time to announce the results of a paternity test LIVE on the Maury Povich Show! But between your tears ("But what about Dad ... er, my ex-Dad ..."), your anger ("how could you do this to me ..."), and your intellectual curiosity ("Does this mean I can get the 8 course tasting menu at Chez Jacques for free?"), the science-nerd part of your mind is wondering exactly how paternity tests work, and how Maury can really claim to have so many decimal places of confidence regarding the result. Read on.

Bioinformatics for Biologists, ed. P. Pevzner and R. Shamir. Published by Cambridge University Press.
© Cambridge University Press 2011.

1.1   What makes you you

You already know the basics about DNA, the famed double helix. You know that it stores your "genetic code" that encodes the genes and proteins that build your body. Of course, your DNA is not exactly the same as anyone else's DNA – even your mom's, since you have two copies of each chromosome, one copy from your mom and one copy from your dad. There are many kinds of DNA differences from person to person, ranging from substitution of a single "base" in the sequence, to insertion, deletion, or rearrangement of a large interval on a chromosome. Numerically, single base substitutions are the most common. Scientists call them "single nucleotide polymorphisms" (SNPs, pronounced "snips"), where the term "polymorphism" means that the substitution is found in only a portion of the human population, while the original base (nucleotide) is found in the remainder. SNPs' abundance makes them a good candidate for use as a "molecular fingerprint" that uniquely identifies each human individual, for paternity tests, forensic tests, etc. For an individual person, only three states are possible for a specific SNP: you either inherited it from both your parents ("homozygous"), from only one of your parents ("heterozygous"), or from neither of your parents ("homozygous normal"). In other words, because you have two copies of each gene, you can only have two, one, or zero copies of a given SNP.

SNPs are extremely interesting scientifically and historically. Some SNPs cause serious diseases such as sickle-cell anemia. For example, β-hemoglobin is a vital component of red blood cells, and helps carry oxygen in the blood. A SNP in the gene encoding β-hemoglobin causes the protein to polymerize into fibers that distort the red blood cell into a sickle-like shape, and damage them. If you inherit a β-hemoglobin gene containing the sickle-cell SNP from *both* your mother and father (i.e. homozygous), you will develop this serious disease. On the other hand, if you inherit one normal copy of the gene (no SNP) from one parent, and one copy containing the SNP from the other parent (i.e. heterozygous), not only does this combination not cause sickle-cell disease, but it actually *protects* you from a completely different disease, malaria (specifically, it reduces your risk of severe malaria by about 10-fold). You will perhaps not be surprised to learn that the sickle-cell SNP appears to have originated in tropical areas of Africa where malaria is common. Scientists believe the sickle-cell SNP is relatively common (despite the fact that it causes sickle-cell disease) because of this protective effect against malaria. Other SNPs cause more moderate but still potent effects on traits such as human personality. For example, serotonin is an important neurotransmitter involved in many aspects of mood and behavior. A number of SNPs in genes affecting serotonin have been shown to significantly change an individual's risk of attempting suicide. Chinese researchers reported that among patients with severe depression, those who were homozygous for one such SNP were three times less likely to attempt suicide, compared with those who were heterozygous or homozygous normal. Since there are over three million common SNPs in the human genome (and an even greater number of less frequent SNPs), an enormous amount of research is ongoing to discover those that play a causal or diagnostic role in human diseases.

Where does a SNP come from? At some moment in the past, a mutation occurred in one person's DNA, either due to ultraviolet light, radiation, or simply the imperfect fidelity of the molecular machinery that copies DNA. This newly created SNP will be passed on to half of that person's descendants on average (which could be a huge number of people, if the population is expanding). Due to random oscillations in the SNP's frequency among successive generations (referred to as "genetic drift"), over time it is increasingly likely either to vanish from the human population, or alternatively become "fixed" in the population (i.e. *everyone* has it). The fact that the SNPs that we detect today haven't reached either of those end points implies that they are relatively recent (in evolutionary terms).

Of course, when a SNP is first created, it isn't created in a vacuum, but in a context of other pre-existing SNPs. In other words, the chromosome on which the new SNP is created *already* contained many SNPs. So at first this SNP is *always* found with that unique fingerprint of SNPs; this is referred to as *genetic linkage*. In successive generations, this linkage will gradually be cut down by the process of *homologous recombination*, in which a matched pair of chromosomes exchange one or more segments. As a result, the SNP will no longer show its original 100% linkage to other SNPs on the entire chromosome, but instead only to neighboring SNPs that are so close to it that no recombination event has yet occurred between them. Over time, recombination events on that chromosome will whittle away these linkages, until eventually the SNPs become no more likely to be found together than expected by random chance. Since recombination is more likely between SNPs that are distant from each other, these associations disappear first. For this reason, the region of SNPs linked to the new SNP will gradually shrink. Thus the size of the "island" of linkage around a given SNP directly tells you how old it is, and the specific SNPs that are linked gives you a "genetic fingerprint" of the person in whom the SNP was first created. Everyone who has that SNP today is descended from that one person.

Think about it. Each one of the three million common SNPs in the human provides a detailed recording of who's related to who, who invaded who, when, etc. Historians have never had such a detailed record of history for each individual before – and it reaches deep into the past, into prehistory. Indeed, some human SNPs are also found in chimpanzees. That means they occurred in an ancestor of both humans and chimps. That's *old*.

1.2 SNPs, forensics, Jacques, and you

That may be fascinating for us science-nerds, but why should Maury Povich care about SNPs? Because SNPs provide an easy and inexpensive way to identify one person's DNA vs. another's, and test relatedness very precisely. Both forensic DNA tests and paternity tests can take advantage of this. And Maury is all over those paternity tests!

Great technology exists for detecting SNPs *en masse*. A single microchip (called a DNA microarray) can detect nearly a million different SNPs simultaneously; a single test machine can run over 750 such microarray samples per week. Only a tiny amount of DNA (200 ng) is required to perform the analysis. Both Rocco and Jacques are good for giving you that amount of their DNA, so your paternity test is a GO. The DNA sample is fragmented into very small pieces (25 to 125 bp), labeled with a fluorescent dye, and placed on the microarray. If a specific SNP is present in the sample, that piece of DNA will bind (base-pair) to a corresponding "probe sequence" on the DNA microarray, which is then scanned with a laser to detect fluorescence at each SNP location on the array. The output signal is simply the amount of fluorescence detected for each SNP. Since each person has two copies of every chromosome (each of which could either have the SNP, or not) the fluorescent signal should cluster into three distinct peaks: little or no fluorescence (indicating that the SNP was absent from both copies); medium fluorescence (indicating the SNP was present on only one copy); and bright fluorescence (indicating its presence on both copies).

If we were performing a forensic DNA test to see whether a suspect's DNA matches a sample obtained from a crime scene, we'd just check whether these fluorescence values matched between the two samples, for every SNP on the array. However, for a paternity test it's a lot more complicated: we don't expect an exact match between your true father and you; you got half your DNA from your mom, and half from your dad. Typically, when you compare your result vs. Jacques' result for a given SNP, there is no definitive interpretation, since most of the possible results are consistent with both him being your father, or not. There are only two clear-cut cases: if Jacques appears to have two copies of a SNP, and you have no copy (or vice versa), he should not be your father. However, these cases are very rare. Moreover, while a typical SNP result may not be interpretable by itself, it does supply useful information on whether he's *likely* to be your father. What we would like to do is develop a computational method that measures the *total evidence* from all the SNPs on the microarray to assess the probability that Jacques is your father.

This is a problem of *statistical inference* – reasoning under uncertainty. It has many angles, but its core principles are both extremely useful and surprisingly simple to learn. Read on.

2 Inference

2.1 The foundation: thinking about probability "conditionally"

Consider the kinds of statements about probability we often hear in the media, such as "the probability of rain is 80%," or "The company's new AIDS diagnostic test is 97% accurate." Mathematicians call these *unconditional probability* statements, which we write as:

$\Pr(H) \equiv$ total probability of event H (over the set of all possible events S).

Using the intuitive concept of probability as the fraction of possible events that meet a particular condition, and indicating "the count of events where H occurred" as $|H|$, this simply becomes

$$\Pr(H) = \frac{|H|}{|S|}.$$

A more sophisticated way to talk about probability is to specify exactly what condition it was measured under. We write a *conditional probability* in the form

$\Pr(H|O) \equiv$ probability that event H occurs in the subset of cases where event O did indeed occur.

Treating these as sets in a "Venn diagram," see Figure 5.1, we write their "intersection" as $H \cap O$. Using this notation, the conditional probability becomes

$$\Pr(H|O) = \frac{|H \cap O|}{|O|}.$$

Following this logic, we can express the "joint probability" that *both* H and O occur, in terms of their separate conditional and unconditional probabilities:

$$\Pr(H \cap O) = \frac{|H \cap O|}{|S|} = \frac{|H \cap O|}{|O|} \frac{|O|}{|S|} = \Pr(H|O)\Pr(O). \tag{5.1}$$

Furthermore, since the order of H, O does not matter for the "intersection" operation (i.e. $H \cap O = O \cap H$), we can equally correctly write the reverse:

$$\Pr(H \cap O) = \Pr(O|H)\Pr(H).$$

Finally, note that our definition of probability inherently sums to one whenever we sum it over the entire set S, as long as our individual "pieces" H do not overlap.

$$\sum_H \Pr(H) = \frac{|H_1|}{|S|} + \frac{|H_2|}{|S|} + \ldots + \frac{|H_n|}{|S|} = \frac{|S|}{|S|} = 1.$$

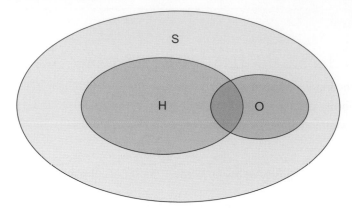

Figure 5.1 A Venn diagram illustrating the conditional probability identity. Each ellipse represents the set of occurrences of a specified event, H or O. The larger ellipse S constitutes the set of all possible events considered in this probability calculation. The intersection $H \cap O$ represents events where both H and O co-occurred.

This property is called "normalization." Applied to a joint probability, it gives another important principle:

$$\sum_H \Pr(H \cap O) = \sum_H \Pr(H|O)\Pr(O) = \left(\sum_H \Pr(H|O)\right)\Pr(O) = \Pr(O).$$

Thus, we can *eliminate* a variable from a joint probability by summing over all possible values of that variable.

2.1.1 The disease test

To understand how this matters for everyday life, let's look at a simple example. A company reports that their new test for a disease is 97% accurate. Table 5.1 shows the raw data, which appear to support this claim. Among patients who do not have disease, the test gives the right answer $960/990 = 97\%$ of the time, and among patients who have disease (a much rarer case), it gives the right answer $9/10 = 90\%$ of the time.

There is just one catch here: these are not the conditional probabilities that a doctor (or patient) cares about! The whole point of the test result (T) is to give information about whether the patient has disease (D); we want to use the *observed* variable T to learn about the *hidden* variable D. Thus the probabilities above ($\Pr(T^-|D^-)$ and $\Pr(T^+|D^+)$) are irrelevant and useless. What we really care about is the converse, the probability that a patient has disease given a positive test result, $\Pr(D^+|T^+)$. And there's the rub: $\Pr(D^+|T^+) = 9/39 = 23\%$. More than three-quarters of the patients

Table 5.1 *A diagnostic disease test*: 1,000 patients were given a diagnostic test that gives either a positive (T^+) or negative (T^-) result, and independently assessed for whether they have the disease (D^+) or not (D^-) by rigorous clinical criteria.

	T^-	T^+	Total
D^+	1	9	10
D^-	960	30	990
Total	961	39	1000

with positive test results do not actually have the disease! This could be a very serious problem, not only because of the stress of patients' being (falsely) told they have the disease, but also because this may subject them to additional expensive and possibly dangerous procedures.

This example illustrates several lessons.

- The "perfect lie": as this example shows, an unconditional probability statement can be both completely misleading and at the same time "factually correct"! The problem with an unconditional probability is that it doesn't tell you what conditions were used to obtain it. What assumptions (sensible or insane) gave rise to this number? You don't know. By choosing different conditions, I can select a number that suits my purposes. As the example demonstrates, even within the strict limits of the correct data, freedom to pick our conditions gives us enough latitude to turn the conclusion upside down! The purpose of conditional probability is to make assumptions explicit.
- Strictly speaking, *every* probability calculation has at least some assumptions. So an unconditional probability statement is really a conditional probability traveling *incognito* – without telling you what its conditions were.
- It is a fatal mistake to confuse one conditional probability with its converse (i.e. $\Pr(X|Y)$ vs. $\Pr(Y|X)$). They are quite different! Once you're aware of this distinction, you will find that people mix up converse probabilities all the time, sometimes due to poor thinking, and sometimes deceptively. When you listen to a politician, newspaper article, advertisement, or anyone else with "something to sell," see if you can catalog all the sins they commit against conditional probability. Remember that "97% test accuracy" may be completely irrelevant to the question that matters – especially if they don't even tell you what conditional probability it represents!

2.2 Bayes' Law

This is all very well, but you may be wondering how this helps us decide whether Jacques is your father. The answer is, conditional probability leads immediately to a simple law for inference. Since (by symmetry) it is equally true that

$$\Pr(H|O)\Pr(O) = \Pr(H \cap O) = \Pr(O|H)\Pr(H).$$

So

$$\Pr(H|O) = \frac{\Pr(O|H)\Pr(H)}{\Pr(O)}. \tag{5.2}$$

This is *Bayes' Law*, and it is inference in a nutshell. It allows us to compute the probability of some *hidden* event H given that some *observable* event O has occurred, provided that we know the converse probability that observation O will occur *assuming* H has occurred. (Intuitively, let's define "observable" as any variable that we can measure directly, with zero uncertainty, and "hidden" as everything else.) For convenience we often replace $\Pr(O)$ by the sum of $\Pr(H \cap O)$ over all possible values of H. Note that this is equivalent to summing the expression that appears in the numerator, and is called "normalizing" the probabilities, since it makes them add up to 1 as probabilities always should.

$$\Pr(H|O) = \frac{\Pr(O|H)\Pr(H)}{\sum_h \Pr(O|h)\Pr(h)}. \tag{5.3}$$

To see how Bayes' Law solves problems, let's look at a simple genetics example.

2.3 Estimating disease risk

A disease is defined as "recessive" if a single copy of the normal gene is sufficient to prevent disease, even if one copy of the genetic variant that causes disease is also present. Say a disease gene has been mapped to the X chromosome. Women have two copies of the X chromosome (they have two female sex chromosomes, XX) whereas men have only one copy (they have one X chromosome and one Y chromosome, XY). For this reason, recessive traits that map to the X chromosome behave differently in men as compared to women. For a man, a single bad copy of the gene (which we will symbolize as x) will give him disease. Such a man will be xY, whereas a woman with one copy of the disease gene (xX) will not develop disease symptoms, because she still has one "good copy" of the gene. Such a woman is referred to as a "disease carrier." Only women with two bad copies of the gene (xx) will show symptoms of the disease.

Consider a woman M who is a disease carrier (xX); she will have no symptoms (which we will symbolize as M^-), but her sons are at high risk for the disease, because they only inherit the X chromosome from their mother (they inherit a male

Y chromosome from their father; only daughters inherit an X chromosome from the father). Specifically, each son S has a 50% probability of inheriting his mother's "bad copy" of the gene (x) and developing disease symptoms, which we will symbolize as S^+.

Let's say a woman comes from a family background where the disease allele x is $\Pr(x) = 0.1$ (i.e. 10%), but shows no symptoms. If she has a single son who is symptom-free (S^-), what is the probability that she is a disease carrier (xX)? We simply apply Bayes' Law:

$$\Pr(xX|S^-) = \frac{\Pr(S^-|xX)\Pr(xX)}{\Pr(S^-|xX)\Pr(xX) + \Pr(S^-|XX)\Pr(XX) + \Pr(S^-|xx)\Pr(xx)}.$$

We know the probabilities of the observations: $\Pr(S^-|xX) = 0.5$, $\Pr(S^-|xx) = 0$, and $\Pr(S^-|XX) = 1$. We also know the probabilities of the woman's genes: $\Pr(XX) = (1 - \Pr(x))^2 = 0.81$, and $\Pr(xx) = \Pr(x)^2 = 0.01$. Thus, without considering any observations, her probability of being a disease carrier is just the remainder, $\Pr(xX) = 1 - 0.81 - 0.01 = 0.18$. Taking into account the observation that her son is symptom-free,

$$\Pr(xX|S^-) = \frac{0.5(.18)}{0.5(.18) + 1(.81) + 0(0.01)} = 0.1. \tag{5.4}$$

Thus, having one disease-free son reduces her probability of being a disease carrier by approximately a factor of 2. (If you want deeper insight into where this number comes from, consider the fact that this outcome (S^-) is twice as likely under the dominant state, XX.) Note that we didn't really need to consider the xx case, since it's completely incompatible with the observation S^-, and thus makes no contribution to the sum.

What if she has a second disease-free son?

$$\Pr(xX|S^-S^-) = \frac{\Pr(S^-S^-|xX)\Pr(xX)}{\Pr(S^-S^-|xX)\Pr(xX) + \Pr(S^-S^-|XX)\Pr(XX)}$$

$$= \frac{0.5(0.5)(.18)}{0.5(0.5)(.18) + 1(1)(.81)} = 0.053.$$

Again the probability has dropped by another factor of 2 (approximately).

What if the woman now has a third son who shows disease symptoms?

$$\Pr(xX|S^-S^-S^+) = \frac{\Pr(S^-S^-S^+|xX)\Pr(xX)}{\Pr(S^-S^-S^+|xX)\Pr(xX) + \Pr(S^-S^-S^+|XX)\Pr(XX)}$$

$$= \frac{0.5(0.5)(0.5)(.18)}{0.5(0.5)(0.5)(.18) + 1(1)(0)(.81)} = 1.$$

A single observation has caused the probability of xX to rocket from 5.3% to 100%, for the simple reason that this observation is impossible under the XX model. Thus

Bayesian inference correctly models even somewhat subtle reasoning processes, which can produce rather dramatic effects like this: a single observation can completely change the entire result. We can see from this example a general principle: a "powerful" observation (one that can change our conclusions dramatically) is one that is highly unlikely under the currently most probable model.

2.4   A recipe for inference

Now that we've seen Bayes' Law in action, we should take stock and try to generalize what we've learned. We can use Bayes' Law as a "recipe" whose parts give us a very clear list of the ingredients necessary for solving any inference problem. Let's take each term of Bayes' Law, give it a name, and state precisely what role it plays in inference:

$$\Pr(H|O) = \frac{\Pr(O|H)\Pr(H)}{\sum_H \Pr(O|H)\Pr(H)}. \tag{5.5}$$

- *What is observed* (O)? The core of inference is distinguishing clearly between hidden variables vs. observed variables. We must be careful not to miscategorize as "observable" quantities that actually are hidden. In general, anything that has uncertainty cannot be considered to be "observable," and should instead be considered hidden.
- *What is hidden* (H)? In science, most things we want to know fall into this "hidden" category; the real question is how to formulate what we want to know as a precise mathematical parameter. This means deciding which aspects of the outward appearance of a problem are extraneous and should be ignored, versus which part(s) are core. And that is the essence of our next ingredient ...
- *What is the likelihood model* $\Pr(O|H)$? In Bayesian inference, the probability of an observation given a hidden state is referred to as a *likelihood*, and the function that allows us to calculate it for a specified pair of observable and hidden variables is a *likelihood model*. Choosing a likelihood model means proposing a process that explains how the observations were produced. A likelihood model usually depends on one or more hidden parameters that shape it. For example, if the observable can only have two possible outcomes (e.g. "rain" vs. "no rain"), one possible model is to assume that each event outcome occurs independently (i.e. whether it rained yesterday has *no* effect on whether it will rain today). This model is called the *binomial probability distribution*, and has only one hidden parameter (usually called θ), the probability of our primary outcome (e.g. the probability that it will rain on any given day). So in this case we would use the binomial distribution as our likelihood

equation, and we would treat θ as the hidden variable whose value we are trying to infer.

- *What is the prior* $\Pr(H)$? We refer to the unconditional probability of H (in the absence of any observations) as its "prior probability." There are two types of priors: those measured directly from previous data sets (as posteriors, see below); and *uninformative* priors. The most common uninformative prior is just a *constant*; in this case, the prior simply cancels from numerator and denominator. However, it should be remembered that priors are important, and that they are one of the major differences between Bayesian inference and other approaches (e.g. maximum likelihood).
- *What is the set of all possible models?* The summation in the denominator must be taken over all possible values of the hidden variable(s).
- *What is the posterior* $\Pr(H|O)$? With all of the above ingredients in hand, we can finally calculate the result, the evidence for a specific model H given the set of observations O. This is called the *posterior probability* of model H.

Paternity inference

So how can we apply all this to Rocco and Jacques' DNA samples to determine which (if either) is your dad? *We just follow the recipe.*

- *What is observed?* The fluorescence signal for each probe on the microarray. Let's call it A for the "candidate dad" sample; B for your DNA sample.
- *What is hidden?* To keep things simple, let's consider only one candidate dad (Rocco or Jacques) at a time. We'll construct two models *dad* and *not-dad*, and calculate their relative posterior probabilities given the observations for that candidate dad. However, there is a bit more to this problem: to calculate these probabilities using SNPs, we also need to determine for each sample how many copies of each SNP it contains. That too is a hidden variable; let's call it $\alpha = 0, 1, 2$ for the "candidate dad", and $\beta = 0, 1, 2$ for you.
- *What is the likelihood model* $\Pr(A|\alpha)$? As we stated before, the fluorescence signal tends to cluster into three distinct peaks, one for each possible value of $\alpha = 0, 1, 2$ (Figure 5.2). Note that the figure represents good separation between the three peaks, which will give stronger paternity results. Bear in mind that for some probes, the three peaks will not be well separated, creating strong uncertainty about the true value of α. Our statistical inference calculation will automatically take this into account in its computation of the evidence.
- *What is the prior* $\Pr(\alpha)$? Say the frequency of the SNP on chromosomes in the general human population is f. Then the chance of getting 2 copies of the SNP is just

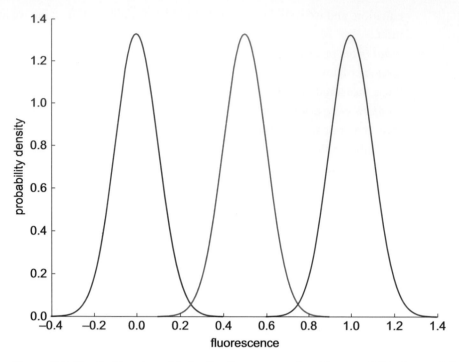

Figure 5.2 The likelihood models for the fluorescence signal for $\alpha = 0$ (blue), $\alpha = 1$ (green), and $\alpha = 2$ (red) for an idealized SNP. As you can see, the fluorescence signal indicates approximately what fraction of the DNA sample contains the SNP.

$Pr(\alpha = 2|f) = f^2$; similarly, the probability of getting 0 copies is $Pr(\alpha = 0|f) = (1 - f)^2$. Consequently the remaining probability $Pr(\alpha = 1|f) = 1 - f^2 - (1 - f)^2 = 2f(1 - f)$.

Next, what should we use as the prior probability $Pr\,(dad)$? Conservatively, your dad could be any adult male on planet Earth, so we can set $Pr(dad) = 1/(3 \times 10^9)$, and $Pr(not\text{-}dad) = 1 - Pr(dad)$.

- *What is the set of all possible models?* There are two possible cases: either the candidate is your dad, or not. For the *not-dad* model, we simply treat α, β as being drawn from the general population, i.e. each just depends on f. For the *dad* model, we make β depend partly on α (because half your DNA comes from your dad). See Figure 5.3 to compare our two models.

Let's consider exactly how the *dad* model modifies our prior for β. For example, if your dad has α copies of the SNP, the chance of getting the SNP from him is $\alpha/2$. Assuming that we don't have any SNP data from your mom, we simply treat her as a member of the general population, i.e. your chance of inheriting a copy of the SNP

(a)
(b)

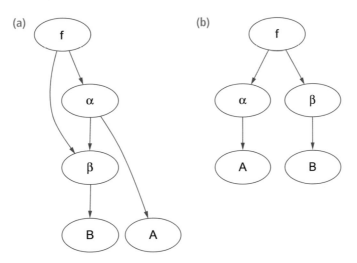

Figure 5.3 Dependency structure of the (a) *dad* model; (b) *not-dad* model.

from her is just f. From this we can immediately infer that your probability of getting $\beta = 2$ copies (i.e. one from your dad, and one from your mom) is just

$$\Pr(\beta = 2|dad, \alpha, f) = \frac{\alpha}{2}f.$$

We can apply the same logic to the $\beta = 0$ case, i.e. your probability of inheriting no copy of the SNP from both your dad and your mom:

$$\Pr(\beta = 0|dad, \alpha, f) = \frac{2 - \alpha}{2}(1 - f).$$

Actually, we're almost done! There is only one more possible case, whose probability we can get by simply subtracting the previous two cases from 1 (after all, the probability of all three cases must sum to 1!):

$$\Pr(\beta = 1|dad, \alpha, f) = 1 - \frac{\alpha}{2}f - \frac{2 - \alpha}{2}(1 - f) = \frac{\alpha}{2} + f - \alpha f.$$

- *What is the posterior* $\Pr(dad|A, B)$? We just follow Bayes' Law, to compute the ratio of the posterior probabilities for the *dad* vs. *not-dad* models. This calculation is easier than it looks. First of all, note that the denominator of Bayes' Law is the same no matter what model you apply it to. For our problem, Bayes' Law gives:

$$\Pr(dad|A, B, f) = \frac{\Pr(A, B|dad, f)\Pr(dad)}{\Pr(A, B|f)}.$$

So if all we want is the "odds ratio" of the posterior probabilities of the two models *dad* vs. *not-dad*, we can just calculate the ratio of the *numerator* of Bayes' Law for the two models:

$$\frac{\Pr(dad|A, B, f)}{\Pr(not\text{-}dad|A, B, f)} = \frac{\Pr(A, B|dad, f)\Pr(dad)}{\Pr(A, B|f)} \frac{\Pr(A, B|f)}{\Pr(A, B|not\text{-}dad, f)\Pr(not\text{-}dad)}$$

$$= \frac{\Pr(A, B|dad, f)\Pr(dad)}{\Pr(A, B|not\text{-}dad, f)\Pr(not\text{-}dad)}.$$

Next, let's look at the likelihood $\Pr(A, B|dad, f)$. We know how to compute a probability that includes the additional variables α, β, i.e.

$$p(A, B, \alpha, \beta|dad, f) = p(A|\alpha)p(B|\beta)p(\alpha|f)p(\beta|dad, \alpha, f).$$

So the obvious question is, how do we get rid of α, β from this probability? That's easy: we just sum over all possible values of $\alpha = 0, 1, 2$, and $\beta = 0, 1, 2$:

$$p(A, B|dad, f) = \sum_{\alpha=0}^{2}\sum_{\beta=0}^{2} p(A, B, \alpha, \beta|dad, f).$$

Plugging in the various probability terms we have

$$\Pr(A, B|dad, f) = \sum_{\alpha=0}^{2}\left(\Pr(\alpha|f)\Pr(A|\alpha)\sum_{\beta=0}^{2}\Pr(\beta|dad, \alpha, f)\Pr(B|\beta)\right)$$

and

$$\Pr(A, B|not\text{-}dad, f) = \left(\sum_{\alpha=0}^{2}\Pr(\alpha|f)\Pr(A|\alpha)\right)\left(\sum_{\beta=0}^{2}\Pr(\beta|f)\Pr(B|\beta)\right).$$

Now we're ready to plug in some data from Jacques and you: the first SNP reading ($A \approx 0.5$, $B \approx 0.5$) indicates $\alpha = 1$ for Jacques and $\beta = 1$ for you (i.e. you both have one copy of the SNP). This result could occur both if Jacques were your father, and if he weren't (you could have gotten this SNP from your mother). But now we can use our probability calculations to weigh the evidence. It turns out to depend strongly on the SNP's frequency in the population (f); see Figure 5.4. At high SNP frequency, the fact that both Jacques and you have the SNP might well just be a coincidence, leading to a *dad/not-dad* ratio of approximately one (i.e. neither model is favored over the other). However, as the SNP frequency becomes smaller, this becomes increasingly unlikely, and gives stronger evidence that Jacques is your father. As you can see from Figure 5.4, the calculations show that at this SNP's known frequency (10%), the data favor the *dad* model by about threefold.

So far we've restricted ourselves to talking about the calculation for a single SNP. But there are a million SNPs on the microarray! Combining the evidence for all the SNPs is very simple. Assuming that our SNP marker set was chosen to be non-redundant (each SNP in the set is independent of the others), we can simply multiply

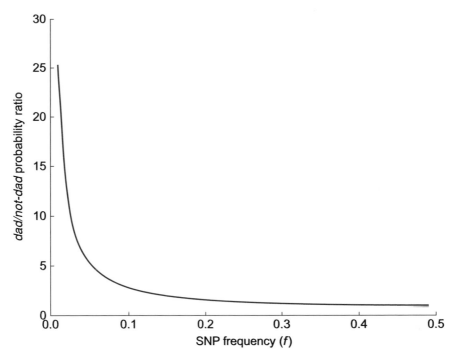

Figure 5.4 Effect of SNP frequency *f* (*x*-axis) on *dad/not-dad* ratio (*y*-axis).

the probabilities computed for each SNP. Even if the evidence from any one SNP is relatively weak, over a million SNPs the total evidence will add up very quickly, to a very big number favoring the correct model and rejecting the incorrect model. Remember that to convince us that the candidate really is your father, the evidence in favor of the *dad* model must be much bigger than the prior odds ratio that we made favor the *not-dad* model (by 3×10^9).

Note that we'll do this analysis separately for Rocco and Jacques. If one of them gets a huge odds ratio in favor of the *dad* model, and the other does not, that would constitute an unambiguous result. Note that there are deeper issues that this calculation does not fully capture; for example, close relatives would also get a favorable odds ratio (because they are more related to you than random), but the result would not be as strong. Additional calculation is required to find the right threshold for distinguishing a true father from a more distant relative.

Note also that we ignored your mother's genetic information in this analysis. We could make it even more accurate, if we included her DNA sample in the calculation as well. This would be very easy to do: we would just make your state (β) depend on your mom's state just like we made it depend on your dad's state (α).

QUESTIONS

(1) What would happen if the fluorescence observations from the "candidate *dad*" (variable A) actually came from your true father's brother? On average, how will the value of $(\Pr A, B \mid dad, f)$ compare with the value expected if the A data really came from your father? On average, how will the value of $\Pr(A, B \mid not\text{-}dad, f)$ compare with the value expected if the A data really came from someone unrelated to your father? What about if the fluorescence observations actually came from your mother?

(2) How exactly would you modify the model to incorporate fluorescence observations (call them variable C) derived from a sample of your mom's DNA? Derive an expression for $\Pr(A, B, C \mid dad, f)$.

(3) How would the model defined in Question 2 handle the case in which the "candidate *dad*" observations (variable A) are actually from your mom's DNA? Specifically, on average, how will the value of $\Pr(A, B, C \mid dad, f)$ compare with the value expected if the A data really came from your father? On average, how will the value of $\Pr(A, B, C \mid not\text{-}dad, f)$ compare with the value expected if the A data really came from someone unrelated to you? How does this compare with the original model presented in the chapter?

PART II

GENE TRANSCRIPTION AND REGULATION

CHAPTER SIX

How do replication and transcription change genomes?

Andrey Grigoriev

From the evolutionary standpoint, DNA replication and transcription are two fundamental processes enabling reliable passage of fitness advantages through generations (in DNA form) and manifestation of these advantages (in RNA form), respectively. Paradoxically, both of these basic mechanisms not only preserve genetic information but also apparently cause systematic genomic changes directly. Here, I show how genome-scale sequence analysis can help identify such effects, estimate their relative contributions, and find practical application (e.g. for predicting replication origins). Visualization of bioinformatics results is often the best way of connecting them to the underlying biological question and I describe the process of choosing the visual representation that would help compare different organisms, genomes, and chromosomes.

 ## Introduction

A species' genome relies on faithful reproduction to reap the benefits of selection. The very fact that the "fine-tuned" genomes of previous generations carrying important fitness advantages can be preserved in the proliferating progeny is the basis of natural selection. That is how we currently understand evolution and life around us, and this grand scheme can operate only under stringent requirements for the precision with which DNA replicates. It is not surprising, therefore, that one observes higher replication fidelity in more complex organisms.

Bioinformatics for Biologists, ed. P. Pevzner and R. Shamir. Published by Cambridge University Press.
© Cambridge University Press 2011.

For the sake of clarity, however, we leave the "more complex organisms" aside for the duration of this chapter. The higher fidelity mentioned above results from many additional processes (including advanced repair) taking place in a cell besides replication. In order to see the inherent properties of one of the key processes in sustaining life, replication and its effects are best observed in simple creatures – bacteria, viruses, and the like.

Having preserved the safe passage of encoded fitness advantages through generations, a way for a species to extract practical value from its genotype is described by the central dogma of molecular biology. Here, transcription represents the first step in the manifestation of selective advantages (conferred by the fidelity of replication), converting them into RNA form. That is followed by the functional manifestation asserted on the protein level via translation, protein folding, etc.

At the level of nucleic acids, both replication and transcription are thus needed to execute the selection. And indeed, they are not commonly viewed as anything else but faithful reproduction machinery, both on the DNA and RNA level. Hence it is perhaps surprising that both of these processes seem to cause significant systematic changes in the genome, even when their enzymatic precision is extremely high and supported by additional sophisticated repair mechanisms. We shall consider the causes and consequences of this paradox.

Interestingly and instructively, evidence for genomic changes induced by replication and transcription comes not from direct biochemical experimentation, but rather from the bioinformatics analysis of sequenced genomes. Such analysis reveals that nucleotide composition of different genomes is linked to their large-scale organization and the specific modes of replication and transcription. We shall see how an organism's "lifestyle" leaves traces in the genome composition in the form of relative nucleotide frequencies and patterns of their change across the chromosomes of modern species.

In what follows, I describe the approaches to detecting such patterns in genomes of different organisms and organelles and how to compare them. More important, however, is the methodology of correct interpretation of the observed features, and here is where our focus shall lie.

2 Cumulative skew diagrams

Scientists had started counting nucleotides in DNA molecules even before the first sequences became available (as exemplified by Chargaff parity rules, discussed later). For example, the *GC content* of a DNA molecule is expressed as a fraction of all nucleotides in the molecule that are either guanines or cytosines (these nucleotides form a base pair with three hydrogen bonds within the double helix). Various properties of DNA have been associated with GC content (higher stability, stronger stacking interactions, etc.), but a detailed discussion of those is beyond the scope of the current chapter. As the sum of G and C nucleotides defines GC content, the difference between total number of G and C nucleotides determines *GC skew* (or *GC strand asymmetry*), which measures cytosine depletion on one strand compared to its complementary strand. Such asymmetry was already observed in the first sequenced genomes (those of viruses), which had appeared with the advent of technologies invented by Sanger and coworkers in the UK and by Maxam and Gilbert in the US.

Let us reproduce some of these results. We first consider a genome of the simian virus and break it into consecutive intervals of, say, 100 basepairs in length (such intervals are called sequence windows). We then calculate differences in the counts of guanine and cytosine in each sequence window and plot these differences vs the window position in the viral genome. We designate [N] for a count of nucleotide N in the window, hence this difference is expressed as [G]–[C]. To avoid the effects of fluctuations we divide it by the GC content within the window and calculate *GC skew*, which we therefore define as the ratio $([G] - [C])/([G] + [C])$.

The skew plot is shown in Figure 6.1a (ignore the b section of the figure for now). Labels on the y-axis are omitted on purpose (except for zero), as we are going to be mainly concerned with the plot shape rather than with the exact values of the skew. The x-axis shows the coordinate of the sequence window expressed as percentage of the genome length, with zero chosen as the start of the sequence file available from GenBank.

It appears that there are more guanines than cytosines (G > C) across some large portions of the genome, and G < C across other large portions. Thus GC skew shows different polarity (or sign, from positive on the left of the plot to negative on the right) over large genome stretches in the SV40 virus. There seems to be a global polarity switch somewhere in the center of this viral genome. It is a circular DNA molecule, so there is another switch (from negative to positive) at the coordinate 100% (or 0%, which is the same). Hence one half of the genome has positive GC skew, while the other half has negative GC skew.

The first sequenced bacterial genome, *Haemophilis influenzae*, also prompted a similar observation, although its plot is somewhat murkier (Figure 6.2a). There also seem to be two global switches of sign of GC skew (one starting a long and predominantly positive stretch of skew, and the other switching it back to negative) and the distance between them is also about 50% of the chromosome length.

One problem with this approach is that it is unclear which of these polarity switches in the middle of the plot of SV40 is actually the global one (where does the long stretch start and end), or what are their coordinates in the genome of *H. influenzae*. Traditional

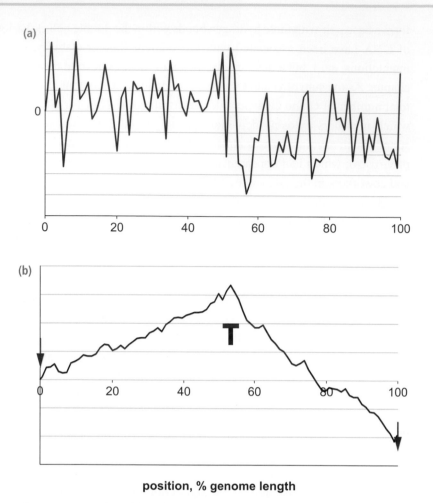

position, % genome length

Figure 6.1 GC skew (a) and cumulative GC skew (b) plots of SV40. As mentioned in the text, *y*-axis values in these and other graphs are omitted on purpose, as the shape of the plots is more important for the purposes of our discussion than the absolute skew values.

techniques of dealing with sequence windows do not really help with the presentation here. Increasing the window size lowers the number of switches, but hides the exact coordinate of the global switch. Smoothing the plot by averaging GC skew in sliding windows does not remove most of the local switches.

In this situation, the solution comes from a numerical integration approach: we could integrate the skew as a function of chromosomal position. In the simplest implementation, it is just a sum of the function values across the thinly sliced adjacent windows (which could be as small as 1 bp). So let us plot *cumulative GC skew* (a cumulative sum of GC skew values we have calculated for individual sequence windows) vs. window

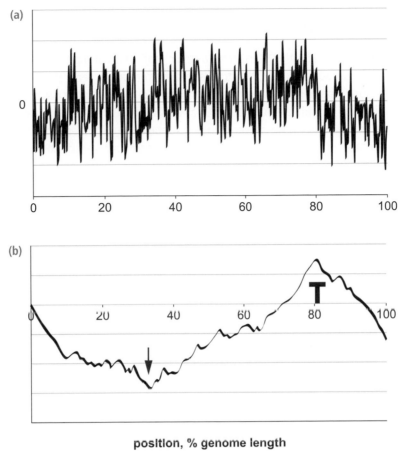

Figure 6.2 GC skew (a) and cumulative GC skew (b) plots of *Haemophilis influenzac.*

coordinate and obtain a graph of an integral (or an antiderivative) of the skew function (Figures 6.1b and 6.2b).

Knowing this integral (almost linear in our case), one easily recognizes the global behavior of the skew itself – it is close to constant on each side of the global switch. A positive skew would then produce a line with positive slope as its integral, while negative skew would produce a line with negative slope. So when cumulative GC skew is plotted for the genomes in question, there is normally a single global maximum and a single global minimum. While not remarkable in terms of calculus, it is striking from the biological point of view: those two points correspond to the terminus and origin of replication (shortened in the literature to *ter* and *ori*, and marked by large T and red arrow on diagrams in Figures 6.1b and 6.2b), respectively. Have a look at Box 6.1 for a refresher on replication and transcription mechanisms and Figure 6.3 for a schematic

Box 6.1 Schematics of replication and transcription

In bacteria and many viruses, replication starts from a single replication origin (middle of the bubble on the right of Figure 6.3) and both parental DNA strands (red) get gradually separated with the bubble growing in both directions. The parental lagging strand forms a duplex with the continuously synthesized nascent leading strand (green) and is thus always in a double-stranded state. The parental leading strand serves as a template for a nascent lagging strand (blue), synthesized as short Okazaki fragments and later ligated into a continuous chain. Hence this template spends some time single-stranded (shown in black).

Transcription also separates the two DNA strands opening a bubble of constant size (on the left of Figure 6.3). However, it is a transient bubble sliding along the transcribed gene in the direction of transcription. The transcribed strand in this process forms a duplex with the nascent mRNA molecule (light blue). The non-transcribed strand (also called "sense strand") remains single-stranded (black) while the bubble is open. As the mRNA is displaced and the bubble moves along, the next fragment of the non-transcribed strand enters a single-stranded state. A gene may occur on either of the two DNA strands and that defines the direction of its transcription. A preponderance of genes on one of the strands would lead to the other strand spending more time single-stranded.

It is important to remember that published DNA genomes are continuous single strands, such as the top strand in Figure 6.3. Hence half of a published sequence of, say, *Escherichia coli* is the leading strand (after the *ori*) and the other half the lagging strand (after *ter* and before *ori*). Clearly, the term "strand" is over-used and this may lead to some confusion.

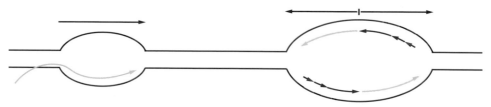

Figure 6.3 Sketch of replication and transcription.

depiction of the replication and transcription bubbles. Pay attention to the differences between leading, lagging, transcribed, and non-transcribed strands.

  ## Different properties of two DNA strands

Cumulative skew plots of three other bacterial genomes – a more exotic linear chromosome of *Borellia burgdorferi* together with the two workhorses of genetics, *Escherichia*

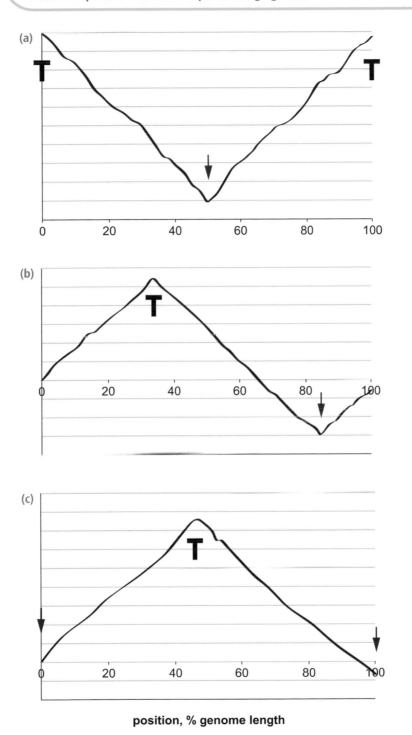

Figure 6.4 Cumulative diagrams of a linear chromosome of *Borellia burgdorferi* (a) and circular chromosomes of *Escherichia coli* (b) and *Bacillus subtilis* (c). Positions of replication termini are shown with a large black T, while a red arrow marks origins. Note that 0% and 100% correspond to the same coordinate on the circular genomes (hence two arrows for *B. subtilis*).

coli and *Bacillus subtilis* – are shown in Figure 6.4, and the vast majority of the nearly 1,000 sequenced bacterial genomes tend to produce very similar graphs. While individual genomes may show peculiar local features, a common global trend of a V-shaped diagram is clearly seen. In every such case, the distance on the x-axis between maximum and minimum of GC skew is about half of the genome length. And in all species where *ori* and *ter* have been detected experimentally, they coincide with the extremities of the species' cumulative plots (not shown here). The global minimum coincides with the *ori*, which means that the genome interval from *ori* to *ter* is G-rich, while the remaining half of a circular chromosome that extends from *ter* to *ori* is C-rich and G-poor. This observation has been generalized, proven experimentally, and is now a widely accepted method of locating *ori* and *ter* in the novel and less-studied microbial genomes.

Such behavior of the skew function means that the minimum and maximum on the graph likely represent the points where global biological properties of the DNA strand change, and that is exactly the case for *ori* and *ter* loci in bacteria: DNA there switches from the leading to the lagging strand, and the mode of synthesis changes, according to the current theories. The global minimum at the *ori* is a start of the leading strand (stretching from *ori* to *ter*), while the lagging strand extends from *ter* to *ori* (on the remaining half of a circular chromosome). One strand undergoes continuous duplication, while Okazaki fragment-driven synthesis takes place on the other strand (leaving it in a single-stranded state as shown in Box 6.1 and Figure 6.3). Such asymmetry could lead to differential accumulation of mutations (and different "mutation pressure") on the two strands.

On the other hand, *ori* and *ter* often mark points in a genome where the prevalent direction of transcription changes. Transcription may also amplify the effects of replication (since leading and transcribed strands would be the same across long genome stretches in many bacterial species). Remarkably, in most bacterial genomes, skew is the strongest when only the third codon positions in genes are taken into account. "Selection pressure" maintaining the gene function by preserving the amino acid sequence through generations is weakest on these codon positions since a mutation there infrequently changes an encoded amino acid. Therefore, mutation pressure may be responsible for the observed skews.

There are multiple hypotheses on the nature of the skews and I recommend to interested readers a thorough review by Frank and Lobry [1]. The most consistent explanation for the effects observed above (and below) is based on spontaneous deamination of C or 5-methylcytosine in single-stranded DNA. This is by far the most frequent mutation that replaces cytosine by uracil (or 5-methylcytosine by thymine) and creates a mismatched basepair T–G. If this mismatch is not repaired, it can lead to pairing the mutated base with A during the next round of replication. Eventually, this would give rise to a relative abundance of G (since C on the other strand is not mutated) and T (since C on this strand is mutated to T) on one strand. Notably, deamination rates rise over 100-fold when DNA is single-stranded.

This does not lead to the situation where all available Cs are replaced by Ts, as further mutagenesis and repair processes continue changing the bases throughout evolution. In fact, AT skew does not always follow in the anti-phase of the GC skew and the behavior of AT skew is much less regular. However, being the most frequent mutation, cytosine deamination seems to shift the equality [C]=[G] consistently towards relative excess of guanine on the DNA strand that spends longer time single-stranded.

This effect is likely a result of two major processes that open the double-stranded DNA (dsDNA): replication and transcription. This effect is observed not only in bacteria but also in archaea, DNA and RNA viruses, and organelles (such as mitochondria).

We look next in more detail at the viral genomes. In all the different schemes of replication and transcription for viruses, one can frequently find surprising correlations with the cumulative skew diagrams of their DNA sequences.

Much like the double-stranded DNA genomes of bacteria (and some archaea), many dsDNA viruses (for example, the human cytomegalovirus) form characteristic V-shapes with global minima near the replication origins. However, it is the other shapes found in cumulative diagrams of viruses that make them very interesting objects for answering our main question: how do transcription and replication change genome composition?

One striking example is the human adenovirus, whose linear dsDNA features two replication origins (one at each end of the genome). Replication leaves the upper strand in Figure 6.5a in a single-stranded state while the lower strand is being duplicated, and then completes the process on the upper strand. This means that the displaced upper strand may be subject to different mutation pressure than the template bottom strand. Assuming a constant speed of replication, mutation pressure will change along the sequence, as the time the displaced part of the upper strand spends single-stranded changes linearly from one end of the molecule to the other. Integration of a linear function results in a second-order polynomial, a parabola.

Remarkably, the GC diagram of human adenovirus type 40 (Figure 6.5b) has a shape very close to parabolic. It points upwards, reflecting a decrease in the skew value from positive to negative along one strand, consistent with the replication mode. The parabolic trendline reaches its global maximum (meaning that the GC skew equals zero) close to the middle of the sequence. Replication may start at either origin, so both strands have a higher GC skew at their respective 5'-ends.

(a)   (b)

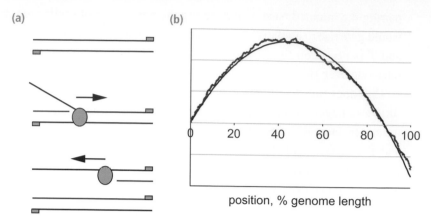

position, % genome length

Figure 6.5 Schema of replication of human adenovirus 40 (a) and its cumulative skew diagram (b). Replication origins are shown as green boxes, replication complex as green circles, newly synthesized DNA strands are in red. The parabolic trendline is shown in (b).

4 Replication, transcription, and genome rearrangements

While connection between mutational patterns and replication seems strong, several papers have reported evidence of mutations caused by the process of transcription. Clearly, transcription by itself would not distinguish between the leading and lagging strand. However, transcription-induced mutations would end up on one strand if bias in gene orientation is strong (e.g. 75% of *B. subtilis* genes are on the leading strand). This could generate the compositional asymmetry between the leading and lagging strand that has been observed in bacterial genomes.

Therefore, replication and transcription may be jointly or separately responsible for the effects observed. As these processes are so different, how do their contributions differ? Using the very same technique but carefully choosing the biological system allows us to address the question. An answer comes from papillomavirus, whose replication and transcription are co-directional in one half of the genome, and opposite in the other half. In other words, the replication is bi-directional, while transcription is unidirectional. If there are separate deamination-driven biases induced by replication and transcription, they should act in concert in one half of its genome, and in the opposite directions in the other half.

If this model is correct, a nearly zero slope on the right of the HPV-1A diagram (Figure 6.6) suggests that a contribution of transcription is comparable to that of replication in papillomavirus. They almost cancel each other out in the region between

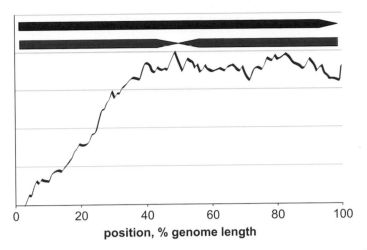

Figure 6.6 Cumulative skew diagram of HPV-16. Blue arrow shows direction of transcription and red arrows depict direction of replication.

50 and 100% of the plot ([G] = 758, [C] = 773), and their combined effects produce significant guanine excess ([G] = 900, [C] = 690) in the other half of the genome.

This leads us to another important consideration. If the integral of a constant value produces a linear plot, why is it sometimes very smooth and sometimes so uneven and jagged (compare *B. subtilis*, Figure 6.4c, and *H. influenzae*, Figure 6.2b)? One explanation is that local irregularities (sequence constraints on amino acid composition, regulatory sequences, etc.) interfere with a global trend. After all, the sequence that we observe now is a snapshot of multiple evolutionary forces acting simultaneously on the same nucleotide positions.

Another explanation is that a sequence inversion would swap the leading and lagging strand and change the skew to its opposite between the borders of the inversion (Figure 6.7). This creates the possibility for deviations from perfect linearity, and it also reverses the direction of transcription for those few genes affected by the inversion. With regard to directionality of transcription and replication this sounds like a chicken and egg question: were genes originally co-directional and inversions have changed that (and introduced jagged skew patterns), or were the genes always divergently transcribed (and thus generated uneven patterns via opposing effects of transcription and replication)?

Furthermore, horizontal transfer of DNA between species and sequence insertions complicates the picture even further. Let us consider an example of a human pathogen, *Helicobacter pylori*, associated with stomach ulcers (Figure 6.8). We can see a familiar

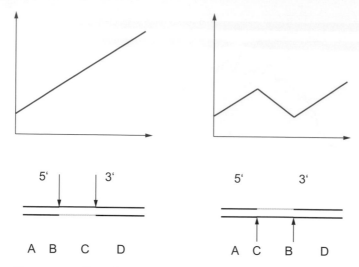

Figure 6.7 Effect of an inversion on the cumulative skew. Schematics of an inversion between two positions B and C is shown, together with the corresponding change in the cumulative skew. As G-rich leading strand fragment BC is replaced by a C-rich lagging strand fragment CB, skew turns from positive to negative over the inverted interval.

V-shaped diagram, featuring a number of inversions and swapped sequences as well as an insertion of a pathogenicity island (most likely, horizontally transferred). Strikingly, in the two strains of this bacterium sequenced about a decade ago the position of the pathogenicity island has remained the same while many other sites in the genome have undergone significant changes, even those in close proximity to the replication origin.

The example of *H. pylori* is also interesting in that we can try and deduce in which of the two strains is the *ori* region more intact (closer to the ancestral strain). Let us consider two facts. First, we note the adjacent positions of the fragments **l, m**, and **n** on the plot in the top diagram versus their scattered and inverted arrangement in the bottom diagram. Second, we note the sharp global minimum in the *ori* region in the top diagram, similar to other bacterial genomes. Logic suggests that the inversions and translocations took place in the strain shown in the bottom diagram, disrupting the original arrangement of the fragments **l, m**, and **n**. Hence the strain shown on top likely features the *ori* organization closest to the ancestral strain, and we were able to infer this purely from the graphical comparison of the cumulative diagrams.

Remarkably, we have not exhausted the value of such comparison in this example. Note where the cumulative skew plot ends in the top and bottom diagrams in Figure 6.8. Following our reasoning, the diagram closest to the ancestral strain (i.e. with fewer rearrangements) ends closer to the *x*-axis. Thus the overall counts of Gs and Cs in

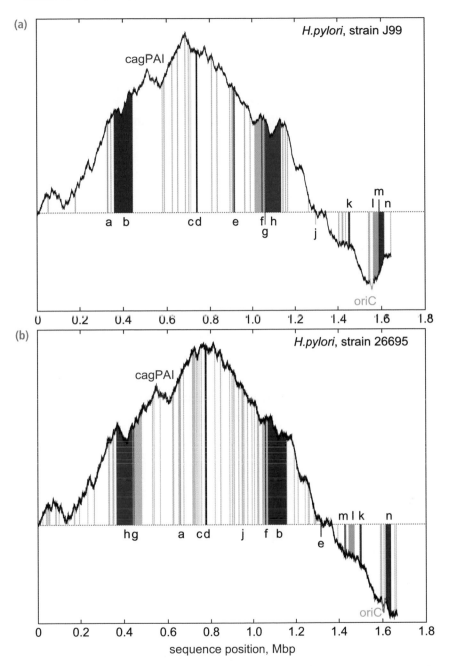

Figure 6.8 Using skew diagrams for compact depiction of genome comparisons between two strains of *Helicobacter pylori*. Colored areas under the curve mark genome rearrangements (designated with letters a–h, j–n). All fragments represent inversions (and, in most cases, translocations), except for the rearrangements designated "a" (only translocation), "j" and "e" (both of which represent reciprocal exchange). A small number of strain-specific genes are not shown; these reside inside larger rearrangements. Note the mirror symmetry of the curve fragments, corresponding to inversions designated by the same letters in the two strains.

Box 6.2 Chargaff parity rules

Counting the numbers of individual nucleotides in the chromosomes was one of the key elements leading to the establishment of DNA structure. There is a well-known Chargaff rule that states that a single strand of a double-stranded DNA molecule contains as many of each of the four nucleotides as there are complementary nucleotides in the second strand. This famous observation paved the way to pairing complementary nucleotides in the DNA structure model.

A later and less-known second Chargaff rule states that a single strand will also contain equal numbers of complementary nucleotides G and C (or A and T). Almost invariably, publications about this rule agree on its rather mysterious origin. There is no mystery, however. If one looks at Figure 6.3, it becomes very clear why Chargaff came to this conclusion when analyzing the *B. subtilis* genome. The right end of the curve lands practically on the x-axis so that the total skew is close to zero (i.e. a total G count is close to that of C).

It is the fact that both stretches of DNA between the origin and terminus in bacterial genomes are of similar length that explains why their contributions to the skew cancel each other out. However, the total skew in many other cases is clearly non-zero; for example, in adenovirus or mitochondrial genomes. Even in bacteria there are clear exceptions. A rearrangement would often be a reason for that, or a horizontal transfer of DNA from another bacterium, as the example of *H. pylori* (Figure 6.8) demonstrates.

that ancestral strain likely were closer to each other. That invites a brief discussion on counting nucleotides through time as a conclusion of this reading (Box 6.2).

DISCUSSION

We have considered here a number of genomes with different schemes of replication and transcription across a variety of organisms. Our computational tool was very simple, yet we could analyze the effects of very fundamental cellular processes. As with many bioinformatics approaches, what counts is not the tool itself, but our ability to interpret its output in the context of a specific biological problem.

Another important point is in making the right choice of the system to study and studying it well. The highly opportunistic nature of viruses apparent in the diverse organization of their small genomes presented us here with many illustrative cases for making conclusions. However, one needs to be patient in order to span that diversity. We must dig through a lot of material in order to interpret correctly even such simple data as nucleotide counts. Luckily, there are plenty of good examples provided by nature (and genome repositories) for us to test our conjectures.

QUESTIONS

(1) For the skew diagrams shown in Figures 6.3a and b, consider a hypothetical large inversion between the coordinates 40% and 60%. What would the resulting diagrams look like?

(2) Now, consider a second, subsequent inversion between the very same coordinates and draw the resulting diagram. What if that second inversion instead took place between the coordinates of 30% and 70%?

(3) Following the logic of the examples in the previous two questions, how can you explain the arrangement of the large colored stripes, designated h and b in the diagrams corresponding to the two strains in Figure 6.8?

REFERENCES

[1] A. C. Frank and J. R. Lobry. Asymmetric substitution patterns: A review of possible underlying mutational or selective mechanisms. *Gene*, 238: 65–77, 1999.

[2] E. Chargaff. Chemical specificity of nucleic acids and mechanism of their enzymatic degradation. *Experientia*, 6: 201–240, 1950.

[3] H. J. Lin and, E. Chargaff. On the denaturation of deoxyribonucleic acid. II. Effects of concentration. *Biochim. Biophys. Acta*, 145: 398–409, 1967.

[4] C. I. Wu and N. Maeda. Inequality in mutation rates of the two strands of DNA. *Nature*, 327: 169–170, 1987.

[5] J. R. Lobry. Asymmetric substitution patterns in the two DNA strands of bacteria. *Mol. Biol. Evol.*, 13: 660–665, 1996.

[6] A. Grigoriev. Analysing genomes with cumulative skew diagrams. *Nucleic Acids Res.*, 26: 2286–2290, 1998.

[7] A. Grigoriev. Genome arithmetic. *Science*, 281: 1923a, 1998.

[8] A. Grigoriev. Strand-specific compositional asymmetries in dsDNA viruses. *Virus Res.*, 60: 1–19, 1999.

CHAPTER SEVEN

Modeling regulatory motifs

Sridhar Hannenhalli

Biological processes are mediated by specific interactions between cellular molecules (DNA, RNA, proteins, etc.). The molecular identification mark, or signature, required for precise and specific interactions between various biomolecules is not always clear, a comprehensive knowledge of which is critical not only for a mechanistic understanding of these interactions but also for therapeutic interventions of these processes. The biological problem we will address here, stated in general terms, is: how do biomolecules accurately identify their binding partners in an extremely crowded cellular environment? An important class of cellular interactions concerns the recognition of specific DNA sites by various DNA binding proteins, e.g. transcription factors (*TF*). Precisely how the TFs recognize their DNA binding sites with high fidelity is an active area of research. While a detailed treatment of this question covers several areas of investigation, we will focus on aspects of the TF–DNA recognition signal that is encoded in the DNA binding site itself. In this chapter we will summarize a number of approaches to model DNA sequence signatures recognized by transcription factor proteins.

1 Introduction

Most biological processes critically depend on specific interactions between biomolecules. A key question in biology is how, in the overly crowded cellular environment, these various interactions are accomplished with high fidelity. Evidence suggests highly developed mechanisms for trafficking, addressing, and recognizing biomolecules within a cell. For instance, brewer's yeast (*Saccharomyces cerevisiae*) feeds on galactose, among other sugars. The yeast needs a mechanisms to sense the presence of galactose in its environment and in response, turn on specific biological

Bioinformatics for Biologists, ed. P. Pevzner and R. Shamir. Published by Cambridge University Press.
© Cambridge University Press 2011.

processes to harness galactose. In the presence of galactose, transcriptional regulator protein GAL4 binds to a specific DNA sequence upstream of several genes, most notably GAL2, involved in galactose metabolism [1]. This entire process, from the sensing of galactose to transmitting information down the signal cascade that culminates in the binding of GAL4 to the GAL2 gene's regulatory sequence and metabolizing galactose, requires many specific interactions between different types of molecules including DNA, RNA, and proteins.

As another example, consider the well-studied JAK-STAT signal transduction pathway which plays a critical role in cell fate decision and immune response in humans. Much like galactose metabolism in yeast, the JAK-STAT system involves sensing specific chemicals outside the cell, transmitting this information across the cell membrane down to the regulatory regions of specific genes, to activate the response system [2]. One can think of such signaling pathways as a relay involving specific interactions starting with the interaction between extracellular chemicals and cell-membrane receptors, culminating in the interaction between transcription factors and DNA in gene regulatory sequences. Questions concerning the specificity of interaction between biomolecules are open in most contexts and are areas of active research.

The problem of interaction specificity could be resolved from first principles if we had two pieces of information, namely the location of an interaction partner and certain identifying features of the partner. For instance, if you were to plan a meeting with a stranger in a large city, you would need to know the approximate meeting location (e.g. corner of 6th and Broad), as well as certain identifying features of this person (e.g. red polka dot suit). A parallel in the cellular environment could be a trans-membrane (location) protein with amino acid sequence HHRHK near the amino terminus (identifying feature). In this example, the identifying feature could also be expressed as a stretch of five positively charged and largely hydrophobic residues. Alternatively, one of the interacting proteins may have a structural feature (the key) which fits a complementary structure on another protein (the lock). These examples provide three different ways of representing the *identifying feature* of the interacting partner, or in other words, these examples are different "models" of the interaction specificity. Based on the different models one can surmise that the task of modeling substrate specificity can be extremely difficult, especially in the realm of proteins. Indeed, the task is complex even for the much simpler case in which the substrate is a nucleic acid molecule (DNA or RNA). While the general principles are common to both proteins and nucleic acids, for the sake of simplicity, we will restrict the exposition to nucleic acids hereafter. In particular, we will discuss the issue of modeling the DNA sites recognized and bound by transcription factors (TF), i.e. transcription factor binding sites (TFBS). To orient the reader, we next provide a brief introduction to transcriptional regulation.

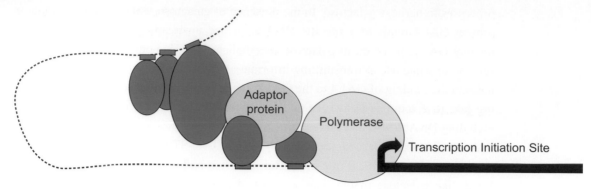

Figure 7.1 Transcription factor proteins (filled ellipses) interact with binding sites (filled rectangles) in the relative vicinity of a gene transcript (black rectangle). The transcription factor binding sites can either be proximal to the transcript (within a few thousand nucleotides) or far (several hundred thousand nucleotides). The interactions between transcription factors is aided by other adaptor proteins. The DNA-bound transcription factors interact with polymerase to regulate transcription.

How much, at what time, and where within an organism any gene product is produced is precisely regulated, and is critical to maintaining all life processes. While the overall regulation of a gene product is executed at various levels – including splicing, mRNA stability, export from nucleus to cytoplasm and translation – much of this regulation is accomplished at the level of transcription. Transcriptional regulation is a fundamental cellular process, and aberrations in this process underlie many diseases [3]. For example, mutations in the *Factor IX* protein is known to cause hemophilia B. Additionally, mutations in the regulatory region immediately upstream of *Factor IX* gene can disrupt the binding of specific TF, which in turn dysregulates the transcription of the gene, thus leading to hemophilia [3]. In eukaryotes, transcriptional regulation is orchestrated by numerous TF proteins. For the most part, TFs regulate gene transcription by binding to specific short DNA sequences in the relative vicinity of the transcription start site of the target gene, and through interactions with each other as well as with the polymerase enzyme. See Figure 7.1 for a schematic.

Precise and specific interaction between the TF and its cognate DNA binding site is a critical aspect of transcriptional regulation. What is the identifying characteristic of the DNA sites recognized by a TF protein? This question remains an open and important one in modern biology. The specific TF–DNA interaction is determined not only by the DNA sequence but also by a number of additional cellular factors. A full description of these determinants is beyond the scope of this chapter. Here we focus on the aspect of TF–DNA interaction that is encoded in the sequence of the DNA binding site itself.

In particular, we will focus on models of TF binding sites. Given several instances of experimentally determined binding sites for a TF, a *model* is a succinct quantitative description of the known binding sites, which not only may provide mechanistic insights into TF–DNA interaction, but also helps identify novel binding sites. Although we have focused our discussion only on TF binding sites, the discussion applies to any DNA signal such as splice sites, polyA sites, and indeed more generally to signals in amino acid sequences. Finally, the signal encoded in the DNA binding site provides only part of the information required for specific interactions with the DNA binding protein. We will conclude with a discussion of additional hallmarks of functional binding sites that can be exploited specifically to identify functional TF binding sites in the genome.

 ## 2 Experimental determination of binding sites

In this section we will briefly summarize the experimental techniques used to determine the DNA binding sites for a specific TF. The sequences obtained from these experiments are then used to construct a model of TF binding. For a detailed review on this topic we refer the reader to [4]. The experimental approaches to binding site determination can be classified as *in vitro* and *in vivo*.

The common *in vitro* techniques include Systematic Evolution of Ligands by EXponential enrichment (SELEX) [5] and protein-binding DNA microarrays [6]. SELEX works as follows. One begins by synthesizing a large library consisting of randomly generated oligonucleotides of fixed length. The solution containing the oligonucleotides is exposed to the TF of interest. Some of the oligonucleotides bind to the TF. The oligomers that are bound by the TF can be separated from the rest (although not perfectly) and a new solution is prepared that is enriched for the bound oligomers. This process of binding to the TF and separating out the bound oligomers is repeated multiple times and in every new round the experimental conditions are varied so that the increasingly stronger binding between the TF and oligomers is favored. Multiple rounds of selection with increasing stringency for the binding results in a solution enriched for oligonucleotides that bind to the TF with high affinity. These oligonucleotides are then cloned and sequenced. In a related experimental techniuqe of protein-binding DNA microarray, the DNA oligomers are immobilized on a glass surface to which a flourescent-labeled TF is exposed. The specific oligomers that bind to the TF of interest are detected through optical signal processing [6]. This approach obviates the need for multiple rounds of enrichment as in SELEX, as well as the need for cloning and sequencing. By their nature, the *in vitro* capture the protein–DNA binding in purified form and in isolation, independent of the other cellular determinants of the binding.

In vivo identification of binding sites is accomplished by two common techniques – *ChIP-chip* and *ChIP-seq*. Both approaches require obtaining the nuclear DNA bound by the TF of interest, followed by DNA digestion, which leaves the TF attached to small stretches of DNA, and then using specific antibody to fish out the TF along with the stretch of DNA bound to it. In the ChIP-chip (Chromatin immunoprecipitation followed by microarray hybridization), the bound DNA is hybridized against a glass array that contains a large set of sequences corresponding to various genomic locations. Thus, the array elements that hybridize to the TF-bound DNA automatically provide the information on the genomic location where the TF binds. In the second technique – ChIP-seq (ChIP followed by high-throughput sequencing) – the microarray hybridization step is replaced by direct sequencing of the TF-bound DNA. The sequences are then mapped to the genome based on sequence similarity. In each of these approaches the TF-bound region is detected with varying resolution, and additional techniques are applied to more precisely map the boundaries of the TF binding sites.

Experimentally determined binding sites are compiled in various databases, most notably TRANSFAC [7] and JASPAR [8]. TRANSFAC is a licensed database which currently includes binding sites for over 1,000 TFs gleaned from the experimental literature. Each individual binding site is assigned a quality score corresponding to the strength of experimental evidence. JASPAR is a freely accessible resource which includes information on ~150 TFs, also curated from experimental literature, and is based on a more stringent set of criteria as compared to TRANSFAC.

3 Consensus

For the rest of the chapter, we will assume that for a given TF we are provided a set of binding sites of a fixed length, and we will focus on the task of modeling these known sites. Therefore, for a transcription factor F, assume that we are given N examples of K bases long DNA sequences bound by F. Denote the N sequences as $X_1, X_2, \ldots, X_N$. Denote the nucleotide base at position j of sequence X_i by $X_{i,j}$, where $X_{i,j} \in \{A, C, G, T\}$. The DNA sequence characteristics that are critical for the protein–DNA interaction have both biological and computational implications. These characteristics should determine the representation of binding specificity. Consider Example 7.1a in which we are provided with 10 experimentally determined binding sites for the yeast TF *Leu3* [9], and each site is 10 nucleotides long.

Example 7.1.

(a)

	1	2	3	4	5	6	7	8	9	10
X_1	C	C	G	G	T	A	C	C	G	G
X_2	C	C	T	G	T	A	C	C	G	G
X_3	C	C	G	C	T	A	C	C	G	G
X_4	C	C	G	G	A	A	C	C	G	G
X_5	G	C	G	G	T	A	C	C	G	G
X_6	C	C	G	T	T	A	C	C	G	G
X_7	C	C	G	C	A	A	C	C	G	G
X_8	C	C	T	G	A	A	C	C	G	G
X_9	G	C	G	G	T	A	A	C	G	G
X_{10}	C	C	G	C	T	A	C	A	G	G

(b)

	1	2	3	4	5	6	7	8	9	10
A	0.0	0.0	0.0	0.0	0.3	1.0	0.1	0.1	0.0	0.0
C	0.8	1.0	0.0	0.3	0.0	0.0	0.9	0.9	0.0	0.0
G	0.2	0.0	0.8	0.6	0.0	0.0	0.0	0.0	1.0	1.0
T	0.0	0.0	0.2	0.1	0.7	0.0	0.0	0.0	0.0	0.0

(c)

A simple and common approach to summarize these known binding sites is called the *consensus* representation in which we create a consensus string of length K and place in position j the *consensus* nucleotide which occurs with the highest frequency at position j in N binding sites. In Example 7.1a, for instance, at position 3 there are 8 Gs and 2 Ts. Thus the consensus at position 3 is G. The consensus sequence of these 10 known examples of binding sites is thus $CCGGTACCGG$. Note that the consensus sequence happens to be the same sequence as X_1.

More formally, given N binding sites, each of length K, let $N_{x,j}$ be the number of binding sites sites having nucleotide x at position j, where $x \in \{A, C, G, T\}$ and

$1 \leq j \leq K$. The normalized frequency of nucleotide x at position j is denoted by $f_{x,j} = (N_{x,j})/N$. Clearly,

$$\sum_{x \in \{A,C,G,T\}} f_{x,j} = 1. \tag{7.1}$$

The consensus sequence of these N binding sites is defined as the K-long nucleotide sequence $C_1 C_2 \cdots C_K$, in which C_j is the nucleotide x that maximizes $f_{x,j}$. The consensus at each position in Example 7.1a is unambiguously defined. However, consider a case where at some position there are 4 Cs, 5 Gs, 1 A and 0 T. In this case, assigning a G as the consensus ignores the fact that nucleotide C is *almost* as likely as G. To address this ambiguity one may use letter S at this position of the consensus string where S represents *strong* bases C and G. Similarly, nucleotides A and G (*purines*) together are represented by letter R. There is an *International Union of Pure and Applied Chemistry* (IUPAC) letter code to denote each combination of nucleotides and which is used to represent consensus in general [10].

Although quite useful for many practical situations, the consensus representation is restrictive as it systematically ignores the rare bases at each position, which might represent biologically important instances of binding sites. Next we discuss the *Position Weight Matrix* representation of binding sites that addresses this specific shortcoming of the consensus model.

4 Position Weight Matrices

The *Position Weight Matrix* (PWM) is currently the most common representation of TF binding sites. Unlike the consensus approach, a PWM captures all observed bases at each position. In its simplest form, a PWM is a probability matrix with 4 rows corresponding to the 4 nucleotide bases and K columns corresponding to each position in the binding site. We will refer to rows 1 through 4 interchangeably as rows A, C, G, T, respectively. The entry corresponding to the jth column (position) and xth row (base) is $f_{x,j}$, defined above as the frequency of x at position j among the binding sites. The PWM corresponding to the binding sites in Example 7.1a is shown in 7.1b.

Note that if there is an insufficient number of known binding sites, i.e. if N is relatively small, then a particular nucleotide base may not be observed at a position. This would result in $f_{x,j} = 0$, which can be interpreted to imply that x is prohibited at position j, even though we know that this is simply due to insufficient sampling of sites and not because of a functional impossibility. A typical solution to deal with this situation is to correct for potentially unobserved data by adding a *prior* (also known as *pseudo count*) to the observed nucleotide counts before computing the frequencies. A simple approach is to add a count of 1 to each observed count, also called the *Laplace prior*. If a Laplace prior is used in Example 7.1a, then the counts in the first column become (*1, 9, 3, 1*) for (*A, C, G, T*), and the first column of the PWM in Example 7.1b becomes (*0.071, 0.644, 0.214, 0.071*). Formally, under the Laplace prior, the frequencies are $f_{x,j} = (N_{x,j} + 1)/(N + 4)$.

There is a quantitative property of a PWM that corresponds to its usefulness in modeling the TF–DNA binding preference. For instance, if the known binding sites for a TF are highly dissimilar to each other, then there is very little knowledge to be gained about the general binding preference. More specifcially, consider a particular column j of a PWM. If each of the 4 nucleotides is equally likely to be observed at that position, i.e. if $f_{x,j} = 0.25$, for each nucleotide base x, then this column conveys no information regarding the binding preference of the TF under consideration. This intuitive notion of information contained in position j of a PWM can be quantified formally using the *Information Content*, which is measured in bits and is defined as

$$I_j = 2 + \sum_{x \in \{A,C,G,T\}} f_{x,j} log_2(f_{x,j}).$$
(7.2)

Note that in the most informative case, when exactly one of the nucleotides, say A, is observed at a position with $f_{A,j} = 1$, $f_{C,j} = 0$, $f_{G,j} = 0$, $f_{T,j} = 0$, then I_j achieves its maximum value of 2 bits.[1] In the other extreme, when all nucleotides are equally likely and $f_{x,j} = 0.25 \ \forall x \in \{A, C, G, T\}$, then I_j achieves its minimum value of 0 bits [11]. One can verify that any other value of probabilities yields a positive information. Example 7.1c shows the *Logo* representation of the motif in Example 7.1b depicting the information content at each position. The x-axis enumerates the binding site positions and the y-axis indicates the information content. The height of each base corresponds to its relative frequency. The figure was generated using the Weblogo tool at *weblogo.berkeley.edu*. For a more detailed discussion on information content and another relative measure called *Relative entropy*, the reader is referred to [12].

While the PWM is a simple, intuitive, and the most commonly used model of TF–DNA interaction, its main drawback is that it assumes independence among different positions in the binding site. Specifically, the preference for a nucleotide at one position has no bearing on the nucleotide preferences at another position. Consider the hypothetical Example 7.2 below which has six binding sites, each four nucleotides long.

---

[1] Here, the value of $0 log_2 0$ is approximated to be 0.

Example 7.2.

X_1	C	G	G	G
X_2	C	G	T	G
X_3	C	G	G	C
X_4	A	T	G	G
X_5	A	T	G	G
X_6	A	T	G	T

In the first column, nucleotides, C and A are equally likely, while in the second column nucleotides G and T are equally likely. Based on this information and assuming independence between these two columns, one would infer that the two binding sites $CGGG$ and $CTGG$ are equally preferred. However, it is more likely that when there is a C at the first position a G is preferred in the second position, and when there is an A at the first position a T is preferred in the second position. In other words, the first and second positions are not independent. A direct experimental measurement of such dependence is laborious. Two specific experimental studies that infer dependence between positions in binding sites can be found in [13] for bacterial Mnt repressor binding sites and in [14] for Egr1 transcription factor binding sites.

5 Higher-order PWM

In Example 7.2, there is likely to be dependence between the first two positions. In this case the preferred binding sites can be better modeled, and thus better predicted, if we consider the first two nucleotides together. For instance, CG and AT are the most likely dinucleotides at the first two positions. In general, if we want to incorporate possible dependencies between nucleotides at every pair of adjacent positions, we can extend the single nucleotide PWM with 4 rows and K columns to a dinucleotide PWM with 16 rows corresponding to all 16 nucleotide combinations and $K - 1$ columns corresponding to all dinucleotide positions. Therefore, in the first column of Example 7.2, the CG and AT dinucleotides will have large frequency values, each "close" to 0.5 each,[2] and all other 14 dinucleotides will have low values, "close" to zero. This dinucleotide-based PWM has also been referred to as the *Position Weight Array* [15, 16]. One can extend the *Position Weight Array* to capture even higher-order dependencies, say among L consecutive nucleotides. This corresponds to enumerating at every position of the binding site the L nucleotides-long sequences starting at the

---

[2] The probabilities will be "close" to 0.5, as opposed to being exactly 0.5, if we add small pseudocounts for the unobserved dinucleotides.

position among all binding sites, i.e. from positions 1 through L, positions 2 through $L + 1$, and so on till positions $K - L + 1$ through K. This results in a PWM with 4^L rows (corresponding to all possible K-long sequences) and $K - L + 1$ columns for any $L \geq 1$, where L represents the number of adjacent nucleotides considered together. This model is equivalent to a *Markov Model* of order $L - 1$, which provides the probability of observing a nucleotide at any position based on the previous $L - 1$ nucleotides. See Figure 7.3b for an example of a first-order Markov Model. The Markov Model is a general statistical tool and is often used to model a variety of molecular sequences.

The main limitation of these higher-order PWMs is a lack of sufficient data, i.e. small values of N. For instance, we cannot reliably infer the preference for a dinucleotide among the 16 possible choices based on only 6 sequences, as in Example 7.2. Moreover, high-order PWMs are still limited in that they do not directly capture the dependence between non-adjacent nucleotide positions, for instance between positions 1 and 3, independent of position 2. In theory, this can be remedied by explicitly enumerating nucleotide combinations for various combinations of positions, although such models suffer from insufficient data to a much greater extent than higher-order PWM models. In the next section we will discuss richer models of TF–DNA binding preferences that attempt to maximize the information captured from the data.

Maximum dependence decomposition

The *Maximum Dependence Decomposition* (MDD) approach, proposed in Genscan [16], explicitly estimates the extent to which the nucleotide at position j depends on the nucleotide at position i. Specifically, MDD estimates the extent to which the nucleotide at position j depends on whether the nucleotide at position i is the consensus (most frequent) nucleotide for that position or a non-consensus nucleotide. For each i all binding site sequences are divided into two groups, C_i and $\overline{C_i}$, depending on whether the nucleotide at position i is the consensus or a non-consensus base, respectively. Within each group the nucleotide frequencies are computed at every position j. For a given position j, the two sets of frequencies are compared using the χ^2 statistic [17]. If position j is independent of position i, then we expect the two sets of nucleotide frequencies to be fairly similar; however, if the two sets of frequencies differ significantly from each other, it would suggest that nucleotide preference at position j depends on the nucleotide at position i. Let f_A, f_C, f_G, and f_T be the normalized frequencies (number of each base divided by the total number of sequences) of the four bases at position j among the sequences in $\overline{C_i}$. Let N be the total number of sequences in C_i. If the four bases were distributed identically in the two sets of sequences C_i and $\overline{C_i}$, then we would expect the number of the four bases at position j among the sequences in C_i to be $N * f_A$, $N * f_C$, $N * f_G$, and $N * f_T$. Let N_A, N_C, N_G, and N_T be the observed number of the four bases at position j among the sequences in C_i. In this context, the χ^2 statistic is defined as:

$$\frac{(N * f_A - N_A)^2}{N * f_A} + \frac{(N * f_C - N_C)^2}{N * f_C} + \frac{(N * f_G - N_G)^2}{N * f_G} + \frac{(N * f_T - N_T)^2}{N * f_T} \quad (7.3)$$

The greater the difference in the two sets of nucleotide frequencies, the higher the value of χ^2 statistic. If the statistic indicates a significant difference[3] between the two frequency distributions then the position j is said to depend on position i. For example, for a set of 20 sequences, if position 1 includes 12 As and 8 Gs, then the consensus C_1 is A. Now for the 12 sequences in which the nucleotide at position 1 is an A, assume that at position 2, 8 have a C and 4 have a T. On the other hand, for the 8 sequences in which the nucleotide at position 1 is a G, at position 2, 7 have a T and 1 has a C. For the sequences with $C_1 = A$, the counts for (A, C, G, T) at position 2 are ($0, 8, 0, 4$), and for the other 8 sequences the nucleotide counts at position 2 are ($0, 1, 0, 7$). Intuitively, the two sets of counts look very different from each other, and the χ^2 statistic formally quantifies this intuition.

Denote the χ^2 statistic quantifying the dependence of position j on position i as $\chi^2(j \mid i)$. The MDD approach proceeds iteratively as follows.

1 Compute $S_i = \sum_{j \neq i} \chi^2(j, i)$ to capture the total dependence on position i.
2 Among all K positions, select position i with the maximum value of S_i, and partition all sequences into two parts based on whether they have C_i or $\overline{C_i}$ at position i.
3 Repeat steps 1 and 2 separately for each of the two sets of sequences obtained in step 2.
4 Stop if there is no significant dependence, or if there is an insufficient number[4] of sequences in the current subset. In either case, construct a standard PWM for the remaining subset of sequences.

Figure 7.2a illustrates the MDD modeling procedure. The above procedure decomposes the entire binding site data set into a tree-like structure. To test whether a given sequence X fits the model, as illustrated in Figure 7.2b, one proceeds down the tree,

---

[3] If there is no real difference between the two frequency distributions then the χ^2 statistic is expected to follow the so-called χ^2 distribution. By comparing the computed χ^2 value to the expected distrbution, one can compute the probability that the two distributions are identical. This probability is called the P-value. If the P-value is small, say below 5%, then we can say that the two distributions are significantly different.

[4] We leave this purposefully vague, as there is no formal rule to define this. Essentially, if the number of remaining sequences is small, say below 5, then it does not pay to further partition them.

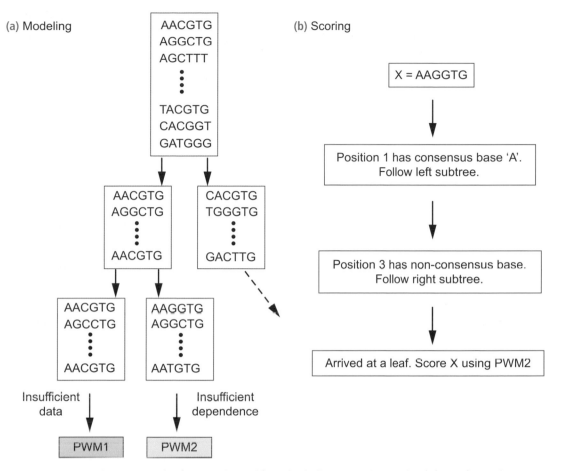

Figure 7.2 The figure, adapted from [16], illustrates the maximal dependency decomposition (MDD) procedure. (a) Modeling. Starting with all binding sites, maximum dependency is detected for position 1 with consensus "A." The sites are then partitioned based on whether or not the nucleotide at position 1 is an "A." Among the sites with "A" in the first position, maximum dependency is detected for position 3 with consensus "C." The sites are further partitioned based on whether or not the nucleotide at position 3 is a "C." The two partitions are not partitioned any further, however, because of either insufficient data or insufficient dependency. The entire MDD model is built following this procedure. (b) Scoring. Given a sequence X, one proceeds down the left subtree because the first base of X is an "A," followed by the right subtree because the third base is not a "C." At this stage, because a leaf is encountered, X is scored using PWM2, corresponding to the current leaf.

where a decision is made at each internal branching point based on whether a specific position of X is a consensus base or not, guiding the search down the appropriate descendent branches of the tree. The search eventually stops at a leaf which corresponds to a PWM, the one that "best" represents the sequence X.

Unlike the *Position Weight Array* mentioned above, which assumes dependence between every pair of adjacent positions, MDD is not restricted to adjacent positions and explicitly evaluates whether there is a statistical dependence between any two positions. However, it is easy to see that MDD requires a large number of sequences.

7   Modeling and detecting arbitrary dependencies

In this section we will discuss a general *Bayesian* approach developed in [18] to model dependencies between arbitrary pairs of binding site positions. In this approach, each of the K binding site positions may depend on any arbitrary set of other positions. This scenario can be best illustrated using a graph structure. Consider a network with K nodes ($s_1, s_2, \cdots, s_K$) corresponding to the positions i through K, where x_i is a random variable representing the nucleotide at position i. We draw an arrow (a directed edge) from node s_i to s_j if the nucleotide at position j depends on the nucleotide at position i; dependence can be determined using the χ^2 statistic. Figure 7.3 shows a few dependency structures for $K = 4$. Consider the simplest case, with 4 nodes and no edges depicted in Figure 7.3a, such that each of the nucleotides is independent, which is precisely the PWM model. In probabilistic terms, the probability of observing a specific binding site $x_1 x_2 x_3 x_4$ is the product of the four independent probabilities, i.e. $P(x_1 x_2 x_3 x_4) = P(x_1)P(x_2)P(x_3)P(x_4)$, where $P(x_i)$ is the entry in the PWM at column i, for nucleotide x_i.

Now consider the dependency shown in Figure 7.3b with three edges. The first position is independent of any other position, while every other position depends on the previous position. In probabilistic terms, $P(x_1 x_2 x_3 x_4) = P(x_1)P(x_2|x_1)P(x_3|x_2)P(x_4|x_3)$, where the notation $P(u|v)$ represents the probability of u conditional on the value of v. This is precisely the first-order Markov Model and is similar to the Weighted Array Matrix model mentioned above. The probability of each nucleotide at the first position is calculated in a fashion identical to that of a PWM. The conditional probabilities can then be derived from the given set of sites in a similar fashion. For instance, if among 10 sequences that have an A at the first position, three have a C at the second position, then $P(x_2 = C \mid x_1 = A) = 0.3$.

Figure 7.3c depicts a more complex dependency structure among the binding site positions. In this case position 2 depends on position 1. Position 3 depends on both positions 1 and 4, while positions 1 and 4 are independent of any other positions. We can write out the probability of observing a DNA sequence $x_1 x_2 x_3 x_4$ as $P(x_1 x_2 x_3 x_4) = P(x_1)P(x_2 \mid x_1)P(x_3 \mid x_1, x_4)P(x_4)$. Similar to the previous case, we can compute the conditional probability $P(x_3|x_1, x_4)$ by computing the fraction of different nucleotides at position 3 for various combinations of dinucleotides at positions 1 and 4. Finally,

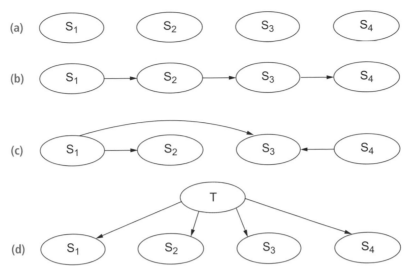

Figure 7.3 The figure illustrates a few possible dependency structures between the binding site positions (adapted from [18]).

Figure 7.3d illustrates a scenario where the nucleotides at the four bases are independent of each other but depend on an extrinsic variable T. For instance, certain TF are known to recognize distinct classes of motifs and the variable T may represent the motif class which in turn determines the nucleotide preferences at the four positions. It is not difficult to see that any arbitrary dependency structure defines a unique model, and given a model, one can precisely estimate the probability of observing a DNA sequence. However, there are a large number of possible dependency structures, and determining all possible dependency structures is not at all trivial. Incidentally, this problem is also encountered in other areas of computational biology, notably when inferring regulatory networks from gene expression data. The issue of searching for the *optimal model* is discussed in more detail in chapter 16 on biological network inference.

 8 # Searching for novel binding sites

The eventual goal of any model of TF–DNA binding is to efficiently and accurately assess whether an arbitrary sequence is likely to bind to the TF, and more generally, to identify potential binding site locations along a long stretch of DNA, possibly an entire genome. For consensus models, the search entails a simple scan of the DNA sequences for a perfect match, or a match with a limited number of mismatches to the consensus sequence. However, in the case of PWMs, detecting the binding sites is less straightforward.

8.1   A PWM-based search for binding sites

Essentially each sequence is assigned a "match" score which represents quantita-
tively its similarity to the PWM. For a PWM, a scoring function can simply be the
product of nucleotide frequencies at each position. For instance, the match score for
CCGGTACCGG (sequence X_1 in Example 7.1a) and using the PWM in Example 7.1b
can be computed as $0.8 \times 1.0 \times 0.8 \times 0.6 \times 0.7 \times 1.0 \times 0.9 \times 0.9 \times 1.0 \times 1.0 =$
0.22. This quantity represents the probability that the sequence confers to, or is gen-
erated by, the PWM. Such a raw score is interpreted (*is this score sufficiently large
to indicate a match of the PWM to the binding site?*) in the context of a specific
background. For instance, a PWM in which, at every position, the bases "C" or "G"
have the highest probability, is expected to achieve a high raw score while searching a
region of the genome that is composed mostly of "C" and "G". In this case, an even
higher raw score should be required.

Various software tools employ different strategies to select a threshold for the raw
score. The MATCH software adapted from [19] employs the following strategy. Let
r denote the raw match score for a PWM for a binding site. The raw score r is
first converted into a percentile score p. If the minimum and maximum achievable
scores by the PWM are r_{min} and r_{max}, then $p = (r - r_{min})/(r_{max} - r_{min})$. MATCH
then searches an input sequence for matches whose percentile score surpasses a user-
defined threshold. The default thresholds are based on a carefully chosen background
to optimize either the false-negative rate, the false-positive rate, or the sum of both
types of errors. Another strategy is to convert the raw score into a P-value, which
estimates the random expectation of observing the raw score (or higher). For instance,
Levy and Hannenhalli use a direct empirical approach. For a PWM, raw scores for
every position on the entire genome (of the species of interest) on either strand are
computed. This empirically estimated background distribution of raw scores provides
a direct way to compute the frequency with which a score of at least r is expected by
chance. If a score of at least r is achieved Q times, then the P-value of this score is
estimated as Q/L, where L is the total length of the genome including both strands
[20]. The other models that incorporate higher-order dependency between positions
can be used to assign a score to novel DNA sequences analogously, and will not be
discussed here.

8.2   A graph-based approach to binding site prediction

In Example 7.1a, it is intuitive that the first sequence $X_1 = CCGGTACCGG$ should
have a high-affinity interaction with the TF, since it is not only known to bind to the
TF, but it is also the consensus sequence. Given a model, we can compute a score
for a sequence indicative of the binding probability or binding affinity. We discussed

above how this score is computed for a PWM. While in Example 7.1a, the consensus sequence happens to be among one of the sequences known to bind the TF, this is often not the case. More problematic and perhaps counterintuitive is the fact that with probabilistic models, such as PWM, a sequence that is not among the known examples may score better than a sequence known to bind the TF. Naughton *et al.* provide a simple illustrative example [21]. Consider three known examples of binding sites for a TF – AAA, AAA, and AGG. If we construct a PWM based on these three sequences, the score for sequence AAG would be $1.0 \times 0.67 \times 0.33 = 0.22$ while the score for AGG will be $1.0 \times 0.33 \times 0.33 = 0.11$. Interestingly, the sequence AAG, which is not known to bind to the TF, has a higher score than the sequence AGG, which is known to bind the TF. The problem is that in order to score a sequence, the probabilistic models use "average" properties of the known sites and not the known sites themselves. To address this shortcoming of probabilistic models, Naughton *et al.* proposed a graph-based approach for scoring a sequence directly from the known binding sites without building an explicit model. The intuition behind their approach is as follows. Assume that we wish to score a sequence X using N distinct sequences known to bind to the TF. Each of the N sequences additively contributes to the score for X, and the individual score contribution is a product of two components. The first component is proportional to the similarity between the sequences X and Y, where Y is one of the N sequences. The second component is proportional to the number of times Y occurs among the known binding sites. Thus the score contribution is high if there is a sequence very similar to X among the known sequences and there are many known instances of this sequence. The details of the precise function used can be found in [21].

 9 ## Additional hallmarks of functional TF binding sites

TF binding sites are typically short (5–15 bp) and various binding sites for a TF can vary substantially. The DNA binding site sequence alone often does not contain sufficient information to explain the specificity with which a TF binds to its cognate binding sites. Thus, on the one hand, there are numerous locations in a genome that harbor DNA sequences strongly matching the TF–DNA binding model, and yet do not seem to bind to the TF in experiments; on the other hand, there are numerous locations experimentally known to be bound to a TF and yet which do not contain any sequences that could be predicted by the TF–DNA interaction model. Therefore, the match to a TF–DNA model, such as a PWM, is only one of the many determinants of functional TF–DNA interactions. There are several other hallmarks of TF binding sites that can be employed to improve the accuracy of binding site identification. Below we briefly mention two such features. Additional determinants of functional TF–DNA interaction are discussed below.

9.1 Evolutionary conservation

Consider a region of the genome that encodes for an important organismal function. Any mutation in this region affecting the specific function may be deleterious to the fitness of the organism and should be purged by evolution. In other words, such a region is likely to be evolving under purifying selection and will thus be conserved across species during evolution. The same principle applies to regulatory regions of the genome that harbor TF binding sites. *Phylogenetic footprints* are non-protein-coding regions of the genome that are highly conserved and are much more likely to be evolving under purifying selection [22]. Due to the recent availability of numerous alignable genome sequences, phylogenetic footprinting has been widely used to identify binding sites [20, 23, 24]. For a detailed review of phylogenetic footprinting we refer the reader to [25]. Although using evolutionary conservation is an effective way to reduce the false-positive rate in binding site prediction, exclusive reliance on conservation is limited for two reasons. First, conserved regions may sometimes be functionally neutral and thus may not harbor an important binding site [26]. Second, several functional binding sites are known not to be conserved, as shown by several studies [27, 28].

9.2 Modular interactions between TFs

Eukaryotic gene regulatory programs achieve complexity through combinatorial interactions among TF. For instance, the expressions of some of the *Drosophila* genes involved in development are regulated through combinatorial interactions among five TF proteins, *Bcd*, *Cad*, *Hb*, *Kr*, and *Kni* [29]. Consistent with the interactions between the TFs, the binding sites for these TF occur in clusters in the regulatory regions of the genes [30]. It seems that binding sites that occur in clusters are more likely to be functional. Thus the prediction of individual binding sites can be improved when subsumed within a search for binding site clusters. Several tools have been developed to detect significant clusters of binding sites in the genome [31, 32]. A cluster of functionally interacting binding sites, typically with multiple instances in the genome (presumably regulating several functionally related genes) is referred to as a *cis-regulatory module* (CRM) [33, 34]. Knowledge of CRMs can aid in accurate identification of individual binding sites [35]. Numerous computational approaches have been proposed to identify CRMs [25, 36–38]. Studies suggest that the binding of a TF to a binding site may depend on the presence or absence of binding sites for other TFs in the relative vicinity [39, 40]. Thus binding sites for a TF can be predicted with greater accuracy if one takes into account the presence/absence of binding sites of specific interacting TF. Binding models have been proposed to exploit such sequence contexts [41, 42].

DISCUSSION

The general problem of accurately identifying transcription factor binding sites is important for a mechanistic understanding of transcriptional regulation. In this chapter we have focused on the narrower problem of modeling the TF–DNA interaction based only on a set of experimentally determined binding site sequences without any other information about the genomic or cellular context. An ideal model should be such that (1) the true DNA binding sites fit the model very well, i.e. the model is *sensitive*, and (2) the DNA sequences that are known not to bind the TF should not fit the model, i.e. the model is *specific*. Moreover, the model should be biologically interpretable. The PWM model, while being simple, does not capture potential dependencies between binding site positions. A full dependence model, on the other hand, is difficult to estimate reliably based only on a small number of exemplar binding sites. Despite the efforts and advances made over the last several years our ability to predict binding sites on a genome scale remains unsatisfactory.

Ultimately, any sequence-based model of TF–DNA interaction does not capture the inherently dynamic cellular state. For instance, how tightly the DNA at any given location on the chromosome is packaged on the nucleosomes, critically determines the TF–DNA interaction and, more generally, transcriptional regulation [43, 44]. It is possible that even a high-affinity binding site may not bind the TF, if the binding site location is tightly wrapped around a nucleosome, which are the basic unit of DNA packaging. Narlikar *et al.* were able to significantly improve the *de novo* motif discovery accuracy by exploiting nucleosome occupancy [45]. Histone modifications can also help identify the condition-specific chromatin structure and can help improve the genome-wide identification of binding sites. Recent application of high-throughput technologies, most notably ChIP-seq [46], have been used to generate genome-wide maps of histone modifications [47–49]. Lastly, post-translational modification states of TF proteins can critically alter the TF–DNA interaction [50]. However, how these modifications affect TF–DNA interaction is not well understood. Improvements in computational modeling of TF–DNA interaction is likely to come from a better biological understanding of these various determinants of TF–DNA interactions coupled with the development of tools that can integrate the heterogeneous information.

QUESTIONS

(1) Consider the following probability matrix representing the DNA binding specificity of a transcription factor.

	1	2	3	4	5
A	0.01	0.10	0.97	0.95	0.50
C	0.03	0.05	0.01	0.01	0.10
G	0.95	0.05	0.01	0.03	0.10
T	0.01	0.80	0.01	0.01	0.30

Calculate the information content (IC) for position 3 and position 5. Briefly explain what information content means and why there is such a difference in this value between positions 3 and 5. In other words, what characteristic of position 5 makes its IC so low, while the IC of position 3 is so high?

(2) What is the consensus binding site for the transcription factor in problem (1)?

(3) Based on the consensus sequence, can you find the most likely binding sites for the TF in the following DNA sequence: ACCAAGTAGATTACTT? Consider both the forward and reverse strands. Now which of these sites is the most likely if you consider the probability matrix above?

(4) Analogous to transcription factors, which bind to DNA, RNA binding proteins (RBP) bind to specific RNA molecules, such as mRNA. They regulate critical aspects of post-transcriptional processing of the mRNA. Much like TF–DNA interaction, RBP–RNA interaction is believed to be specific. What aspects of the target mRNA are likely to be important for specific RBP–RNA interaction?

REFERENCES

[1] J. M. Huibregtse, P. D. Good, G. T. Marczynski, J. A. Jaehning, and D. R. Engelke. Gal4 protein binding is required but not sufficient for derepression and induction of gal2 expression. *J. Biol. Chem.*, 268: 22219–22222, 1993.

[2] D. Hebenstreit, J. Horejs-Hoeck, and A. Duschl. Jak/stat-dependent gene regulation by cytokines. *Drug News Perspect.*, 18: 243–249, 2005.

[3] J. Villard. Transcription regulation and human diseases. *Swiss Med. Wkly*, 134: 571–579, 2004.

[4] L. Elnitski, V. X. Jin, P. J. Farnham, and S. J. Jones. Locating mammalian transcription factor binding sites: A survey of computational and experimental techniques. *Genome Res.*, 16: 1455–1464, 2006.

[5] C. Tuerk and L. Gold. Systematic evolution of ligands by exponential enrichment: RNA ligands to bacteriophage T4 DNA polymerase. *Science*, 249: 505–510, 1990.

[6] M. L. Bulyk. Protein binding microarrays for the characterization of DNA-protein interactions. *Adv. Biochem. Eng. Biotechnol.*, 104: 65–85, 2007.

[7] V. Matys, O. V. Kel-Margoulis, E. Fricke, *et al.* TRANSFAC and its module TRANSCOMPEL: Transcriptional gene regulation in eukaryotes. *Nucleic Acids Res.*, 34: D108–D10, 2006.

[8] A. Sandelin, W. Alkema, P. Engstrom, W. W. Wasserman, and B. Lenhard. JASPAR: An open-access database for eukaryotic transcription factor binding profiles. *Nucleic Acids Res.*, 32: D91–D94, 2004.

[9] X. Liu and N. D. Clarke. Rationalization of gene regulation by a eukaryotic transcription factor: Calculation of regulatory region occupancy from predicted binding affinities. *J. Mol. Diol.*, 323: 1–8, 2002.

[10] A. Cornish-Bowden. Nomenclature for incompletely specified bases in nucleic acid sequences: Recommendations 1984. *Nucl. Acids Res.*, 13: 3021–3030, 1985.

[11] T. D. Schneider, G. D. Stormo, L. Gold, and A. Ehrenfeucht. Information content of binding sites on nucleotide sequences. *J. Mol. Biol.*, 188: 415–431, 1986.

[12] G. D. Stormo. DNA binding sites: Representation and discovery. *Bioinformatics*, 16: 16–23, 2000.

[13] T. K. Man, J. S. Yang, and G. D. Stormo. Quantitative modeling of DNA-protein interactions: Effects of amino acid substitutions on binding specificity of the MNT repressor. *Nucl. Acids Res.*, 32: 4026–4032, 2004.

[14] M. L. Bulyk, P. L. Johnson, and G. M. Church. Nucleotides of transcription factor binding sites exert interdependent effects on the binding affinities of transcription factors. *Nucl. Acids Res.*, 30: 1255–1261, 2002.

[15] M. Q. Zhang and T. G. Marr. A weight array method for splicing signal analysis. *Comput. Appl. Biosci.*, 9: 499–509, 1993.

[16] C. Burge and S. Karlin. Prediction of complete gene structures in human genomic DNA. *J. Mol. Biol.*, 268: 78–94, 1997.

[17] M. J. Campbell and D. Machin. *Medical Statistics: A Commonsense Approach*. 3rd edn. Wiley, Chichester 2002.

[18] Y. Barash, G. Elidan, N. Friedman, and T. Kaplan. Modeling dependencies in protein-DNA binding sites. In: Proceedings of the Seventh Annual International Conference on Research in Computational Molecular Biology, Berlin, Germany. ACM Press, New York, 2003, 28–37.

[19] K. Quandt, K. Frech, H. Karas, E. Wingender, and T. Werner. Matind and matinspector: New fast and versatile tools for detection of consensus matches in nucleotide sequence data. *Nucl. Acids Res.*, 23: 4878–4884, 1995.

[20] S. Levy and S. Hannenhalli. Identification of transcription factor binding sites in the human genome sequence. *Mamm. Genome*, 13: 510–514, 2002.

[21] B. T. Naughton, E. Fratkin, S. Batzoglou, and D. L. Brutlag. A graph-based motif detection algorithm models complex nucleotide dependencies in transcription factor binding sites. *Nucl. Acids Res.*, 34: 5730–5739, 2006.

[22] D. A. Tagle, B. F. Koop, M. Goodman, *et al.* Embryonic epsilon and gamma globin genes of a prosimian primate (*Galago crassicaudatus*). Nucleotide and amino acid sequences, developmental regulation and phylogenetic footprints. *J. Mol. Biol.*, 203: 439–455, 1988.

[23] W. W. Wasserman and J. W. Fickett. Identification of regulatory regions which confer muscle-specific gene expression. *J. Mol. Biol.*, 278: 167–181, 1998.

[24] X. Xie, J. Lu, E. J. Kulbokas, *et al.* Systematic discovery of regulatory motifs in human promoters and 3′ UTRS by comparison of several mammals. *Nature*, 434: 338–345, 2005.

[25] W. W. Wasserman and A. Sandelin. Applied bioinformatics for the identification of regulatory elements. *Nat. Rev. Genet.*, 5: 276–287, 2004.

[26] M. A. Nobrega, Y. Zhu, I. Plajzer-Frick, V. Afzal, and E. M. Rubin. Megabase deletions of gene deserts result in viable mice. *Nature*, 431: 988–993, 2004.

[27] E. T. Dermitzakis and A. G. Clark. Evolution of transcription factor binding sites in mammalian gene regulatory regions: Conservation and turnover. *Mol. Biol. Evol.*, 19: 1114–1121, 2002.

[28] E. Emberly, N. Rajewsky, and E. D. Siggia. Conservation of regulatory elements between two species of *Drosophila*. *BMC Bioinformatics*, 4: 57, 2003.

[29] D. Niessing, R. Rivera-Pomar, A. La Rosee, *et al.* A cascade of transcriptional control leading to axis determination in *Drosophila*. *J. Cell. Physiol.*, 173: 162–167, 1997.

[30] B. P. Berman, Y. Nibu, B. D. Pfeiffer, *et al.* Exploiting transcription factor binding site clustering to identify cis-regulatory modules involved in pattern formation in the *Drosophila* genome. *Proc. Natl Acad. Sci. U S A*, 99:757–762, 2002.

[31] M. Rebeiz, N. L. Reeves, and J. W. Posakony. Score: A computational approach to the identification of cis-regulatory modules and target genes in whole-genome sequence data. Site clustering over random expectation. *Proc. Natl Acad. Sci. U S A*, 99: 9888–9893, 2002.

[32] S. Sinha, E. Van Nimwegen, and E. D. Siggia. A probabilistic method to detect regulatory modules. *Bioinformatics*, 19 Suppl. 1, I292–I301, 2003.

[33] M. Z. Ludwig, N. H. Patel, and M. Kreitman. Functional analysis of eve stripe 2 enhancer evolution in *Drosophila*: Rules governing conservation and change. *Development*, 125: 949–958, 1998.

[34] H. Bolouri and E. H. Davidson. Modeling DNA sequence-based cis-regulatory gene networks. *Dev. Biol.*, 246: 2–13, 2002.

[35] O. Hallikas, K. Palin, N. Sinjushina, *et al.* Genome-wide prediction of mammalian enhancers based on analysis of transcription-factor binding affinity. *Cell*, 124: 47–59, 2006.

[36] J. W. Fickett and W. W. Wasserman. Discovery and modeling of transcriptional regulatory regions. *Curr. Opin. Biotechnol.*, 11: 19–24, 2000.

[37] S. Hannenhalli. Eukaryotic transcriptional regulation: Signals, interactions and modules. In N. Stojanovic (ed.) *Computational Genomics*. Horizon Bioscience, Norfolk, 2007, 55–82.

[38] S. Hannenhalli. Eukaryotic transcription factor binding sites – Modeling and integrative search methods. *Bioinformatics*, 24: 1325–1331, 2008.

[39] A. Hochschild and M. Ptashne. Cooperative binding of lambda repressors to sites separated by integral turns of the DNA helix. *Cell*, 44: 681–687, 1986.

[40] S. Lomvardas and D. Thanos. Nucleosome sliding via TBP DNA binding in vivo. *Cell*, 106: 685–696, 2001.

[41] D. Das, N. Banerjee, and M. Q. Zhang. Interacting models of cooperative gene regulation. *Proc. Natl Acad. Sci. U S A*, 101: 16234–16239, 2004.

[42] L. Wang, S. Jensen, and S. Hannenhalli. An interaction-dependent model for transcription factor binding. In: *Lecture Notes in Computer Science*. Volume 4023. Springer, Berlin/Heidelberg, 2005, 225–234.

[43] W. Reik. Stability and flexibility of epigenetic gene regulation in mammalian development. *Nature*, 447: 425–432, 2007.

[44] M. M. Suzuki and A. Bird. DNA methylation landscapes: Provocative insights from epigenomics. *Nat. Rev. Genet.*, 9: 465–476, 2008.

[45] L. Narlikar, R. Gordan, and A. J. Hartemink. A nucleosome-guided map of transcription factor binding sites in yeast. *PLoS. Comput. Biol.*, 3: e215, 2007.

[46] P. J. Park. Chip-seq: Advantages and challenges of a maturing technology. *Nat. Rev. Genet.*, 10: 669–680, 2009.

[47] A. Barski, S. Cuddapah, K. Cui, *et al.* High-resolution profiling of histone methylations in the human genome. *Cell*, 129: 823–837, 2007.

[48] D. E. Schones, K. Cui, S. Cuddapah, *et al.* Dynamic regulation of nucleosome positioning in the human genome. *Cell*, 132: 887–898, 2008.

[49] E. Birney, J. A. Stamatoyannopoulos, A. Dutta, *et al.* Identification and analysis of functional elements in 1 genome by the encode pilot project. *Nature*, 447: 799–816, 2007.

[50] M. Neumann and M. Naumann. Beyond ikappabs: Alternative regulation of nf-kappab activity. *FASEB J.*, 21: 2642–2654, 2007.

CHAPTER EIGHT

How does the influenza virus jump from animals to humans?

Haixu Tang

As shown by the 2009 Swine Flu outbreak, the influenza epidemics are often caused by human-adapted influenza viruses originally infecting other animals. The influenza viruses infect host cells through the specific interaction between the viral hemagglutinin protein and the sugar molecules attached to the host cell membrane (called *glycans*). The molecular mechanism of the host switch for Avian influenza viruses was thus believed to be related to the mutations that occurred in the viral hemagglutinin protein that changed its binding specificity from avian-specific glyans to human-specific glycans. This theory, however, is not fully consistent with the epidemic observations of several influenza strains. I will introduce the bioinformatics approaches to the analysis of *glycan array* experiments that revealed the glycan structural pattern recognized by the hemagglutinin from viruses with different host specificities. The glycan motif finding algorithm adopted here is an extension of the commonly used protein/DNA sequence motif finding algorithms, which works for the trees (representing glycan structures) rather than strings (as protein or DNA sequences).

1 Introduction

The recent outbreak of "swine flu" is not the first flu pandemic (i.e. the spread of an infectious disease in the human population across a large region) in human history. Three worldwide outbreaks of influenza flu occurred in the twentieth century, in 1918, 1957, and 1968, respectively. "Spanish flu" is known as the most deadly natural

Bioinformatics for Biologists, ed. P. Pevzner and R. Shamir. Published by Cambridge University Press.
© Cambridge University Press 2011.

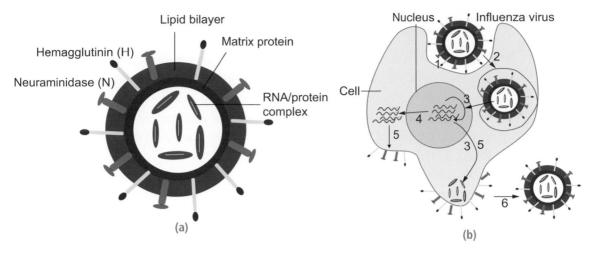

Figure 8.1 A schematic illustration of (a) the structure of the influenza virus; and (b) the infection process of the influenza virus. The virus contains a lipid bilayer attached by two kinds of membrane proteins, the hemagglutinin and the neuraminidase, and an inner layer of matrix proteins. The virus infects epithelial cells of the host respiratory systems in six steps (see text for details).

disaster, which swept around the world in 1918 and killed about 50–100 million people. Although the number of deaths in the subsequent pandemics were less significant – it is estimated that the 1957 and 1968 pandemics killed approximately one million people each, whereas the 2009 pandemic killed more than 18,000 people worldwide according to the statistics of the World Health Organization – the death rate remains similar and is comparable to that of the seasonal flu. It was not until the 1930s that the cause of the influenza was found to be a virus. To date, three types of influenza virus were discovered (A, B, and C, respectively), among which influenza A is responsible for the regular influenza outbreaks.

All influenza viruses belong to one family of RNA viruses (Orthomyxoviridae) that has RNA (ribonucleic acid) as their genetic materials. The influenza virion is a globular particle (Figure 8.1a) with a diameter of about 100 nm. The surface of the virion is protected by a lipid bilayer, the same component as the plasma membrane, which is derived from the plasma membrane of its host cell. Two kinds of membrane proteins are attached on the viral surface, i.e. ~500 copies of hemagglutinin (also called the "H" protein) and ~100 copies of neuraminidase (also called the "N" protein). The influenza virion carries eight RNA molecules consisting of genes encoding the H and N proteins, the matrix proteins and the nucleoproteins. Within the lipid bilayer, the RNA molecules were further protected by another layer of matrix proteins and many copies of nucleoproteins associated with them.

The influenza virus infects epithelial cells of the host respiratory systems. The whole infection process involves six steps (Figure 8.1b).

1 The virus binds to the epithelial cells through the interaction between the hemagglutinin and the glycans[1] attached to glycoproteins on the host cell surface.
2 The virus is swallowed up by the host cell (a process called *endocytosis*).
3 Fusion of the viral membrane with the vesicle membrane releases the content of the virus into the cytosol, and the viral RNAs enter the nucleus of the cell where the RNAs will be reproduced.
4 Fresh copies of viral RNAs enter the cytosol.
5 Some viral RNA molecules in the cytosol act as messenger RNA to be translated into the proteins for the new virus particles, while other viral RNA molecules are assembled into the core of the new virus particles.
6 The new virus buds off from the membrane of the host cell, aided by the neuraminidase encoded by the virus RNAs.

It is clear that the two viral surface proteins, the hemagglutinin and the neuraminidase, play essential roles in the infection process of the influenza viruses. The hemagglutinin acts as the "initiator" that recognizes and captures the target cells, whereas neuraminidase acts as the "terminator" that releases the fresh virus from the host cells. Not surprisingly, these two proteins became the primary targets for the design of antiviral drugs and effective vaccines against influenza.[2] For the same reason, influenza viruses are usually classified into subtypes based on the sequence divergence of their hemagglutinin (H) and neuraminidase (N) genes. A total of 16 types of H genes and 9 types of N genes are known to date. A majority of severe pandemics of human influenza were caused by the H1N1 (including the 2009 "Swine Flu") and the H3N2 viruses.

Since the discovery of influenza viruses, thousands of influenza virus strains have been collected. The analysis of their genetic materials (i.e. the RNA molecules) has shown that the flu pandemics occur when the virus acquires a new variant of the genes encoding the H or N proteins. Where did these "new" variants come from? In many cases, domestic animals appear to be the source. In fact, influenza viruses can infect not only humans, but also domestic animals such as pigs (causing "Swine Flu"), horses, chickens, ducks, and some wild birds (causing "Avian Flu"). Although most influenza viruses can only infect either humans or another animal, some animal flu viruses have jumped from animals to humans, which has caused several major flu outbreaks. The

---

[1] The carbohydrates (sugars) linked to other molecules (such as proteins or lipids) are called glycans in biochemistry.
[2] For instance, the antiviral drugs Oseltamivir (trade name *Tamiflu*) and Zanamivir (trade name *Relenza*) that slow down the spread of influenza are both effective inhibitors of the neuraminidase.

H2 viruses that appeared in 1957 and the H3 viruses that appeared in 1968 originated from Avian Flu viruses, whereas the 2009 "Swine Flu" pandemic was caused by a new H1N1 influenza virus that circulated in pigs.

Now a fundamental biological problem arises: how can influenza viruses jump from animals to humans? As we mentioned briefly above, the molecular mechanism for influenza viruses to *recognize* its appropriate target cell involves the specific interaction between the hemagglutinin and glycans on the surface of the host cell. Hence, a straightforward model may explain the host switch of influenza viruses, which is based on three hypotheses: (1) structurally distinct glycans are present on the surface of animal and human cells; (2) hemagglutinin proteins can recognize these subtle structural distinctions; and (3) some mutations occurring on hemagglutinin of animal viruses result in the switch of its binding specificity from animal glycans to human glycans. To study the validity of this model and, more importantly, to characterize the subtle glycan structural features that can be recognized by influenza viruses, the glycan array technique is used to assay the binding affinity of hemagglutinins on various glycan structures. In this chapter, we will introduce the bioinformatics concept for the analysis of glycan array experimental data in an attempt to elucidate the distinct features that are recognized by human viruses but not by animal viruses.

The rest of the chapter is organized as follows. We will first introduce the molecular basis of the host switch of influenza viruses, then we will briefly describe the glycan array experiments for characterizing hemagglutinin binding specificity, and finally we will introduce the computational approach to the glycan array data analysis. We will conclude the tutorial by discussing some specific aspects of the bioinformatics topics related to glycobiology.

2 Host switch of influenza: molecular mechanisms

Although DNA and proteins have garnered most of the attention in modern molecular cell biology, other classes of biomolecules are no less important. Carbohydrates (or sugars) were well studied in biochemistry for their roles as the structural molecules and in cellular metabolisms. Recent advancement in the research of glycans, a field called *glycobiology*, however, has concentrated on their relatively new roles as signaling molecules. All cells carry a dense coating of covalently linked sugar chains (called glycans or oligosaccharides) on their outer surface, which modulate a large variety of interactions between the cell and other cells in a multicellular organism, or between organisms, e.g. between host and viral or parasite cells. The initial step for the infection

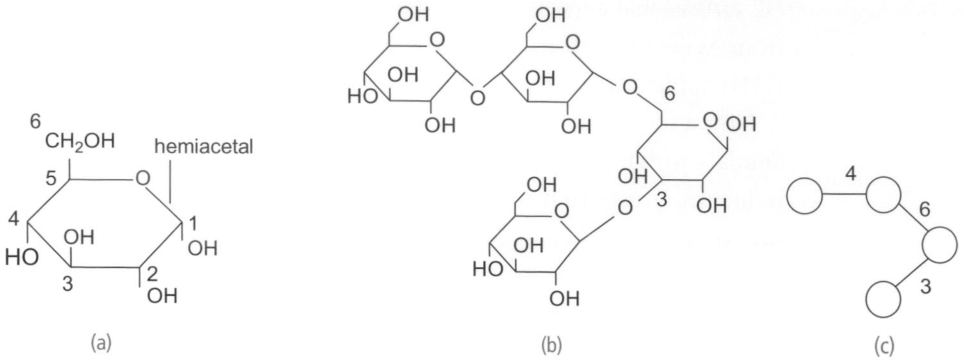

Figure 8.2 The structure of glycans. (a) The cyclic structure of a glucose; (b) the structure of a tetraglucose, consisting of four glucoses with a bifurcation branching of 1–3 and 1–6 linkages; (c) the tree representation of the tetraglucose.

of influenza viruses, in which hemagglutinin proteins on the virus surface interact with the glycans on the host cell surface, is an example of these cell communication processes.

2.1 Diversity of glycan structures

The study of the biological functions of glycans has advanced relatively slower than the study of proteins or nucleic acids, for two reasons. First, glycans exhibit more complex structures than proteins and nucleic acids, and the complexity is not due to their compositions. There are only a limited number of building blocks, called *monosaccharides*, in glycans, of which those common ones found in higher animal glycans are listed in Table 2.1. Each monosaccharide is a small carbohydrate, and contains six carbon atoms that can be numbered as the organic chemistry nomenclature such that the hemiacetal carbon is referred to as C1 (Figure 8.2a). Two monosaccharides react and form a glycosidic bond between the C1 group of one monosaccharide and the alcohol group of the other while releasing a water molecule. Depending on which alcohol group participates in the reaction, there are four different types of glycosidic bonds, called 1–2, 1–3, 1–4, and 1–6 linkages.[3] A monosaccharide can be linked to more than one monosaccharide at a time (by covalent bonds called *glycosidic bonds*) and form branching structures. As a result, a general form of a glycan can be represented by a labeled tree,[4] in which each monosaccharide is represented by

---

[3] Since the reductive carbon atom in sialic acids are labeled as the *second* carbon, three possible linkages of sialic acid residues are classified as 2–3, 2–4, 2–6 linkages, respectively.

[4] Mathematically, a *tree* is a graph with no cycles, in which each node has zero or more *children* nodes and at most one *parent*. The nodes having no child are called the *leaf* nodes. The only node in a tree with zero parent

Table 8.1   Symbolic representations of common monosaccharides

Symbols[1]	Monosaccharide residues and abbreviations
○	Hexoses, e.g. galactose (Gal), glucose (Glc), and mannose (Man)
□	*N*-acetylhexosamines (HexNAc), e.g. N-acetylglucosamine (GlcNAc) and *N*-acetylegalactosamine (GalNAc)
◇	Sialic acids, e.g. *N*-acetylneuraminic acid (Neu5Ac) and *N*-glycolylneuraminic acid (Neu5Gc)
◆	Uronic acids, e.g. iduronic acid (IdoA) and glucuronic acid (GlcA)
△	Deoxyhexoses, e.g. fucose (Fuc)
★	Pentoses, e.g. xylose (Xyl)

[1] Each symbol represents a class of monosaccharides with the same atomic compositions (i.e. the same chemical formula) but different chemical *configurations*, referred to as the *isomers*, e.g. the galactose and glucose. Isomers are distinguished by different colors in the glycan representation (as shown in Figure 8.4).

a symbol (see Table 2.1 for the list of such symbols) and each glycosidic bond is represented by an edge. The number of branches of the tree is bounded by 4, because there are at most 4 glycosidic bonds that can be formed by one monosaccharide. In higher animals, there are usually two branches (two glycosidic bonds). We say the structure of a glycan is known when not only its monosaccharide sequence but also its whole branching structure and all linkage types are characterized. Second, glycans are synthesized through a template-free and step-wise process. The complex glycosylation machinery that assembles monosaccharides into oligosaccharides consists of hundreds of proteins. More importantly, to carry out biological functions, glycans are often attached to other classes of biomolecules, such as proteins and lipids, forming different *glycoconjugates*. In higher animals, the synthetic glycoconjugates can be classified according to the biomolecules they are attached to. A glycoprotein is a glycoconjugate in which one or more glycans are covalently attached to a protein through *N*-linked or *O*-linked glycosylations (Figure 8.3a). Most glycoproteins are anchored on the plasma membrane, with the glycans oriented toward the extracellular side. Many of these glycans act as the specific receptors for various kinds of viruses, bacteria, and parasites, including the influenza viruses.

is called the *root* node. The *depth* of a node is defined as the length (i.e. the number of edges) of the path from the node to root. A *subtree* of a tree is defined as the tree consisting of a subset of *connected* nodes in the original tree. A *complete* subtree is then defined as a subtree consisting of a node and all its descendents (children, children of children, etc.). Both the nodes and edges in a tree can be *labeled*. For example, the nodes in a glycan tree are labeled by the monosaccharide residues, and the edges in a glycan tree are labeled by the linkage type.

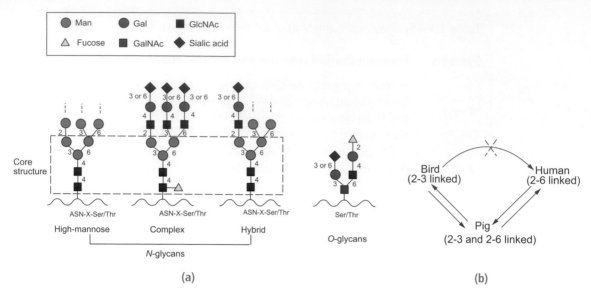

Figure 8.3 Glycan receptors and the host switch of influenza viruses. (a) Schematic representions of glycans attached to proteins. The *N*-linked (or *N*-) glycosylation occurs at an asparagine residue within the sequence pattern of Asn-X-ser/Thr (NXS/T), where *N* can be any amino acid residue but proline. All *N*-glycans share a common pentasaccharide core structure (with two GlcNAc and three Man residues), and can be further divided into three main classes: high-mannose-type, complex-type, and hybrid-type, based on the monosaccharide sequences extended from the core structure. The extended sequence of the high-mannose-type *N*-glycans contains only mannose residues in all their branches, whereas the extended sequence of the complex-type *N*-glycans alternates between GlcNAc and Gal residues (called the lactosamine repeats) and terminates with sialic acid or fucose residues, and the hybrid-type *N*-glycans contain some branches of high-mannose-type extended sequences, and some branches of complex-type extended sequences. The *O*-glycan (or *O*-) glycosylation occurs via the linkage between a GalNAc and a serine or threonine residue on the protein and can be extended into a large variety of oligosaccharides. The complex- or hybrid-types of *N*-glycans and *O*-glycans may contain sialic acids or fucoses as *terminal* residues, referred to as the sialylated and fucosylated glycans, respectively. The sialylated glycans are the ligands of the influenza hemagglutinins. (b) Molecular mechanisms for the host switch of influenza virus strains. The hemagglutinin of human influenza viruses have a binding preference for 2–6 linked sialylated glycans, whereas the hemagglutinin of avian viruses have a binding preference for 2–3 linked sialylated glycans. The respiratory epithelial cells of pigs express both 2–3 linked and 2–6 linked sialylated glycans, and thus can be infected by both human and avian influenza viruses. A new pandemic influenza strain might arise from the mix of the gene segments from the avian and human viruses that infect the same host (e.g. pigs).

2.2   Molecular basis of the host specificity of influenza viruses

A notable property of the glycans attached to the animal cell membranes is that they are of great *microheterogeneity*, i.e. there exist many different glycans on the cell surface, of which some share similar structures. Accordingly, unlike the protein–protein interaction that involves two or more specific proteins, glycan binding proteins often interact with a class of glycans that have a common structural pattern. The influenza hemagglutinin is a well-studied viral glycan-binding protein that specifically binds to sialylated glycans. The specificity of this interaction for different subtypes of influenza viruses varies substantially. Human influenza viruses bind only to cells expressing glycans of 2–6 linked sialic acids (to galactoses), whereas the other animal influenza viruses also bind to 2–3 linked sialic acids. Further investigation shows that this linkage preference is caused by a single mutation occurring in the hemagglutinin gene. This finding seems to be consistent with many observations related to the host specificity and switches of influenza viruses. Indeed, the 2–6 linked sialylated glycans are abundant in human respiratory epithelia, whereas the respiratory epithelia of the birds mainly express 2–3 linked sialylated glycans. The respiratory epithelia of some animals (e.g. pig) have receptors with both 2–3 linked and 2–6 linked sialylated glycans. According to the *vessel theory* of influenza pandemics (Figure 8.3b), pigs can act as the intermediate host on which the genetic materials from human and avian viruses are mixed, resulting in new pandemic strains that retain the ability to transmit within the human population, but are sufficiently different to reduce the efficiency of the host's immune response. It was hypothesized that both the 1957 H2N2 and the 1968 H3N2 pandemic strains arose from this mechanism.

The correlation between the transmission efficiency and the hemagglutinin–glycan binding specificity was observed on some influenza virus strains (e.g. the highly pathogenic human 1918 viruses). However, several cases were found to be inconsistent with this theory. For instance, switching hemagglutinin binding specificity of one human influenza virus (SC18) from 2–6 to 2–3 resulted in a virus strain (AV18) that is supposed to be transmissable in birds according to the theory, but is not in practice. Two experimentally collected H1N1 strains both show a mixed 2–3/2–6 binding specificity; however, one strain (NY18) does not transmit efficiently in the human population, whereas the other (Tx91) does. Finally, some chimeric H1N1 strains with increased binding affinity to 2–6 linked sialylated glycans actually spread less efficiently than the original strains in human and pig populations. All these results suggest a more complicated scenario of the host switch of influenza viruses.

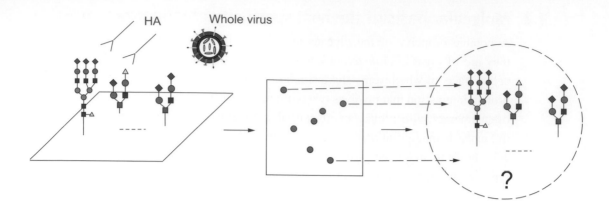

Figure 8.4 Elucidation of glycan structural determinants for a glycan binding protein (e.g. the viral hemagglutinin) through the glycan array technology. To characterize the binding specificity of a glycan binding protein (GBP) to various glycans, a library of synthetic glycans are printed onto the surface of a microarray slide, on which each spot represents a specific glycan. The GBP–glycan interaction can then be detected by incubating the slides with labeled GBPs (e.g. the hemagglutinins), and identifying the glycans corresponding to spots with signals. The identified glycans that potentially bind to the GBP can be used to characterize the glycan structural pattern recognized by the GBP, known as the glycan motif finding problem.

2.3   Profiling of hemagglutinin–glycan interaction by using glycan arrays

Until recently, the analysis of specificity of influenza hemagglutinins relied on virus-based assays, such as the competetive binding of glycoproteins (associated with glycans of great microheterogeneity) to the immobilized viruses. Although these assays demonstrated that the specificity of viral hemagglutinins is more complex than the recognition of 2–3 or 2–6 linked glycans, they were relatively low-throughput and were only optimized to certain virus strains. The development of glycan array technology enabled the study of the interaction between glycan binding proteins and glycans in a high-throughput manner. A glycan array comprises a library of synthetic (thus structurally known) glycans that are automatically printed on a glass slide (Figure 8.4). To investigate the specificity of influenza hemagglutinins, one can design a library of hundreds of glycans containing sialic acids, with various linkage, such as 2–3 or 2–6 linked. Therefore, the array provide an opportunity to simultaneously assay the interaction between hemagglutinins and hundreds of its potential glycan ligands. The subset of glycans can then be detected that interact with hemagglutinin proteins on a specific influenza virus strain (Figure 8.4). Note that the interaction assay can be conducted by using either the whole virus or recombinant hemagglutinin, which can be detected by fluorescent antibodies that bind to it.

Glycan array experiments report a group of structurally known glycans as the potential ligands of hemagglutinin proteins. The next question is what structural pattern these glycans share that can be recognized by the hemagglutinin. For example, since we have known the hemagglutinin proteins from a human influenza virus strain recognize 2–6 linked sialylated glycans, we anticipate that all detected glycans binding to human viral hemagglutinin should contain 2–6 linked sialic acids as terminal residues. Our expectation of the structural pattern actually goes beyond that. We want to investigate, besides the specifically linked sialic acid, whether there exist other common structure patterns among the detected glycan ligands. This leads to the formulation of the glycan motif binding problem, which attempts to identify a common structural pattern from a given set of glycans.

3 The glycan motif finding problem

The glycan motif finding problem resembles the well-studied DNA sequence motif finding problem. A DNA motif is defined as a DNA sequence pattern of some biological significance, e.g. the binding sites of a transcription factor (TF). The pattern is usually short (i.e. 5–20 bp long) and is known to recur in the regulatory regions of a number of genes. Given a set of DNA sequences (regulatory regions), the motif finding attempts to find overrepresented motifs. The input to the DNA motif finding problem can be retrieved from various resources, ranging from the comparative analysis of multiple genomes (i.e. the orthologous gene clusters) to the high-throughput genomics data from a single genome, such as gene microarray analysis (to find co-expressed genes that are likely co-regulated by the same TFs), Chromatin Immunoprecipitation (ChIP) (to find the genomic segment that a TF binds to), or protein binding arrays.

Depending on the representation of the DNA motifs, DNA motif finding algorithms can be roughly divided into three categories. The word-based methods assume that the DNA motif is a short sequence of some fixed length l (also called an l-tuple, e.g. TATAAA) that recur in the input sequences as the *exact* same copy. The consensus methods use a similar assumption, except that they allow some variation from the "consensus" motif. Finally, the profile methods employ sequence profiles (also called position weight matrix, PWM) to represent DNA motifs, which is a $4 \times l$ matrix ($l =$ the motif length) with each column representing the frequency of four nucleotides at each motif position. The word-based methods are simple to implement. For a fixed word length l, one needs to test whether each l-tuple in the input sequence is overrepresented or not. In contrast, consensus-based and profile-based methods need to apply sophisticated probabilistic algorithms (for details see Chapter 7). The overrepresentation of an l-tuple can be measured by a simple statistical test on the counts of the l-tuple in the DNA sequences. Given N input DNA sequences of the same length L, denote n as the number of sequences containing a specific l-tuple. What is the probability for an l-tuple to be observed in a random DNA sequence of length L? Since there are in total 4^l l-tuples in DNA sequences and they occur at equal probability in a random DNA sequence, each l-tuple has the equal probability of $(L - l + 1)/4^l$, and the *expected* number of sequences containing the l-tuple, denoted as n_e, is then $(N \times (L - l + 1))/4^l$. The greater n is than n_e, the more probable that an l-tuple is "overrepresented" in the input DNA sequences. The significance of the l-tuple can be measured by its probability of being observed n times in N random DNA sequences, which can be derived by using probability theory, or using simulation experiments [2].

Below, we introduce a similar approach to the glycan motif finding problem, in which we assume the glycan motif (the structural pattern recognized by GBPs, e.g. the hemagglutinin) is a *treelet*. Given a labeled tree, an l-treelet is a tree with l nodes that is a *subgraph* of the tree.[5] The glycan motif finding problem is then transformed to the search for overrepresented treelets in a given set of N glycan trees that can be solved by a *treelet counting* approach (Figure 8.5a). in two independent steps:

1 enumerate all l-treelets in each of N input glycan trees and count the number of trees (among N input glycan trees) that contain it as a subgraph, defined as the l-treelet *occurence*;

2 determine if an l-treelet is overrepresented in the set of input glycan trees based on its occurrence.

The enumeration of all l-treelets in a glycan tree can be achieved by a *recursive* algorithm. Denote $S(T, l)$ as the set of l-treelets in a tree T. In some special cases (or the *boundary* cases), $S(T, l)$ can be obtained directly. For instance, if T has fewer than l nodes, there is no l-treelet in T, or $S(T, l) = \emptyset$, where $\emptyset$ designates an empty set; if T has exactly l nodes, it has one and only one l-treelet that is the whole tree T, or $S(T, l) = T$; and finally, because the 1-treelet should contain only one node, $S(T, 1)$ should be the set of nodes in T. However, in general, $S(T, l)$ needs to be obtained recursively. Consider $S(T, l, v)$ as the set of l-treelets in T rooted by the node v. Obviously, $S(T, l)$ is the union of $S(T, l, v)$ for all nodes in T (or $S(T, l) = \cup_{v \in T} S(T, l, v)$). Assume the root of T (denoted as r) has n direct children

---

[5] A treelet is a subgraph of a tree if and only if both the topology and the node/edge labels match. Notably, a treelet of a tree is formally defined in graph theory as a *subtree* of tree (see Figure 8.5a for examples).

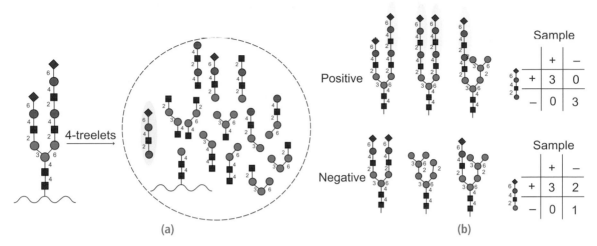

Figure 8.5 Glycan motif finding problem. (a) Enumerating 4-treelets in a complex-type N-glycan. All 4-treelets appear once in the glycan tree. The highlighted 4-treelet was found to be overrepresented in the human viral hemagglutinin binding glycans detected by glycan array experiments. (b) Determining if a treelet is overrepresented in a positive $(+)$ sample of glycans rather than a negative $(-)$ sample, derived from a glycan array experiment (see text for details). The occurrence of a treelet in a sample is defined as the number of glycans in the sample containing this treelet. A treelet is overrepresented if it occurs more frequently in the positive sample than in the negative sample, which can be conducted by constructing a 2×2 *contigency* table. For a specific treelet, the first row (denoted as $+$) in the table displays its occurrences in the positive and negative samples, respectively, whereas the second row (denoted as $-$) displays the number of glycans in the positive and negative samples that do not contain it. Intuitively, the treelet shown in the top table is more likely overrepresented in the positive sample than the treelet shown in the bottom table. The significance of the overrepresentation for a treelet can be obtained by a Fisher's exact test, as described in the text.

$(n \leq 4$ for glycan trees) (denoted as $v_1, v_2, ..., v_n$). We denote the complete subtrees of T that are rooted by $v_i (i = 1, 2, ..., n)$ as T_{v_i}. Any l-treelet of T is either rooted by r or is an l-treelet in one of the complete subtrees T_{v_i}. If we have obtained the set of k-treelets for each of these complete subtrees (for $k = 1, 2, ..., l$), i.e. $S(T_{v_i}, k)$, we can then construct the set of l-treelets of T by the union of several non-intersected sets: (1) the set of l-treelets in T_{v_i}, i.e. $S(T_{v_i}, l)$; and (2) the set of l-treelet rooted by r. The second set can be computed by enumerating the possible combination of n treelets with a total number of $l - 1$ nodes, each rooted by one v_i (thus a member of $S(T_{v_i}, k, v_i)$). The recursion continues until it reaches a boundary case.

After obtaining all l-treelets in a given set of glycan trees, the next step is to determine, for each of these treelets, if it occurs in a significantly large subset of trees.

At a first glance, we can devise a method similar to the one we use to compute the significance of the DNA l-tuples. For each of the input trees i, we can also count the total number of l-treelets it contains, denoted as k_i.[6] If we assume the input glycan tree is randomly chosen, then the expected number of trees containing any l-treelet is the same and equal to $(\sum_i k_i)/t^l$, where t is the total number of monosaccharides observed in the glycan trees (≈ 6). Unfortunately, this approach has a strong drawback. Glycans have regular structures and cannot be assumed to be random sequences, because they are synthesized through a series of reactions. For example, all glycans share the same core structure consisting of five monosaccharide residues (Figure 8.3a). As a result, overrepresented l-treelets detected by this method may correspond to the recurrent glycan structures rather than the structural pattern recognized by hemagglutinin.

To address this issue, we need to adopt a different approach. Consider all M glycans printed on the glycan array. If an l-treelet is not overrepresented in the glycans binding to hemagglutinin, it should occur in a *proportional* number of glycans in the set of glycans binding to hemagglutinin and the set of glycans not binding to hemagglutinin (Figure 8.5). To test whether a specific l-treelet is overrepresented in the first set in comparison to the second, we can employ a Fisher's exact test on a 2×2 contingency table [3].[7] Assume that there are N glycans detected to bind to hemagglutinin, and $M - N$ glycans not. For each l-treelet i, we count the number of glycans containing it in these two sets, denoted as n_i^+ and n_i^-, respectively. Then the four cells of the contingency table are n_i^+ and n_i^- (the first row), and $N - n_i^+$ and $M - N - n_i^-$ (the second row). Fisher showed that, if the l-treelet is not overrepresented in the hemagglutinin-binded glycans, the probability of obtaining these values follows a hypergeometric distribution,

$$P = \frac{\frac{M!}{n_i^+! n_i^-! (N - n_i^+)! (M - N - n_i^-)!}}{\frac{M!}{(n_i^+ + n_i^-)!(M - n_i^+ - n_i^-)!} \times \frac{M!}{(M-N)! N! N}}. \tag{8.1}$$

Note that in the equation, the nominator computes the number of possible ways to configure the M glycans into 4 groups so that each group consists of the number of glycans as the number in the 4 cells in the 2×2 contingency table (i.e. n_i^+, n_i^-, $N - n_i^+$, and $M - N - n_i^-$, respectively), and the denominator computes the number of possible ways to configure M glycans into 4 cells so that the sum of the numbers in two rows and two columns are kept as the sums in the contingency table. The probability can be used to measure the significance of an l-treelet – the treelet is significantly overrepresented in the hemagglutinin-bound glycans if the probability is small (e.g. < 0.01).

---

[6] Note that k_i is determined not only by the number of nodes in the tree i, but also its topology. Therefore, k_i needs to be obtained for each input tree separately.

[7] In statistics, a contingency table is used to display the frequency of two or more variables in a matrix format.

The last question is how to choose an appropriate size of the treelet (i.e. l) to search for. In fact, we can use different sizes, e.g. $l = 2, 3, 4, ...$, and report the overrepresented l-treelet for each l. In practice, the search is limited to a certain size (e.g. ≤ 5 monosaccharide residues) because the hemagglutinin–glycan binding interface does not likely extend beyond that size. In the bioinformatics studies of the glycan array data, two glycan motifs were found to be overrepresented in the glycans binding to human viral hemagglutinins, including the 2–6 linked disaccharide (Sia–Gal), and a linear oligosaccharide of four residues (GlcNAc–Gal–GlcNAc–Gal) with specific linkages (as shown in Figure 8.5a). The first result is consistent with the known binding preference of human influenza viruses, whereas the second is new and indicates that human influenza viruses may prefer to bind to N-glycans containing a long branching with more than one lactosamine repeat (GlcNAc–Gal). This finding led to a new model for the host preference of influenza viruses through hemagglutinin–glycan interaction, which has been also supported by other evidence [4].

DISCUSSION

A majority of important bioinformatics algorithms are developed to analyze sequences because the two most important biomolecules, proteins and nucleic acids, are linear molecules and can be represented as sequences. Glycans, on the other hand, have branching structures and should be represented as labeled trees. Nevertheless, many algorithms designed for proteins and nucleic acids can be extended to the analysis of glycans.

QUESTIONS

(1) The host switch for influenza viruses is caused by the altered binding specificity of viral hemagglutinin proteins, which, from an evolutionary perspective, is an effect of adaptive selection on the viral hemagglutinin genes when the viruses jump from the population of their original host (e.g. avian) to the population of a new host (e.g. human). To characterize the adaptively selected residues on viral hemagglutinin proteins, we have collected a set of viral hemagglutinin protein sequences (Figure 8.6a), some of which are from avian viruses (cluster 1) and the others are from human viruses (cluster 2).

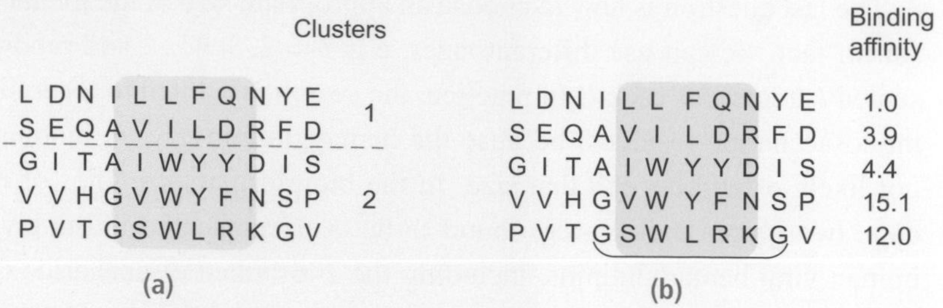

Clusters		Binding affinity

```
        Clusters                              Binding
                                              affinity
   L D N I L L F Q N Y E                 L D N I L L  F Q N Y E     1.0
   S E Q A V I L D R F D    1            S E Q A V I  L D R F D     3.9
   G I T A I W Y Y D I S                 G I T A I W  Y Y D I S     4.4
   V V H G V W Y F N S P    2            V V H G V W  Y F N S P     15.1
   P V T G S W L R K G V                 P V T G S W  L R K G V     12.0

             (a)                                  (b)
```

Figure 8.6 A schematic example for characterizing key residues involved in the alteration of glycan binding specificity of viral hemagglutinin proteins. (a) A set of viral hemagglutinin protein sequences are collected and multi-aligned. These sequences can be partitioned into two clusters: the first two sequences are from avian viruses and the remaining three sequences are from human viruses. (b) Each of the proteins is assayed for human-specific glycans and its (average) binding affinity is measured. Note: the residues within the conserved regions are highlighted in gray areas.

(a) Devise a method to predict the key amino acid residues involved in the binding specificity alteration of viral hemagglutinin.

(b) Assume each of these proteins has been assayed by glycan array experiments to human-specific glycans and its (average) binding affinity has been measured. Using these data, devise a method to predict the key residues involved in the binding specificity alteration.

(2) In order to elucidate the glycan pattern that a hemagglutinin protein recognizes, each putative glycan motif (represented by an l-treelet) is evaluated to determine if it is overrepresented in the glycans binding to hemagglutinin in comparison to the set of glycans not binding to hemagglutinin by a Fisher's exact test. This method can be extended to characterize the glycan binding pattern of other glycan-binding proteins. However, some glycan-binding proteins may recognize multiple (e.g. two) glycan motifs that are similar to each other. In this case, any individual glycan motif may not show high statistical significance when being evaluated using the statistical method described in this chapter. Explain why this may happen, and devise a computational method to address this issue.

(3) Given two independent samples of observations, Wilcoxon's rank-sum test is a non-parametric statistical hypothesis test to assess if they have equally large (or small) values [13]. To compute it, we first rank the observations from both samples together. Then the rank-sum test U is defined as,

$$U = R_1 - \frac{n_1(n_1 + 1)}{2}$$

where R_1 is the sum of ranks of the observations in the first sample and n_1 is the number of observations in the first sample, respectively. Note, U can be equivalently defined on the observations in the second sample (for details see [13]).

In this chapter, when we evaluate the overrepresentation of glycan motifs, we assume the glycans on the glycan array can be partitioned into two sets: one (positive) set of glycans binding to hemagglutinin and the other (negative) set not binding to the hemagglutinin. In practice, what we obtain from a glycan assay is the binding affinity between each glycan on the array and the hemagglutinin, and the positive and negative glycans are partitioned based on an empirical threshold: glycans with binding affinity above the threshold are assigned to be positive, and the other glycans are assigned to be negative. To avoid an arbitrary chosen threshold, devise a statistical method based on Wilcoxon's rank-sum to evaluate the overrepresentation of glycan motifs.

FURTHER READING

I recommend an excellent respective article by H. Nicholls [5] for those who are interested in the biology of influenza viruses. Those who are interested in glycobiology should refer to the encyclopedia of glycobiology, *Essentials of Glycobiology*, by A. Varki *et al.* [6], or a more concise textbook, *Introduction to Glycobiology*, by M. E. Taylor and K. Drickamer [7]. I skipped many details regarding the diversity of the chemical structure of glycans (e.g. their stereochemical configurations) that can be found in these books.

The rapid advancement of glycobiology benefited from the development of high-throughput technologies, in particular, glycan array and mass spectrometry. Mass spectrometry (MS) is a complementary high-throughput technology to glycan array, and can be used to infer the composition and structure of glycans in biological samples. To learn more about these techniques, one can refer to recent reviews [8, 9].

The treelet counting approach introduced in this chapter for glycan array data analysis was first developed by R. Sasisekharan and colleagues from Massachusetts Institute of Technology [4]. More sophisticated algorithms for pattern recognition in glycan structures were reviewed by K. Aoki-Kinoshita in her recent book [10] and an advanced tutorial [11].

The binding preferences of influenza viral hemagglutinin are supported by different analytical methodologies – the glycan array approach is just one of them. For instance, MS analysis has shown a substantial diversity, as well as predominant expression of long oligosaccharide branch (with multiple lactosamine repeats) 2–6 linked sialylated glycans in the human upper respiratory epithelial cells, which is consistent with the motif finding results from glycan array data [4]. Another line of evidence was from the 3-dimensional structure simulation of hemagglutinin–glycan interactions. A class of structural bioinformatics approach called *molecular dynamics* can be used to elucidate the energy profile of hemagglutinin–glycan interaction, and thus characterize the substructures of glycans (monosaccharide residues) that contribute to the binding specificity. This kind of study can also predict the mutations in hemagglutinin that are responsible for the change of its glycan binding preference [12].

REFERENCES

[1] M. F. Berger, A. A. Philippakis, A. M. Qureshi, *et al*. Compact, universal DNA microarrays to comprehensively determine transcription-factor binding site specificities. *Nat. Biotechnol.*, 24:1429–1435, 2006.

[2] J. van Helden, B. Andrei, and J. Collado-Vides. Extracting regulatory sites from the upstream region of yeast genes by computational analysis of oligonucleotide frequencies. *J. Mol. Biol.*, 281:827–842, 1998.

[3] A. Agresti. A survey of exact inference for contingency tables. *Statist. Sci.*, 7:131–153, 1992.

[4] A. Chandrasekaran, A. Srinivasan, R. Raman, *et al*. Glycan topology determines human adaptation of avian H5N1 virus hemagglutinin. *Nat. Biotechnol*, 20:107–113, 2008.

[5] H. Nicholls. Pandemic influenza: The inside story. *PLoS Biol.*, 4:e50, 2006.

[6] A. Varki, R. D. Cummings, J. D. Esko, *et al*. *Essentials of Glycobiology*. 2nd edn. Cold Spring Harbor Laboratory Press, New York, 2009.

[7] M. E. Taylor and K. Drikamer. *Introduction to Glycobiology*. Oxford University Press, Oxford, 2006.

[8] J. Stevens, O. Blixt, J. C. Paulson, *et al*. Glycan microarray technologies: Tools to survey host specificity of influenza viruses. *Nat. Rev. Microbiol.*, 4:857–864, 2006.

[9] A. Dell and H. R. Morris. Glycoprotein structure determination by mass spectrometry. *Science*, 291:2351–2356, 2001.

[10] K. Aoki-Kinoshita. *Glycome Informatics: Methods and Applications*. Chapman & Hall/CRC Press, 2009.

[11] K. F. Aoki-Kinoshita. An introduction to bioinformatics for glycomics research. *PLoS Comput. Biol.*, 4:e1000075, 2008.

[12] E. I. Newhouse, D. Xu, P. R. Markwick, *et al*. Mechanism of glycan receptor recognition and specificity switch for avian, swine, and human adapted influenza virus hemagglutinins: A molecular dynamics perspective. *J. Am. Chem. Soc.*, 131:17,430–17,442, 2009.

[13] W. J. Conover. *Practical Nonparametric Statistics*. 2nd edn. John Wiley & Sons, 1980, 225–226.

PART III

EVOLUTION

CHAPTER NINE

Genome rearrangements

Steffen Heber and Brian E. Howard

Genome rearrangements are one of the driving forces of evolution, and they are key events in the development of many diseases. In this chapter, we focus on a selection of topics that will provide undergraduate students in bioinformatics with an introduction to some of the key aspects of genome rearrangements and the algorithms that have been developed for their analysis. We do not attempt to provide a comprehensive overview of the history or the results in this field. Our presentation is in many parts inspired by the textbook *An Introduction to Bioinformatics Algorithms* by Neil Jones and Pavel Pevzner [1], by lectures from Anne Bergeron [2] and Julia Mixtacki [3], and by several reviews of genome rearrangements and the associated combinatorial and algorithmic topics [4–7]. We will begin with a brief review of the basic biology related to this topic.

1   Review of basic biology

The *genome* of an organism encodes the blueprint for its proteins and ultimately determines that organism's developmental and metabolic fate. Genetic information is stored in double-stranded *deoxyribonucleic acid (DNA)* molecules. Each individual DNA strand is a long sequence of the nucleotides adenine, cytosine, guanine, and thymine, which are commonly referred to using the letters A, C, G, and T. In each strand, the fifth carbon atom of each ribose molecule in the sugar–phosphate backbone is attached to the third carbon atom of the next ribose molecule (Figure 9.1a). However, the two strands are oriented in opposite directions. One strand proceeds in the forward,

Bioinformatics for Biologists, ed. P. Pevzner and R. Shamir. Published by Cambridge University Press.

5′ to 3′ direction, and the other one in reverse, from 3′ to 5′. Both strands are complimentary in the sense that an A nucleotide in one strand pairs with a T nucleotide in the other strand, and a G nucleotide in one strand pairs with a C nucleotide in the other. Therefore, the nucleotide sequence in one strand determines a complementary sequence in the other strand, and the two sequences are in reverse complementary orientation.

Genomes are partitioned into organized structures called *chromosomes* (Figure 9.1b). A chromosome can either be linear or circular. Linear chromosomes have regions of repetitive DNA at their ends called *telomeres*, which protect the chromosomes from damage and from fusing to each other. Each chromosome contains multiple *genes*, or stretches of DNA that are responsible for encoding proteins or functional RNAs. We can label each gene with an orientation dependent on the strand (forward or reverse) on which it is located. To simplify matters, we will assume that each gene appears exactly once in the genome and that consecutive genes are well separated from one another by an intergenic region. If we substitute integers for genes and encode the location of a gene on either the forward or reverse strand by a sign, a chromosome can then be represented as a linear or circular sequence of signed integers (Figure 9.1c). However, in real genomes, several copies of a gene might sometimes exist, and genes can be nested or overlap each other. In these cases, a more flexible genome representation is required.

Even genomes of closely related individuals, for example parents and their children, differ slightly from one another. These differences become more distinct if we compare genomes from different species. A large portion of genetic differences are caused by *point mutations*, in which only one nucleotide is changed at a time. Point mutations include *substitutions*, where one nucleotide is exchanged for another, as well as *insertions* and *deletions*, where individual nucleotides are added or removed.

In contrast to point mutations, *genome rearrangements* are mutations that affect multiple nucleotides of a genome simultaneously. A genome rearrangement occurs when one or two chromosomes break and the fragments are reassembled in a different order. Here, we assume that breakpoints only occur between genes – since, in most cases, a breakpoint inside a gene will compromise the gene function and cause the affected organism to die. (Exceptions to this rule do exist.) The result of a genome rearrangement is a new genome sequence that has a modified gene order, but which does not differ from the original genome in nucleotide composition. Rearrangements can cause dramatic differences in gene regulation and can have a strong effect on the phenotype of an organism. Genome rearrangements are therefore of fundamental importance for understanding chromosomal differences between organisms, and they have been linked to important diseases, including cancer [8]. Figure 9.2 illustrates some of the most common types of genome rearrangements.

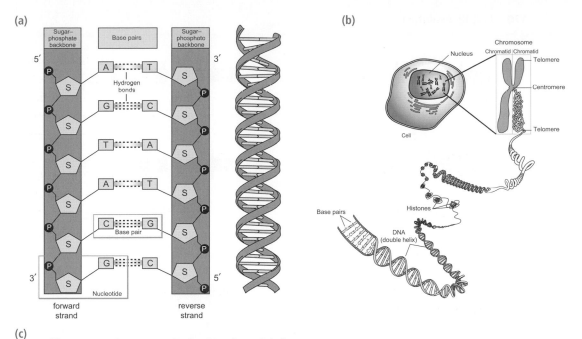

(c)

Homo sapiens, part of mitochondrial genome

Bombyx mori, part of mitochondrial genome

Replace gene names by integers
ND3 →1, ND4L →2, ND4 →3, ND5 →4, ND6 →5, CYTB →6, RNS →7, RNL →8, ND1→9, ND2 →10
and gene orientation by '+' and '−' signs

Homo sapiens: (1   2   3   4   −5   6   7   8   9   10)
Bombyx mori: (1   −4   −3   −2   5   6   −9   −8   −7   10)

Figure 9.1 Basic biology. (a) Nucleotide base pairing and strand orientation result in reverse
complementary sequences. The "forward" direction is called the 5′ direction, and the reverse
direction is the 3′ direction. Each individual nucleotide also has a 5′ and 3′ end, and the 3′ end
of each consecutive nucleotide can only bind to the 5′ end of the next nucleotide. (b) Higher
levels of DNA organization. Figures 9.1a and 9.1b are taken, modified, and printed with the
permission of the National Human Genome Research Institute (NHGRI), artist Darryl Leja.
(c) Example of rearranged genomes (modified from [2]). Shown are part of the mitochondrial
genome of *Homo sapiens* (human) and *Bombyx mori* (silkworm). Each arrow represents a
single gene; for example, "CYTB" stands for cytochrome b. The direction of the arrow indicates
which strand, forward or reverse, the gene resides on. If we encode gene names by integers
and gene orientation by signs, we can represent the genome parts by signed permutations.

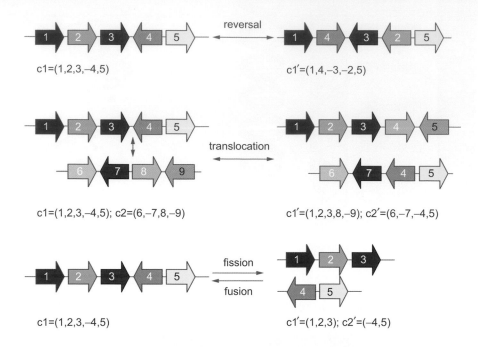

Figure 9.2 Four important types of genome rearrangements: reversal, translocation between chromosomes, and fusion and fission (special cases of translocation). The directions of the large arrows indicate gene orientation on the forward or reverse strand.

Reversals (sometimes also called *inversions*) are one important type of genomic rearrangement. A reversal occurs when a segment of a chromosome is excised and then reinserted in the opposite direction with forward and reverse strands exchanged. As a result, the gene order and orientation for any genes within this segment is reversed. In Figure 9.1c you can observe the effect of reversals. For example, the segment containing the genes RNS, RNL, and ND1 in the human mitochondrial genome appears reversed in the mitochondrial genome of the silkworm. What other reversals can you find in this example?

If we ignore signs and replace genes with characters, genome rearrangements are similar to a familiar word puzzle: *anagrams*. An anagram is a word or phrase formed by rearranging the characters of another word or phrase. For example, the phrase "eleven plus two" can be rearranged into the new phrase "twelve plus one." As with rearranged genomes, the meaning of an anagram might be quite different from the original, for example, "forty-five" can be rearranged into "over fifty." To check if two phrases are anagrams of each other, we can draw a character dot-plot, a matrix where the axes are labeled by the phrases, and a dot is printed at position (i,j) if the ith character

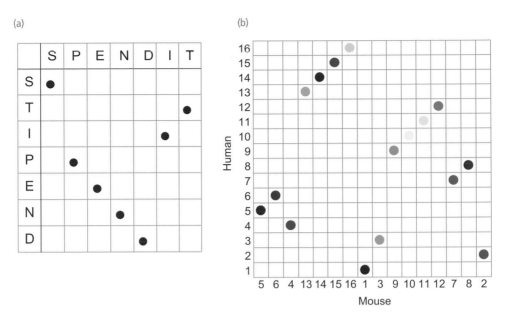

Figure 9.3 Dot-plot examples. (a) Character dot-plot of the anagram pair "stipend" and "spend it." (The space character is ignored.) (b) Genome dot-plot of human and mouse X-chromosome.

of phrase one occurs at position j in phrase two (Figure 9.3). If the two phrases are anagrams, and if no character occurs more than once, then there should be exactly one dot in each column and row.

Distance metrics and the genome rearrangement problem

Evolutionary changes such as point mutations and genome rearrangements can be used to define a variety of useful *distance metrics* between sequences. For example, assume that you are given two *homologous* gene sequences, A and B, that originate from the same ancestral gene, C. (Genes in different organisms are called homologous if they originate from the same gene in a common ancestor.) Using a given set of edit operations, the minimum number of changes necessary to transform sequence A into sequence B defines the *edit distance*, d_{edit}, between A and B. Accordingly, the fewer changes one needs to transform one sequence into the other, the more similar the two sequences are.

(a)   S1 = AGCTT   S2 = AGCCTG   S3 = ACAG

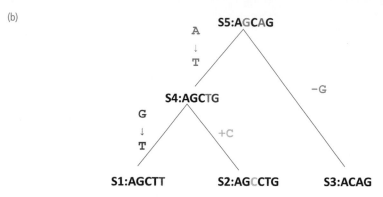

d_{edit}(S1,S2) = 2   AGCTT $\xrightarrow{\text{1) T}\rightarrow\text{G}}$ AGCTG $\xrightarrow{\text{2) insert C}}$ AGCCTG

d_{edit}(S1,S3) = 3   AGCTT $\xrightarrow{\text{1) T}\rightarrow\text{G}}$ AGCTG $\xrightarrow{\text{2) T}\rightarrow\text{A}}$ AGCAG $\xrightarrow{\text{3) delete G}}$ ACAG

d_{edit}(S2,S3) = 3   AGCCTG $\xrightarrow{\text{1) delete C}}$ AGCTG $\xrightarrow{\text{2) T}\rightarrow\text{A}}$ AGCAG $\xrightarrow{\text{3) delete G}}$ ACAG

(b)

Figure 9.4 Edit distance. (a) Edit distances and the corresponding sequence changes. (b) Evolutionary tree that uses a minimum number of point mutations (nucleotide change G->T (red), A->T (blue), insertion + C (yellow), deletion − G (green)) to explain the data. The sequences S4 and S5 are hypothetical because we cannot observe these ancestral sequences.

Computing the edit distance using point mutations is similar to solving the popular word puzzle where you are given a start word and a target word, and your goal is to successively change, add, or delete characters until the target word is reached. Here is an example for the pair "spices" and "lice":

spices → slices → slice → lice.

In general, finding the *minimum* number of necessary transformations is a difficult problem. Often, there are many possible alternative transformation sequences, for example:

spices → spice → slice → lice.

Moreover, even if you are given a feasible transformation sequence, it may be difficult to decide if this sequence is optimal.

Figure 9.4 shows a few examples of how edit distance can be computed for related DNA sequences. In biology, assuming that the minimum number of changes reflects the true evolutionary distance (*parsimony assumption*), the edit distance can be used to compute sequence alignments, and to infer evolutionary relationships among species.

As with point mutations, biologists have used genome rearrangements for measuring the similarity between genomes, and for reconstructing evolutionary relationships. Dobzhansky and Sturtevant pioneered this type of research by analyzing reversals in polytene chromosomes of the fruit fly *Drosophila pseudoobscura* [9]. Polytene chromosomes often occur in the salivary glands of fly larvae. They originate from multiple rounds of chromosome replication (without cell division) where the individual replicated DNA molecules remain fused together. Having multiple genome copies in an individual cell allows the larval tissue to increase the cell volume, and to have a higher rate of transcription. The resulting giant chromosomes are much larger than normal chromosomes and show a pattern of chromosomal bands that correlates with large chromosomal regions. By comparing the chromosomal bands of giant chromosomes with a light microscope, genome rearrangements can be detected; however, no information about the orientation of genes or genomic markers can be inferred. Dobzhansky and Sturtevant demonstrated that there are multiple reversals present in strains of flies inhabiting different geographic regions, and that these reversals can be used to construct a phylogeny of the analyzed fly strains [9]. Figure 9.5 shows a sketch of the original data set and the corresponding phylogenetic relationships.

In order to infer the evolutionary tree displayed in Figure 9.5, Dobzhansky and Sturtevant were faced with what computer scientists now call the *genome rearrangement problem*: given a pair of genomes, find the shortest sequence of rearrangements that transform one genome into the other. Similar to the edit distance defined above, this minimum number of rearrangements also defines a distance metric between genomes, and can therefore be used to infer phylogenetic relationships between species.

Dobzhansky and Sturtevant's original data set consisted of only a few genetic loci, but the recent availability of a large number of fully sequenced genomes gives us access to hundreds of genes in hundreds of genomes. This causes serious problems. Think about how long it would take you just to read 100 gene names aloud. How long would it then take you to find a sequence of reversals that transforms one genome with 100 genes into another genome? If you have found a reversal sequence, how can you be sure the problem cannot be solved with fewer reversals? These challenges have motivated computer scientists to design algorithms for analyzing genome rearrangement data, and, as a result, many different computational approaches now exist. In the following, we will discuss several of these approaches, which vary according to the distance metrics they use and the types of genomic operations they allow, such as signed and unsigned reversals, translocations, and double-cut-and-join operations.

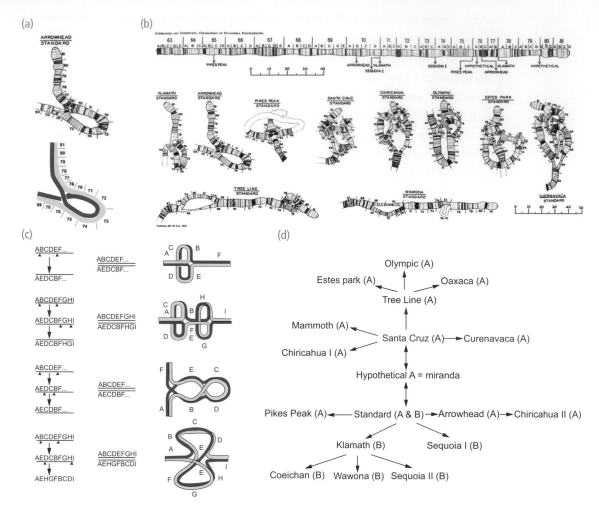

Figure 9.5 Dobzhansky's data. (a) In chromosome three of *Drosophila pseudoobscura* several genome rearrangements exist. For example, the *Standard* arrangement and the *Arrowhead* arrangement differ by an inversion of the chromosomal segment 70–76, highlighted in part (b) of this figure. This inversion results in a loop structure that is formed during the pairing of homologous chromosomes in meiosis of Standard–Arrowhead heterozygotes. (b) Configurations observed in the third chromosome in various inversion heterozygotes. (c) Schematic representation of the pairing of chromosomes differing in a single or a double inversion. Above: a single inversion; second from above: two independent inversions. (d) Phylogeny of the gene arrangements in the third chromosome of *Drosophila pseudoobscura*. Any two arrangements connected by an arrow in the diagram differ by a single inversion. Figures are taken from [9] and printed with the permission of the Genetics Society of America.

3 Unsigned reversals

In a very simple version of the genome rearrangement problem, we will assume that both genomes consist of the same set of genes, that we do not have any information about the orientation of the genes, and that only reversals can occur. These assumptions are motivated by Dobzhansky and Sturtevant's experiment where, due to the limited resolution of light microscopes, only the order of chromosomal markers could be observed, but not their orientation. To formally represent the problem, and to make the data more amenable to computational analysis, we encode the two genomes as permutations of unsigned integers. Let us start with a toy example. Assume that you are given the gene order of 6 genes along a chromosome in two fly species; for example $\pi_1 = (1\ 5\ 3\ 2\ 4\ 6)$ in species 1 and $\pi_2 = (5\ 3\ 2\ 4\ 6\ 1)$ in species 2. Since, in this experiment, the gene orientation cannot be observed, the encoding does not include a sign $(+\text{ or }-)$. Assuming that both genomes originated from a common ancestor but have been modified by genome rearrangements, we would like to learn how to transform gene order 1 into gene order 2 using a sequence of reversals since gene orientation is unobservable, we will use *unsigned reversals*, which reverse the order of the affected genes, but do not change their orientation. For example, in gene order π_1, a reversal of the interval delimited by genes 3 and 4 will result in the new gene order $(1\ 5\ 4\ 2\ 3\ 6)$. To standardize the presentation, we rename the genes such that permutation π_2 becomes the identity permutation, i.e. we replace:

$$5 \rightarrow 1' \quad 3 \rightarrow 2' \quad 2 \rightarrow 3' \quad 4 \rightarrow 4' \quad 6 \rightarrow 5' \quad 1 \rightarrow 6'.$$

After renaming, we obtain order $\pi_1' = (6'\ 1'\ 2'\ 3'\ 4'\ 5')$ and order $\pi_2' = (1'\ 2'\ 3'\ 4'\ 5'\ 6')$. This procedure simplifies our problem without essentially changing it – the label change can easily be reversed. Our original problem can now be stated as a *genome sorting problem*: given an input permutation (π_1') find a minimum number of reversals d_{rev} that transforms the input permutation into the identity permutation (π_2'). To simplify the presentation, we will drop "$'$" in the remainder of this discussion.

A simple, mechanistic procedure to find a sequence of reversals that can transform any permutation, π, into the identity consists of iteratively locating the element, i, in π and moving it via a reversal to its correct location, with i increasing from 1 to $n - 1$ (see Algorithm 1).[1] In the following, $\pi[j] = i$ denotes that the element i is at position j in π, and $\pi \bullet r(i, j)$ indicates an unsigned reversal of $\pi[i \dots j]$.

---

[1] This and the subsequent algorithm BreakpointReversalSort were taken from the textbook *An Introduction to Bioinformatics Algorithms* [1] and were first described in the seminal paper by John Kececioglu and David Sankoff [10].

For example, in the renamed π_1 above, the element $i = 6$ is at position $j = 1$, so $\pi_1[1] = 6$.

Algorithm 1: GREEDYREVERSALSORT (π)

```
1   for i ← 1 to n − 1
2       j ← position of element i in π (i.e. π[j] = i)
3       if j ≠ i
4           π ← π • r(i, j)
5           output π
6       if π is the identity permutation
7           return
```

For the example above, this algorithm will result in the following sequence, where the individual reversals have been underlined:

$$(6\,1\,2\,3\,4\,5) \rightarrow (1\,6\,2\,3\,4\,5) \rightarrow (1\,2\,6\,3\,4\,5) \rightarrow (1\,2\,3\,6\,4\,5) \rightarrow (1\,2\,3\,4\,6\,5) \rightarrow (1\,2\,3\,4\,5\,6).$$

For any pair of permutations π_1 and π_2, this procedure will always find a sequence of reversals that transforms permutation π_1 into permutation π_2; however, it will not always find the minimum number of reversals. In our example there exists a shorter sequence of only two reversals:

$$(6\,1\,2\,3\,4\,5) \rightarrow (6\,5\,4\,3\,2\,1) \rightarrow (1\,2\,3\,4\,5\,6). \tag{9.1}$$

Is it possible to find an even shorter sequence of reversals? In this example, it is easy to verify that there is no shorter solution. However, in general, determining if a given rearrangement scenario is of minimum length is quite difficult. An exhaustive search through all possible sequences of reversals will always find the solution of minimum length, but due to the large search space and the corresponding running time, this approach is not practical. You might think that maybe a better algorithm will do the job, but it has been shown that the genome sorting problem is NP-hard [11]. This implies that, so far, no one has found an algorithm that remains efficient for growing permutation sizes, and that, unless P = NP, no such algorithm can exist. Unfortunately, many computer scientists believe that P $\neq$ NP. On the other hand, even if there is no efficient way to compute an optimal solution, an *approximation algorithm* might still allow the swift discovery of a useful, suboptimal solution. Trading exactness for efficient running time, these algorithms are not guaranteed to find a shortest possible reversal sequence; however, often it is possible to ensure that the resulting approximation is not too far off from an optimal solution, and for many applications this might be good enough. Later, we will describe such an algorithm (Algorithm 2: BreakpointReversalSort).

To find a lower bound for the number of reversals necessary for sorting a permutation, we extend the input permutations by the artificial elements 0 and $n + 1$ at either end. You can interpret these markers as telomeres. In the extended permutation, we call a pair of neighboring elements *adjacent* if they occur consecutively in the target permutation, i.e. in our setting, if the elements correspond to consecutive integers. (Remember that we assume that, after relabeling, the target permutation is the identity.) Otherwise, the pair is called a *breakpoint*. The identity permutation is the only permutation without breakpoints. Let $b(\pi)$ denote the number of breakpoints in permutation π. Since a single reversal can eliminate, at most, two breakpoints, we can derive a simple lower bound for the minimum number of reversals necessary to sort an input permutation π:

$$d_{rev} \geq \left\lceil \frac{b(\pi)}{2} \right\rceil \tag{9.2}$$

where the *ceiling function* $\lceil x \rceil$, denotes the smallest integer greater than or equal to x.

In our example, this bound immediately answers the question of whether there is a shorter transformation sequence than the one given in Equation (9.1). Since $\lceil \frac{b(\pi)}{2} \rceil = 2$, there cannot be any shorter transformation. You might be tempted to suggest a sorting algorithm where every step removes two breakpoints; however, you will soon find that there are permutations for which no single reversal will reduce the number of breakpoints. For instance, try this example: (0 1 5 6 7 2 3 4 8 9).

Although it is not always possible to remove a breakpoint with a single reversal, we can guarantee that within two reversals at least one breakpoint will be eliminated. This can be shown by introducing the notion of *strips*: a strip is an interval between successive breakpoints. In the above example, we have the strips: [0, 1], [5, 6, 7], [2, 3, 4], and [8, 9]. A strip is called *decreasing* if the elements in this interval occur in decreasing order; otherwise, it is called *increasing*. Single element strips will be called decreasing, except for the strips [0] and [n + 1], which will be called increasing. If a permutation π has a decreasing strip, then there exists a reversal that decreases the number of breakpoints by at least one. Assume k is the smallest right border of any decreasing strip. This implies that the element $k - 1$ is at the right border of an increasing strip, followed by a breakpoint. Assume further that in π the element $k - 1$ is followed by the element y and that the element k is followed by the element x (also a breakpoint). If the element $k - 1$ is to the right (left) of k, then the reversal of the interval $x, ..., k - 1 (y, ..., k$, respectively) will remove at least one breakpoint. (The reversal will remove *two* breakpoints if x and y are adjacent.) The following two sketches indicate the relative location of k, $k - 1$, x, and y before and after performing the reversal; a breakpoint is indicated by a "|" symbol.

$k - 1$ to the right of k: $\quad (... \ k \mid \underline{x \ ... \ k - 1} \mid y \ ...) \rightarrow (... \ k \ k - 1 \ ... \ x \mid y \ ...)$

$k - 1$ to the left of k: $\quad (... \ k - 1 \mid \underline{y \ ... \ k} \mid x \ ... \ ) \rightarrow (... \ k - 1 \ k \ ... \ y \mid x \ ...).$

If the permutation π only has increasing strips, we can generate a decreasing strip by reversing one strip, and reduce the number of breakpoints with the second reversal. This motivates the following algorithm.

Algorithm 2: BREAKPOINTREVERSALSORT (π)
1 **while** $b(\pi) > 0$
2   **if** π has a decreasing strip
3   Choose reversal r that minimizes $b(\pi \bullet r)$
4   **else**
5   Choose a reversal r that flips an increasing strip in π
6      $\pi \leftarrow \pi \bullet r$
7   **output** π
8 **return**

How many iterations does this algorithm need to sort an arbitrary input permutation? As long as there are decreasing strips in the permutation, each iteration will decrease the number of breakpoints by at least one. When there is no decreasing strip, the algorithm will reverse an increasing strip without decreasing the number of breakpoints. This creates a decreasing strip and guarantees the existence of a reversal that will reduce the number of breakpoints in the next iteration. Therefore, this algorithm guarantees that during two consecutive reversals at least one breakpoint is removed. Although we cannot guarantee that this procedure will find the minimum number of reversals necessary to sort the permutation, we can argue that the constructed solution will not use more than four times the minimum number of reversals. To see this, assume that we are given an input permutation with $b(\pi)$ breakpoints. We know that any algorithm will need at least $\frac{\lceil b(\pi) \rceil}{2}$ reversals for sorting π – possibly more. The above algorithm will need at most $2b(\pi)$ reversals, which is at most $\frac{2b(\pi)}{\lceil \frac{b(\pi)}{2} \rceil} \leq 4$ times the optimal number of reversals.

4 Signed reversals

While Dobzhansky and Sturtevant could only observe the relative order of a few genetic markers (chromosome bands) with their light microscope, nowadays completely sequenced genomes offer a much higher resolution. The location of genes can be pinned down to individual nucleotides, and we can also learn about each gene's orientation, i.e. their location on one of the two complementary DNA strands. The latter information, in particular, is extremely useful for designing efficient algorithms

(a)   (b)   (c)

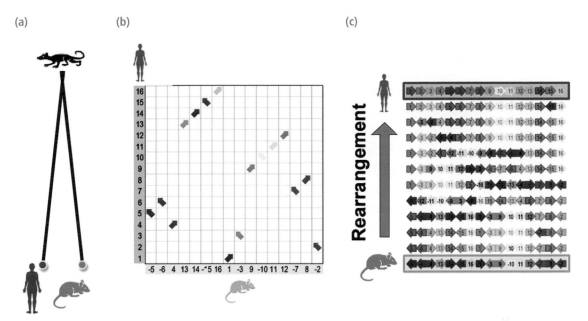

Figure 9.6 Reversal scenario, human and mouse. (a) Human and mouse are descendants from a common evolutionary ancestor. (b) Synteny blocks, which are groups of genes or genomic markers present in both organisms with an evolutionarily conserved order, are used as the basic input elements for various rearrangement algorithms. A genomic dot-plot of the synteny blocks in human and mouse reveals that the human and mouse X-chromosomes are permutations of one another. (c) A series of 10 reversals transforms the mouse X-chromosome into the human X-chromosome.

to find optimal rearrangement scenarios. However, despite the higher level of resolution in sequenced genomes, reconstructing genome rearrangement scenarios is more complicated than you might expect. Identifying the corresponding (homologous) gene pairs in different organisms itself is not easy, and there are many processes such as point mutations, horizontal gene transfer, deletions, and expanding repeat families that complicate this task even further. Moreover, even if we know the correct gene order and orientation in two completely sequenced genomes, this does not suffice to infer the precise location and extent of all genome rearrangement events since, for example, rearrangements in an intergenic region between two consecutive genes are overlooked.

To overcome these problems, researchers do not focus solely on genes, but start from a dense set of genomic anchors – short genomic substrings that are derived from both genes and intergenic regions and that can be uniquely mapped to both genomes. These anchors are filtered and clustered in order to identify groups of anchors with an evolutionarily conserved order. (See [12] for the details of this procedure.) The resulting groups are called *synteny blocks*, and they are the basic input elements for rearrangement algorithms. In the following, we will represent synteny blocks by integers and their orientation (strand) by a "+" or "−" sign, as we did previously for genes. Under this notation, genomes correspond to *signed* permutations, and a reversal will now not only reverse the order of the involved elements, but also simultaneously flip the sign of each affected element.

Figure 9.6 shows a genomic dot-plot comparing human and mouse X-chromosomes. A series of reversals transforms the mouse X-chromosome into the human X-chromosome. Although the inclusion of orientation information may at first seem to complicate the problem, it turns out that this additional constraint allows the design of efficient genome rearrangement algorithms. While the computation of the unsigned reversal distance is an NP-hard problem, signed reversal distances can be computed using an $O(n)$ time algorithm [11, 13]. The details of these algorithms and their variations are beyond the scope of this presentation, and the interested reader is referred to the following thorough overview [7].

5 DCJ operations and algorithms for multiple chromosomes

So far, we have only considered rearrangements that affect a single chromosome. However, many genomes consist of multiple chromosomes, and genome rearrangements like translocation, and fusion and fission (special types of translocations, see Figure 9.2) affect two different chromosomes simultaneously. Hannenhalli and Pevzner [14] were the first to propose a polynomial-time algorithm for computing the multi-chromosomal genome rearrangement distance, d_{HP}, which counts the minimum number of reversals and/or translocations necessary to sort two genomes that consist of multiple linear chromosomes. This algorithm essentially caps and concatenates all chromosomes, and sorts the resulting artificial "super-chromosome" via signed reversals. The algorithm is quite complex, requiring multiple parameters, and it has been revised several times [15–17]. An implementation is provided on the GRIMM web server (http://grimm.ucsd.edu/GRIMM/, [15]).

The DCJ model is an alternative rearrangement model introduced by Yancopoulos and colleagues [18]. This model computes the distance metric, d_{DCJ}, using the Double-Cut-and-Join (or DCJ) genome rearrangement operations. Like Hannenhalli and Pevzner's approach, the DCJ genome rearrangement algorithms are efficient, but they are also relatively easy to implement. Our description here follows the presentation of Anne Bergeron and colleagues [19, 20]. Once again, a gene and its orientation are represented by a signed integer. The genes of a genome are grouped into chromosomes, which can either be linear, in which case both telomeres are represented by the special symbol "o," or circular without a telomere. For example, consider a genome consisting of a linear chromosome $c1 = (1\ -2\ 3\ 4)$ and a circular chromosome $c2 = (5\ 6\ 7)$. In the DCJ model, this genome is represented as $c1 = (o\ 1\ -2\ 3\ 4\ o)$ and $c2 = (5\ 6\ 7)$. The DCJ genome rearrangement operations act on the intergenic regions between consecutive genes, or between a gene and a neighboring telomere. A DCJ operation breaks one or two intergenic regions (possibly on different chromosomes), and joins the resulting open ends. To describe this operation elegantly, we will replace each positively oriented gene g by an interval $[-g,+g]$ and each negatively oriented gene $-g$ by $[+g,-g]$, where $+g$ and $-g$ represent the gene ends (often also denoted as $5'$ and $3'$ gene ends). In addition, we represent each telomere by the special character "o" which has no orientation (see Figure 9.7). An intergenic region, also known as an *adjacency*, can now be encoded by its unordered pair of neighboring gene ends, or by an unordered pair consisting of one gene end and a telomere symbol. In addition, we also allow "special" adjacencies $\{o,o\}$ consisting of two telomere symbols. These adjacencies do not actually correspond to a known biological structure, but simplify the representation of certain DCJ transformations. In our example, c1 has the adjacencies $\{o,-1\}$, $\{1,2\}$,$\{-2,-3\}$,$\{3,-4\}$,$\{4,o\}$ and c2 has the adjacencies $\{5,-6\}$, $\{6,-7\}$, $\{7,-5\}$. Knowing all adjacencies of a genome is equivalent to knowing the original gene order and orientation. Simply start with any adjacency and extend to the left and right, matching adjacencies until a telomere is reached (in the case of a linear chromosome), or an already chosen gene is encountered (in the case of a circular chromosome). Repeat this procedure until all adjacencies have been used and you have reconstructed the genome.

A DCJ operation "breaks" two intergenic regions (adjacencies) and rearranges the fragments. Formally, this corresponds to replacing a pair of adjacencies $\{a,b\}$ and $\{c,d\}$ by $\{a,d\}$ and $\{c,b\}$, or $\{a,c\}$ and $\{b,d\}$. Here, the variables a, b, c, and d represent different (signed) gene ends or telomeres; for telomeres we assume "$+o$" $=$ "$-o$." A special case of this operation occurs when one of the adjacencies is $\{o,o\}$. In this case we get the rearrangement, $\{a,b\}$ $\{o,o\}$ $\leftrightarrow$ $\{a,o\}$ $\{b,o\}$, which corresponds to replacing the adjacency $\{a,b\}$ by the pair of adjacencies $\{a,o\}$ and $\{b,o\}$.

The DCJ operations can be used to implement a variety of different types of genome rearrangements, including reversals, translocations, chromosome fusion and fission, transpositions, and block exchanges. For example, if we apply a DCJ operation that replaces $\{1,2\}$ and $\{3,-4\}$ by $\{1,3\}$ and $\{2,-4\}$ in the above chromosome $c1$, we obtain the rearranged chromosome $c1' = (o\ 1\ -3\ 2\ 4\ o)$. In this case, the DCJ rearrangement corresponds to a signed reversal of genes 2 and 3 (Figure 9.7b). If we apply the DCJ operation that replaces $\{1,2\}$ and $\{3,-4\}$ by $\{1,-4\}$ and $\{2,3\}$, the rearrangement excises the chromosomal interval $[2,-3]$ and transforms it into a new circular chromosome (Figure 9.7c), resulting in $c11' = (o\ 1\ 4\ o)$ and $c12' = (2,\ -3)$. If we break the

(a)

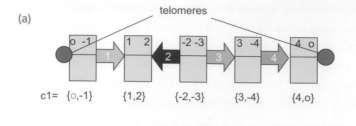

c1= {o,-1}   {1,2}   {-2,-3}   {3,-4}   {4,o}

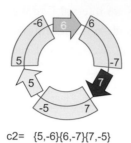

c2= {5,-6}{6,-7}{7,-5}

(b)

DCJ 1: {a,b} {c,d} ⟶ {a,c} {b,d}

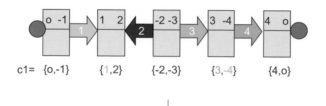

c1= {o,-1}   {1,2}   {-2,-3}   {3,-4}   {4,o}

{1,2}{3,-4} ⟶ {1,3}{2,-4}

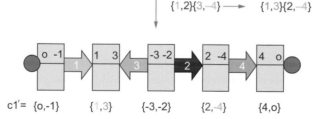

c1'= {o,-1}   {1,3}   {-3,-2}   {2,-4}   {4,o}

Figure 9.7 Double-Cut-and-Join (DCJ) operations. (a) Encoding of one linear and one circular chromosome using the adjacency notation described in the text. Adjacencies are depicted by orange boxes. (b–d) DCJ operations can be used to implement a variety of different types of genome rearrangements. Panel (b) illustrates how a DCJ operation can be employed to implement a signed reversal of genes 2 and 3. In panel (c), genes 2 and 3 are excised from the chromosome resulting in one linear and one circular chromosome. Panel (d) shows the transformation of a circular chromosome into a linear chromosome using a DCJ operation.

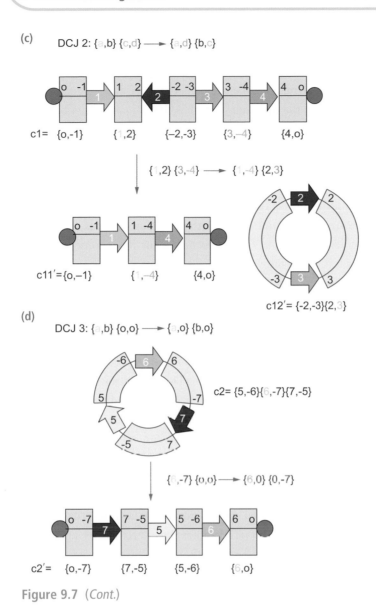

Figure 9.7 *(Cont.)*

adjacency $\{6,-7\}$ of the circular chromosome $c2$ and replace it by $\{6,o\}$ and $\{o,-7\}$ we obtain the linearized chromosome $c2' = (o\ 7\ 5\ 6\ o)$ shown in Figure 9.7d. Similar to the above Hannenhalli and Pevzner distance, the *DCJ distance*, d_{DCJ}, is defined as the minimum number of DCJ rearrangement operations necessary to transform one genome into another. Since the DCJ distance has several other rearrangement types available in addition to the reversals and translocations of the Hannenhalli and Pevzner distance, we get $d_{DCJ} \leq d_{HP}$.

One major advantage of the DCJ model is the availability of simple graph algorithms that transform one genome into another. As an example we describe in the following the algorithm DCJSORT that was originally presented by Bergeron and colleagues [17]. Assume that you are given two genomes, A and B, containing the same set of n genes. We define the adjacency graph $AG(A,B) = (V,E)$, a bipartite graph where V contains one vertex for each adjacency of genome A and one vertex for each adjacency of genome B. In the following we will refer to the set of vertices derived from genome A and B as V_A and V_B, respectively. Each gene, g, defines two edges, one connecting the adjacencies of A and B where +g occurs as a gene border, the other connecting the adjacencies where –g occurs. The idea of algorithm DCJSORT is to find and apply a sequence of DCJ operations to genome A that reduces, in each step, the number of adjacencies of genome B that do not occur in genome A. If there are no such adjacencies left, the resulting genomes are identical and a sequence of DCJ operations that transforms genome A into genome B has been found.

DCJSORT operates in three phases. In phase one, the adjacency graph $AG(A,B)$ is constructed. In phase two, the algorithm searches for adjacencies $\{p,q\}$ in genome B where the corresponding (single) vertex $w=\{p,q\} \in V_B$ of $AG(A,B)$ is incident to a pair of vertices $u1 = \{p,l\} \in V_A$ and $u2 = \{q,m\} \in V_A$ (corresponding to two adjacencies in genome A). The algorithm applies the DCJ operation that replaces $\{p,l\}$ and $\{q,m\}$ by $\{p,q\}$ and $\{l,m\}$ to genome A and updates the adjacency graph correspondingly. This increases the number of shared adjacencies between target genome B and the transformed genome A by at least one. When no such adjacencies remain, it can be concluded that *if* there are still adjacencies in genome B that do not appear in the transformed genome A, then these adjacencies are incident to only one vertex $u = \{p,l\} \in V_A$, and these adjacencies therefore include telomeres. In this case, each incident vertex $u = \{p,l\} \in V_A$ corresponds to the two adjacencies $\{p,o\}$ and $\{o,l\}$. In phase three, DCJSORT handles these vertices by applying a DCJ operation that replaces the adjacency $\{p,l\}$ with $\{p,o\}$ and $\{o,l\}$ and updates the adjacency graph correspondingly. See Figure 9.8 for an example. This simple algorithm finds a sequence of DCJ operations of minimum length $d_{DCJ}(A,B)$ that transforms genome A into genome B. Moreover, let C denote the number of cycles, and I the number of paths with an odd number of edges in $AG(A,B)$. We have $d_{DCJ}(A,B) = n - C - I/2$. For a proof, as well as further details about an implementation with $O(n)$ worst-case running time, the reader is referred to [19, 20].

Algorithm 3: DCJSORT (A,B)

1   Generate adjacency graph AG(A, B) of A and B
2   **for each** adjacency {p, q} with p, q≠o in genome B **do**
3   let u={p,l} be the vertex of A that contains p
4   let v={q,m} be the vertex of A that contains q

(a)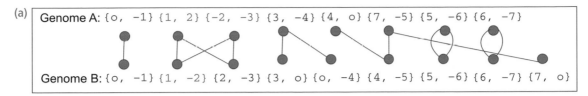

(b)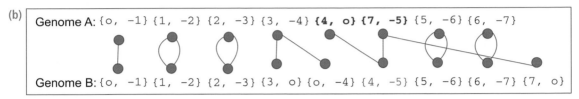

(c)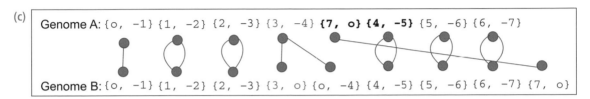

(d)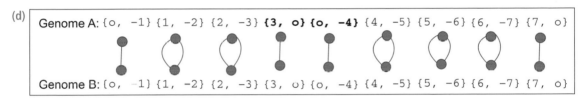

Figure 9.8 DCJSORT transforms genome A: (o1 − 2 3 4 o)(5 6 7) into genome B: (o 1 2 3 4 o)(o 5 6 7 o). Phase one (panel a): The adjacency graph is generated. Phase two (panels b and c): {1, 2}{−2, −3} → {1, −2}{2, −3} and {4, o}{7, −5} → {7, o}{4, −5}. Phase three (panel d): {3, −4} → {3, o}{o, −4}. The affected adjacencies are marked red.

```
 5      if u ≠ v then
 6          replace vertices u and v in A by {p,q} and {l,m}
 7          update edge set
 8      end if
 9   end for
10   for each telomere {p,o} in B do
11      let u = {p,l} be the vertex of A that contains p
12      if l ≠ o then
13          replace vertex u in A by {p,o} and {o,l}
14          update edge set
15      end if
16   end for
```

DISCUSSION

Genome rearrangements are an important natural engine of genetic variation and are therefore critical for a deep understanding of evolution, and the origin of many important diseases, including cancer. Simultaneously, rearrangements are also an interesting application field for demonstrating basic principles of algorithm design, providing students with an opportunity to learn how to model genome rearrangements, to apply and analyze genome sorting algorithms, and to compare exact and approximate solutions to the problem.

While the first studies of genome rearrangements were performed using low-resolution marker maps from giant chromosomes in fruit flies, rapid advancements in sequencing technology have now made it possible to compare the entire genomes of hundreds of organisms. Motivated by this data avalanche, we investigate the performance of various approaches to solving genome rearrangement problems. Beginning with an analogy to familiar recreational word games, we demonstrate how one can describe and model genome rearrangements using permutations. We show that transforming one genome into another is similar to the classic problem of computing the edit distance between two homologous sequences, or, equivalently, of computing an optimal alignment. Throughout the chapter, we proceed to introduce a series of increasingly complex distance metrics and genome transformation operations, illustrating how these choices influence the resulting genome sorting algorithms.

Interestingly, the computational complexity of rearrangement algorithms is very different depending on how exactly the problem is modeled. While it is quite simple to find a sequence of rearrangements that transforms one chromosome into another, for unsigned reversals, finding the shortest such sequence is NP-hard and might take a long time for large genomes [11]. This provides a natural motivation for developing approximation algorithms. On the other hand, for signed reversals, the problem can be solved exactly in linear time [13]. Furthermore, the same approach can also be generalized to multi-chromosomal genomes, although the resulting algorithms are rather difficult to understand and implement [14–16]. The alternative DCJ model uses an extremely flexible genome rearrangement operation that acts on multi-chromosome genomes and the corresponding algorithms for finding optimal DCJ rearrangement sequences are both simple and efficient [18–20]. Together, these varied approaches to the genome rearrangement and sorting problem illustrate an intimate connection between biological data, mathematical modeling, and the design of efficient and practical computer algorithms – a theme that has become increasingly important in many areas of modern biology.

QUESTIONS

(1) Describe the similarities and differences between a word transformation scenario and a point mutation scenario.

(2) Describe the similarities and differences between word anagrams and genome rearrangements.

(3) Can you transform the word "stipend" into "spend it" using unsigned reversals? You can ignore the space character in this example.

(4) Can you find a permutation without any decreasing strip where the number of breakpoints can be reduced by a reversal?

(5) Can you find a DCJ operation that implements the rearrangements shown in Figure 9.2?

REFERENCES

[1] N. C. Jones and P. A. Pevzner. *An Introduction to Bioinformatics Algorithms*. MIT Press, Cambridge, MA, 2004.

[2] A. Bergeron. Applications of Genome Rearrangements. http://acim.uqam.ca/~anne/INF4500/Rearrangements.ppt.

[3] J. Mixtacki. Double cut-and-join and related operations in genome rearrangement. http://ows.molgen.mpg.de/2006/lectures/mixtacki.pdf.

[4] S. Hannenhalli and P. A. Pevzner. Towards a computational theory of genome rearrangements. Computer science today: Recent trends and developments. *Lecture Notes in Computer Science*, 1000:184–202, 1995.

[5] D. Sankoff and J. H. Nadeau, eds. *Comparative Genomics: Empirical and Analytical Approaches in Gene Order Dynamics, Map Alignment and the Evolution of Gene Families*. Kluwer Academic Press, Dordrecht, 2000.

[6] M. Blanchette. Evolutionary puzzles: An introduction to genome rearrangement. *Lecture Notes in Computer Science*, 2074:1003–1011, 2001.

[7] G. Fertin, A. Labarre, I. Rusu, E. Tannier, and S. Vialette. *Combinatorics of Genome Rearrangements*. MIT Press, Cambridge, MA, 2009.

[8] P. Stankiewicz and J. R. Lupski. Genome architecture, rearrangements and genomic disorders. *Trends Genet.*, 18(2):74–82, 2002.

[9] T. Dobzhansky and A. H. Sturtevant. Inversions in the chromosomes of *Drosophila pseudoobscura*. *Genetics*, 23(1):28–64, 1938.

[10] J. Kececioglu and D. Sankoff. Exact and approximation algorithms for the inversion distance between two permutations. *Algorithmica*, 13:180–210, 1995.

[11] A. Caprara. Sorting permutations by reversals and eulerian cycle decompositions. *SIAM J. Discrete Math.*, 12(1):91–110, 1999.

[12] P. A. Pevzner and G. Tesler. Genome rearrangements in mammalian evolution: Lessons from human and mouse genomes. *Genome Res.*, 13:37–45, 2003.

[13] D. A. Bader, B. M. Moret, and M. Yan. A linear-time algorithm for computing inversion disctance between signed permutations with an experimental study. *J. Comput. Biol.*, 8(5):483–491, 2001.

[14] S. Hannenhalli and P. A. Pevzner. Transforming men into mice: Polynomial algorithm for genomic distance problem. In: *36th Annual IEEE Symposium on Foundations of Computer Science* (FOCS), 1995, 581–592.

[15] G. Tesler. Efficient algorithms for multichromosomal genome rearrangements. *J. Comput. Syst. Sci.*, 65(3):587–609, 2002.

[16] M. Ozery-Flato and R. Shamir. Two notes on genome rearrangements. *J. Bioinf. Comput. Biol.*, 1(1):71–94, 2003.

[17] G. Jean and M. Nikolski. Genome rearrangements: A correct algorithm for optimal capping. *Inform. Process. Lett.*, 104:14–20, 2007.

[18] S. Yancopoulos, O. Attie, and R. Friedberg. Efficient sorting of genomic permutations by translocation, inversion and block interchange. *Bioinformatics*, 21(16):3340–3346, 2005.

[19] A. Bergeron, J. Mixtacki, and J. Stoye. A unifying view of genome rearrangements. *Algorithms in Bioinformatics, 6th International Workshop*, WABI, 2006, 163–173.

[20] A. Bergeron, J. Mixtacki, and J. Stoye. A new linear time algorithm to compute the genomic distance via the double cut and join distance. *Theoret. Comput. Sci.*, 410:5300–5316, 2009.

CHAPTER TEN

Comparison of phylogenetic trees and search for a central trend in the "Forest of Life"

Eugene V. Koonin, Pere Puigbò, and Yuri I. Wolf

The widespread exchange of genes among prokaryotes, known as horizontal gene transfer (HGT), is often considered to "uproot" the Tree of Life (TOL). Indeed, it is by now fully clear that genes in general possess different evolutionary histories. However, the possibility remains that the TOL concept can be reformulated and remains valid as a statistical central trend in the phylogenetic "Forest of Life" (FOL). This chapter describes a computational pipeline developed to chart the FOL by comparative analysis of thousands of phylogenetic trees. This analysis reveals a distinct, consistent phylogenetic signal that is particularly strong among the Nearly Universal Trees (NUTs), which correspond to genes represented in all or most of the organisms analyzed. Despite the substantial amount of apparent HGT seen even among the NUTs, these gene transfers appear to be distributed randomly and do not obscure the central tree-like trend.

 1 ## The crisis of the Tree of Life in the age of genomics

The Tree of Life (TOL) is one of the dominant concepts in biology, starting from the famous single illustration in Darwin's *Origin of Species* to twenty-first century undergraduate textbooks. For approximately a century, beginning with the first, tentative trees published by Haeckel in the 1860s and up to the foundation of molecular evolutionary

Bioinformatics for Biologists, ed. P. Pevzner and R. Shamir. Published by Cambridge University Press.
© Cambridge University Press 2011.

analysis by Zuckerkandl, Pauling, and Margoliash in the early 1960s, phylogenetic trees were constructed on the basis of comparing phenotypes of organisms. Thus, by design, every constructed tree was an "organismal" or "species" tree; that is, a tree was assumed to reflect the evolutionary history of the corresponding species. Even after the concepts and early methods of molecular phylogeny had been developed, for many years, it was used simply as another, perhaps, particularly powerful and accurate approach to the construction of species trees. The TOL concept remained intact, with the general belief that the TOL, at least in principle, would accurately represent the evolutionary relationships between all lineages of cellular life forms. The discovery of the universal conservation of rRNA and its use as the molecule of choice for phylogenetic analysis pioneered by Woese and coworkers [1, 2] resulted in the discovery of a new domain of life, the archaea, and boosted the hopes that the definitive topology of the TOL was within sight.

However, even before the era of complete genome sequencing and analysis, it has become clear that in prokaryotes some common and biologically important genes have experienced multiple exchanges between species known as horizontal gene transfer (HGT); hence the idea of a "net of life" as an alternative to the TOL. The advances of comparative genomics have revealed that different genes very often have distinct tree topologies and, accordingly, that HGT appears to be the rule rather than an exception in the evolution of prokaryotes (bacteria and archaea) [3–5].

It seems worth mentioning some remarkable examples of massive HGT as an illustration of this key trend in the evolution of prokaryotes. The first case in point pertains to the most commonly used model of microbial genetics and molecular biology, the intestine bacterium *Escherichia coli*. Some basic information on the genome of *E. coli* and other sequenced microbial genomes is available on the website of the National Center for Biotechnology Information at the National Institutes of Health (http://www.ncbi.nlm.nih.gov/sites/entrez?db=genome). The most well-studied laboratory isolate of *E. coli* on which most of the classic experiments of molecular biology have been performed is known as K12. The K12 genome encompasses 4,226 annotated protein-coding genes (there is always uncertainty as to the exact number of the genes in a sequenced genome, for instance, because it remains unclear whether or not some small genes actually encode proteins; however, the estimate suffices for the present discussion). Several other sequenced genomes of laboratory *E. coli* strains possess about the same number of genes. In contrast, genomes of pathogenic strains of *E. coli* are typically much larger, with one strain, O157:H7, encoding 5,315 annotated proteins. The nucleotide sequences of the shared genes in all strains of *E. coli* are identical or differ by just one or two nucleotide substitutions. In a stark contrast, the differences between the genomes of laboratory and pathogenic strains concentrate in several "pathogenicity islands" that comprise up to 20% of the genome. The pathogenicity islands encompass genes typically involved in bacterial pathogenesis such as toxins, systems for their secretion, and components of prophages. One can imagine that the pathogenicity islands were present in the ancestral *E. coli* genome but have been deleted in K12 and other laboratory strains. However, the gene contents of the islands differ dramatically between the pathogenic strains, so that in three-way comparisons of *E. coli* genomes only about 40% of the genes are shared typically. Thus, the only possible conclusion is that the pathogenicity islands spread between bacterial genomes via rampant HGT, conceivably driven by selection for survival and spread of the respective bacterial pathogens within the host organisms.

The second example involves apparent large-scale HGT across much greater evolutionary distances, namely, between the two "domains" of prokaryotes, bacteria and archaea [1, 2]. The distinction between these two distinct domains of microbes was established by phylogenetic analysis of rRNA sequences and the sequences of other conserved genes, and has been supported by major distinctions between the systems of DNA replication and the membrane apparatus of the respective organisms. Comparative analysis of the first few sequenced genomes of bacteria and archaea supported the dichotomy between the two domains: most of the protein sequences encoded in bacterial genomes show the greatest similarity to homologs from other bacteria and cluster with them in phylogenetic trees, and the same pattern of evolutionary relationships is seen for archaeal proteins. However, the analysis of the first sequenced genomes of hyperthermophilic bacteria, *Aquifex aeolicus* and *Thermotoga maritima*, yielded a striking departure from this pattern: the protein sets encoded in these genomes were shown to be "chimeric," i.e. they consist of about 80% typical bacterial proteins and about 20% proteins that appear distinctly "archaeal," by sequence similarity and phylogenetic analysis. The conclusion seems inevitable that these bacteria have acquired numerous archaeal genes via HGT. In retrospect, this finding might not appear so surprising because bacterial and archaeal hyperthermophiles coexist in the same habitats (e.g. hydrothermal vents on the ocean floor) and have ample opportunity to exchange genes. Similar chimeric genome composition, but with reversed proportions of archaeal and bacterial genes, has been subsequently discovered in mesophilic archaea such as *Methanosarcina*.

Beyond these and related observations made by comparative genomics of prokaryotes, HGT is thought to have been crucial also in the evolution of eukaryotes, especially as a consequence of endosymbiotic events in which numerous genes from the genome of the ancestors of mitochondria and chloroplasts have been transferred to nuclear genomes [6]. These findings indicate that no single gene tree (or any group of gene trees) can provide an accurate representation of the evolution of entire genomes; in other words, the results of comparative genomics indicate that a perfect TOL fully reflecting the evolution of cellular life forms does not exist. The realization that HGT is a major evolutionary phenomenon, at least among prokaryotes, led to a crisis of the TOL concept which is often viewed as a paradigm shift in evolutionary biology [4].

Of course, the inconsistency between gene phylogenies caused by HGT, however widespread, does not alter the fact that all cellular life forms are linked by an uninterrupted tree of cell divisions (*Omnis cellula e cellula* according to the famous motto of Rudolf Virchow) that goes back to the earliest stages of evolution and is violated only by endosymbiosis events that were key to the evolution of eukaryotes but not prokaryotes. Thus, the difficulties of the TOL concept in the era of comparative genomics concern the TOL as it can be derived by the phylogenetic analysis of multiple genes and genomes, an approach often denoted "phylogenomics," to emphasize that phylogenetic studies are now conducted on the scale of complete genomes. Accordingly, the claim that HGT "uproots the TOL" means that extensive HGT has the potential to completely decouple molecular phylogenies from the actual tree of cells. However, such decoupling has clear biological connotations given that the evolutionary history of genes also describes the evolution of the encoded molecular functions. In this chapter, the phylogenomic TOL is discussed with such an implicit understanding.

The views of evolutionary biologists on the evolving status of the TOL in the age of comparative genomics span the entire spectrum of positions from: (i) persisting denial of the major importance of HGT for evolutionary biology; to (ii) "moderate" overhaul of the TOL concept; to (iii) genuine uprooting, whereby the TOL is declared obsolete [7]. The accumulating data on diverse HGT events are quickly making the first "anti-HGT" position plainly untenable. Under the intermediate moderate approach, despite all the differences between the topologies of individual gene trees, the TOL still makes sense as a representation of a central trend (consensus) that, at least in principle, could be elucidated through a comprehensive comparison of trees for individual genes [8]. By contrast, under the radical "anti-TOL" view, rampant HGT eliminates the very distinction between the vertical and horizontal transmission of genetic information, so the TOL concept should be abandoned altogether in favor of some form of a network representation of evolution [7].

This chapter describes some of the methods that are used to compare topologies of numerous phylogenetic trees and the results of the application of these approaches to the analysis of approximately 7,000 phylogenetic trees of individual prokaryotic genes that collectively comprise the "Forest of Life" (FOL). This set of trees does gravitate to a single tree topology, suggesting that the "TOL as a central trend" concept is potentially viable.

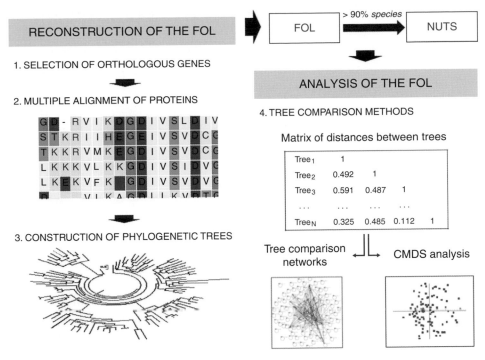

Figure 10.1 The bioinformatic pipeline for the analysis of the Forest of Life.

 ## The bioinformatic pipeline for analysis of the Forest of Life

The realization that, owing to widespread HGT, the evolutionary history of each gene is in principle unique brings the emphasis on phylogenomics; that is, genome-wide comparative analysis of phylogenetic trees. This task depends on a bioinformatic pipeline which leads from protein sequences encoded in the analyzed genomes to a representative collection of phylogenetic trees (Figure 10.1). The pipeline consists of several essential steps: (1) selection of genes for phylogenetic analysis, (2) multiple alignment of orthologous protein sequences, i.e. amino acid sequences of proteins encoded by "the same" gene from different organisms (in evolutionary biology, such genes are usually called orthologs), (3) construction of phylogenetic trees, (4) calculation of the distances between trees and construction of a tree distance matrix, (5) clustering and classification of trees on the basis of the distance matrix. Obviously, this pipeline incorporates a variety of computational methods, and it is impractical to present all of them in detail within a relatively short chapter. However, a brief outline of these methods is given below. The current collection of complete microbial genomes includes over 1,000 organisms

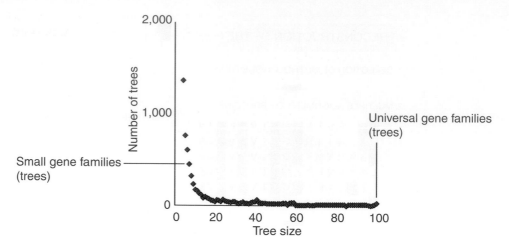

Figure 10.2 The distribution of the trees in the FOL by the number of species.

(http://www.ncbi.nlm.nih.gov/genomes/MICROBES/microbial_taxtree.html), so it is impractical to use them all for phylogenetic analysis as it quickly becomes prohibitively computationally expensive with the increase of the number of species. Therefore, the FOL was analyzed using a manually selected representative set of 100 prokaryotes [9].

The great majority of orthologous gene clusters include a relatively small number of organisms. In the set of clusters selected for phylogenomic analysis of the FOL, the distribution of the number of species in trees showed exponential decay, with only about 2,000 out of the approximately 7,000 clusters including more than 20 species (Figure 10.2). The truly universal gene core of cellular life is tiny and continues to shrink as new genomes are sequenced, owing to the loss of "essential" genes in some organisms with small genomes and to errors of genome annotation. Among the trees in the FOL, there were about 100 Nearly Universal Trees (NUTs), i.e. trees for gene families represented in all or nearly all analyzed organisms; almost all NUTs correspond to genes encoding proteins involved in translation and transcription [9]. The NUTs were analyzed in parallel with the complete set of trees in the FOL.

Before constructing a phylogenetic tree, the sequences of orthologous genes or proteins need to be aligned, i.e. all homologous positions have to be identified and positioned one under another to allow subsequent comparative analysis of the sequences. For large evolutionary distances, as is the case between many members of the analyzed set of 100 microbial genomes, trees are constructed using multiple alignments of protein sequences (Figure 10.1).

Once the sequences of orthologous proteins are aligned, the construction of phylogenetic trees becomes possible. Many diverse approaches and algorithms have been developed for building phylogenetic trees. There is no single "best" phylogenetic method that would be optimal for solving any problem in evolution, but in general the highest quality of phylogenetic reconstruction is achieved with maximum likelihood methods that employ sophisticated probabilistic models of gene evolution [10].

The construction of the trees (about 7,000 altogether) provides for an attempt to identify patterns in the FOL and address the question of whether or not there exists a central trend among the trees that perhaps could be considered an approximation of a TOL. To perform such an analysis, it is necessary first to build a complete, all-against-all matrix of the topological distances between the trees; obviously, this matrix is a big, approximately 7,000 × 7,000 square table in which each cell contains a distance between two trees.

So how does one compare phylogenetic trees and how are the distances in the matrix calculated? Comparison of trees is much less commonly used than phylogenetic analysis per se, but in the age of genomics, it is rapidly becoming a mainstream methodology. Essentially, what is typically compared are the topologies (that is, the branching order) of the trees, and the distance between the topologies can be captured as the fraction of the tree "splits" that are different (or common) between two compared trees (Figure 10.3). An additional idea implemented in the method for tree topology comparison illustrated in Figure 10.3 is to take into account the reliability of the internal branches of the tree, so that the more reliable branches contribute more than the dubious ones to the distance estimates. The reliability or statistical support for tree branches is usually estimated in terms of the so-called bootstrap values that vary from 0 (no support at all) to 1 (the strongest support). In the Boot Split Distance (BSD) method for tree topology comparison illustrated in Figure 10.3, the contribution of each split is weighted using the bootstrap values.

3 Trends in the Forest of Life

3.1 The NUTs contain a consistent phylogenetic signal, with independent HGT events

Figure 10.4 represents the NUTs as a network in which the edges are drawn on the basis of the topological distances between the trees (see the preceding section and Figure 10.3). Clearly, the topologies of the NUTs are highly coherent, so that when a relatively short distance of 0.5 is used as the threshold to draw edges in the network, almost all the nodes in the network are connected (Figure 10.4). In 56% of the NUTs, representatives of the two prokaryotic domains, archaea and bacteria, are perfectly separated, whereas the remaining 44% of the NUTs showed indications of HGT between archaea and bacteria. Of course, even in the 56% of the NUTs that showed no sign of

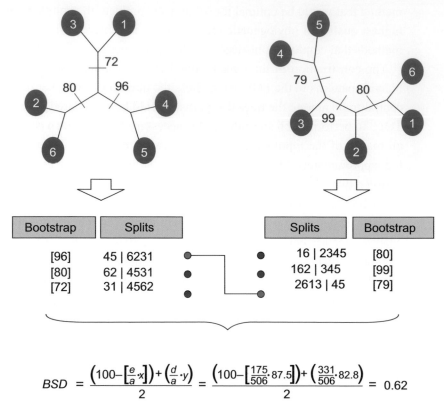

Figure 10.3 Comparison of phylogenetic tree topologies. Identical (equal) splits are shown by connected green circles, and different splits are shown by red circles. Bootstrap values are shown as percent. The Boot Split Distance (BSD) between the trees was calculated using the formula shown in the figure. The designations are:

$e = \sum$ Bootstrap of equal splits
$d = \sum$ Bootstrap of different splits
$a = \sum$ Bootstrap of all splits
$x = $ Mean Bootstrap of equal splits
$y = $ Mean Bootstrap of different splits

interdomain gene transfer, there were many probable HGT events within one or both domains, indicating that HGT is indeed common, even in this group of nearly universal genes.

To analyze the structure of a distance matrix between any objects, including phylogenetic trees, researchers often use so-called multidimensional scaling that reveals clustering of the compared objects. Cluster analysis of the NUTs using the Classical MultiDimensional Scaling (CMDS) method shows lack of significant clustering: all

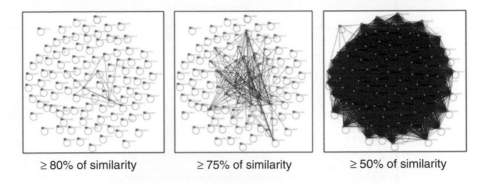

≥ 80% of similarity   ≥ 75% of similarity   ≥ 50% of similarity

Figure 10.4 The network of similarities among the NUTs. Each node denotes a NUT, and nodes are connected by edges if the topological similarity between the respective trees exceeds the indicated threshold (in other words, if the distance between these trees is sufficiently low). The circular arrows show that each node is connected with itself.

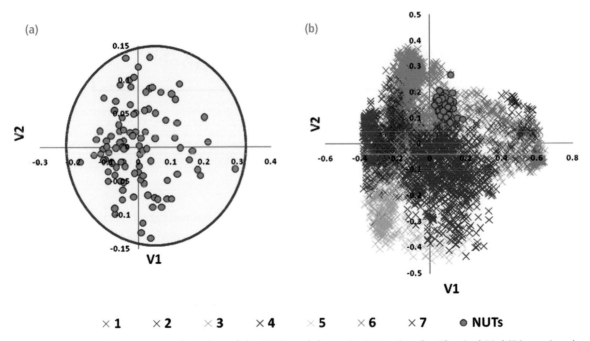

Figure 10.5 Clustering of the NUTs and the entire FOL using the Classical MultiDimensional Scaling (CMDS) method. (a) The best two-dimensional projection of the clustering of the 102 NUTs in a 30-dimensional space. (b) The best two-dimensional projection of the clustering of 3,789 largest trees from the FOL in a 669-dimensional space. The seven clusters are color-coded and the NUTs are shown by circles.

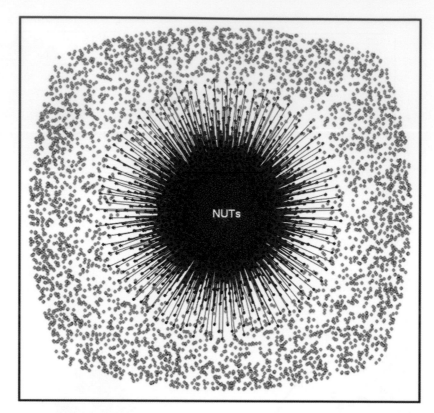

Figure 10.6 The FOL network and the NUTs. The figure shows a network representation of the 6,901 trees in the FOL. The 102 NUTs are shown as red circles in the middle. The NUTs are connected to trees with similar topologies: trees that show at least 50% of similarity with at least one NUT are shown as purple circles and are connected to the NUTs. The rest of the trees are denoted by green circles.

the NUTs formed a single, unstructured cloud of points (Figure 10.5a). This organization of the tree space is best compatible with random deviation of individual NUTs from a single, dominant topology, mostly as a result of HGT but also in part due to random errors of the tree construction procedure. The results of this analysis indicate that the topologies of the NUTs are scattered within a close vicinity of a consensus tree, with the HGT events distributed at least approximately randomly, a finding that is compatible with the idea of a "TOL as a central trend."

3.2 The NUTs versus the FOL

The structure of the FOL was analyzed using the CMDS procedure, with the results being very different from those seen with the NUTs: in this case, seven distinct clusters of trees were revealed (Figure 10.5b). The clusters significantly differed with respect to the distribution of the trees by the number of species, the partitioning of archaea-only and bacteria-only trees, and the functional classification of the respective genes [9]. Notably, all the NUTs formed a compact group within one of the clusters and were roughly equidistant from the rest of the clusters (Figure 10.5b). Thus, the FOL seems to contains several distinct "groves" of trees with different evolutionary histories. The critical observation is that all the NUTs occupy a compact and contiguous region of the tree space and, unlike the complete set of the trees, are not partitioned into distinct clusters by the CMDS procedure (Figure 10.5a). Moreover, the NUTs are, on average, highly similar to the rest of the trees in the FOL as shown in Figure 10.6. Taken together, these findings suggest that the NUTs collectively could represent a central trend in the FOL.

DISCUSSION: THE TREE OF LIFE CONCEPT IS CHANGING, BUT IS NOT DEAD

Prokaryotic genomics revealed the wide spread of HGT in the prokaryotic world and is often claimed to "uproot" the TOL [4]. Indeed, it is now well established that HGT spares virtually no genes at some stages in their history [5], and these findings make obsolete a "strong" TOL concept under which all (or the substantial majority) of the genes would tell a consistent story of genome evolution (the species tree, or the TOL) when analyzed using appropriate data sets and methods. However, is there any hope of salvaging the TOL as a statistical central trend [8]? Comprehensive comparative analysis of the "forest" of phylogenetic trees for prokaryotic genes outlined here suggests a positive answer to this crucial question of evolutionary biology [9].

This analysis results in two complementary conclusions. On the one hand, there is a high level of inconsistency among the trees comprising the FOL, owing primarily to extensive HGT, a conclusion that is supported by more direct observations of numerous likely transfers of genes between archaea and bacteria. However, there is also a distinct signal of a consensus topology that was particularly strong among the NUTs. Although the NUTs show a substantial amount of apparent HGT, these transfers seem to be distributed randomly and did not obscure the vertical signal. Moreover, the topologies of the NUTs are quite similar to those of numerous other trees in the FOL, so although the NUTs cannot represent the FOL completely, this set of largely consistent, nearly universal trees is a good candidate for representing a central trend.

QUESTIONS

(1) Do the phylogenetic trees for all genes in a genome possess the same topology?

(2) Is it possible to detect a common central trend in a genome-wide analysis of tree topologies?

(3) What are the biological functions of genes that are nearly universally conserved among cellular life forms?

REFERENCES

[1] N. R. Pace, G. J. Olsen, and C. R. Woese. Ribosomal RNA phylogeny and the primary lines of evolutionary descent. *Cell*, 45: 325–326, 1986.

[2] C. R. Woese. Bacterial evolution. *Microbiol. Rev.*, 51: 221–271, 1987.

[3] T. Dagan, Y. Artzy-Randrup, and W. Martin. Modular networks and cumulative impact of lateral transfer in prokaryote genome evolution. *Proc. Natl. Acad. Sci. U S A*, 105: 10039–10044, 2008.

[4] W. F. Doolittle. Phylogenetic classification and the universal tree. *Science*, 284: 2124–2129, 1999.

[5] J. P. Gogarten and J. P. Townsend. Horizontal gene transfer, genome innovation and evolution. *Nat. Rev. Microbiol.*, 3: 679–687, 2005.

[6] T. M. Embley and W. Martin. Eukaryotic evolution, changes and challenges. *Nature*, 440: 623–630, 2006.

[7] W. F. Doolittle and E. Bapteste. Pattern pluralism and the Tree of Life hypothesis. *Proc. Natl. Acad. Sci. U S A*, 104: 2043–2049, 2007.

[8] Y. I. Wolf, I. B. Rogozin, N. V. Grishin, and E. V. Koonin. Genome trees and the Tree of Life. *Trends Genet.*, 18: 472–479, 2002.

[9] P. Puigbo, Y. I. Wolf, and E. V. Koonin. Search for a Tree of Life in the thicket of the phylogenetic forest. *J. Biol.*, 8: 59, 2009.

[10] J. Felsenstein. *Inferring Phylogenies*. Sinauer Associates, Sunderland, MA, 2004.

CHAPTER ELEVEN

Reconstructing the history of large-scale genomic changes: biological questions and computational challenges

Jian Ma

In addition to point mutations, larger-scale structural changes (including rearrangements, duplications, insertions, and deletions) are also prevalent between different mammalian genomes. Capturing these large-scale changes is critical to unraveling the history of mammalian evolution in order to better understand the human genome. It also has profound biomedical significance, because many human diseases are associated with structural genomic aberrations. The increasing number of mammalian genomes being sequenced as well as recent advancement in DNA sequencing technologies are allowing us to identify these structural genomic changes with vastly greater accuracy. However, there are a considerable number of computational challenges related to these problems. In this chapter, we introduce the ancestral genome reconstruction problem, which enables us to explain the large-scale genomic changes between species in an evolutionary context. The application of these methods to within-species structural variation and disease genome analysis is also discussed. The target audience of this chapter is advanced undergraduate students in biology.

Bioinformatics for Biologists, ed. P. Pevzner and R. Shamir. Published by Cambridge University Press.
© Cambridge University Press 2011.

1 Comparative genomics and ancestral genome reconstruction

1.1 The Human Genome Project

The Human Genome Project (HGP) is one of the greatest scientific achievements in the twentieth century. In 2001, the draft of the human genome was completed. The human genome has been sequenced in high quality in terms of accuracy and coverage (i.e. the proportion of sequenced bases). One may ask the question: does this mean that we have almost successfully understood our genomes? Unfortunately, this is not the case. We have only scratched the surface of this question, and we are actually at the very beginning of this long scientific journey. Much effort and investigation is still needed to understand how genes and the genome contribute to the complex cellular functions of our body.

1.2 Comparative genomics

During evolution, negative (or purifying) selection causes genomic sequences that yield functional products to evolve more slowly than the neutral expectation. Therefore, an important approach to identify the functional sequences in the human genome is to compare it with the genomes of other species and search for conserved regions. Since the HGP, a number of other mammalian genome sequencing projects have been completed, including mouse, rat, dog, chimpanzee, rhesus macaque, opossum, and cow. The genome sequences from these mammals have greatly advanced the study of mammalian comparative genomics [1]. Scientists have developed various computational methods to compare these sequenced genomes to select candidate functional regions to further test in the laboratory. More mammalian species are planned to be sequenced.

Besides conserved regions, these mammalian genomes have also provided us with a great opportunity to elucidate dramatic genomic differences between species. For example, Figure 11.1 shows the large-scale chromosomal differences between human and mouse. The sequences in the mouse genome are colored according to their similarity counterparts (or homology) in the human genome. We can observe that a stretch of DNA in human can be scattered into different places in mouse. The figure illustrates about 100 large homologous pieces between human and mouse. In other words, if we cut the human genome into these pieces, we can rearrange them to make a genome similar to that of the mouse genome. We now know that these differences are caused by chromosomal changes that happened in the past, some time after the human and mouse diverged (approximately 80 million years ago (MYA)). However, can we

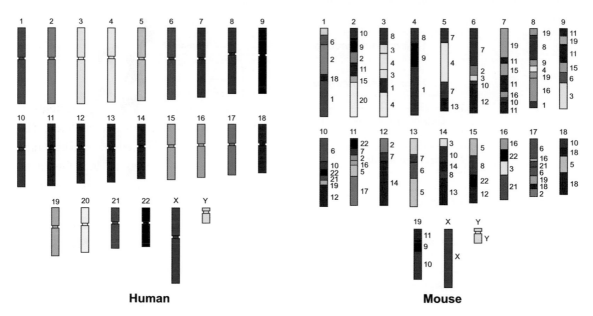

Figure 11.1 This figure illustrates the genomic differences between mouse and human. There are about 100 homologous segments (i.e. the segments in human and mouse share common ancestry) in total illustrated here. The colors and corresponding numbers next to the mouse chromosomes indicate the human counterparts. Figure adapted from the original figure courtesy of Lawrence Livermore National Laboratory.

determine when these changes happened? Did they happen on the human lineage after human mouse divergence or on the mouse lineage?

In fact, if we compare only the human and mouse genomes, we cannot answer this question. Since they both evolved from a common ancestor, more species are needed to determine when the genomic rearrangements happened after human and mouse diverged. Figure 11.2 illustrates mammalian evolution. The phylogenetic tree shows the evolutionary relationships between human and some representative mammalian species, from the closest relative chimpanzee (divergence time 4–5 MYA), to platypus, which shares a mammalian common ancestor with human approximately 160 MYA. We are particularly interested in the changes in molecular evolution along the branch toward modern human, because those genomic innovations may greatly contribute to distinguishing human from other mammalian species. Hence, systematic comparative genomic analysis will shed light on one of the most exciting questions in science – how did we become human?

We know that the differences between mammalian genomes in Figure 11.2 are the result of evolutionary changes after their divergence from their common ances- tor. For example, almost all placental mammals share a common ancestor, called the Boreoeutherian common ancestor. Over the last 100 million years, that ancestor's

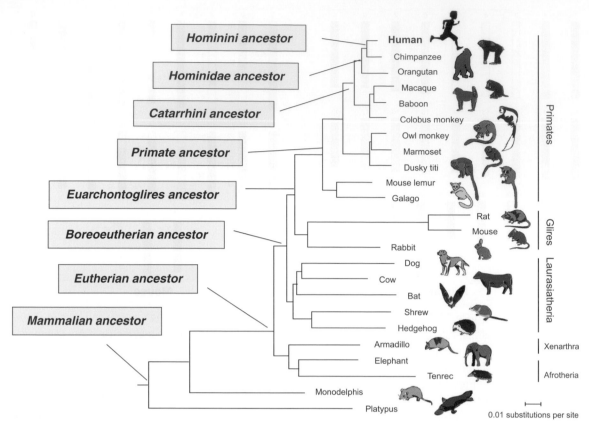

Figure 11.2 The phylogeny of mammalian species. Modified from figure 1 in [2] with the relationship among Boreoeutheria, Xenarthra, and Afrotheria adjusted based on [3].

descendants have evolved into a complex array of different placental mammals – about 5,000 currently living on the planet. As the result of speciation events and many significant changes in each lineage, we see remarkable differences among living placental mammals, both genetic and morphological. If we could somehow obtain the genome of those ancestral species at the precise moment of speciation for each branch in the phylogenetic tree in Figure 11.2, we would be able to compare two genomes on both sides (one ancestor and one descendant) and determine exactly what happened during a particular period of time in mammalian evolution. That would be incredibly exciting, since this unraveled trajectory would tell us how the human genome reached its present state of evolution. Sadly, although new technologies allow us to get DNA from specimens of some relatively recent ancient species, e.g. Neanderthal [4] and woolly mammoth [5], we cannot directly obtain DNA sequences older than a million years. However, the mammalian genomes already sequenced and the additional diverse set of mammalian species that will be sequenced in the future give us an alternative approach.

NM_177028	A			V			G			W			V			I			F			A		
Boreoeutherian	G	C	C	G	T	G	G	G	C	T	G	G	G	T	C	A	T	C	T	T	T	G	C	C
euArc	G	C	C	G	T	G	G	G	C	T	G	G	G	T	C	A	T	C	T	T	T	G	C	C
primate	G	C	T	G	T	G	G	G	C	T	G	G	G	T	C	C	T	C	T	T	T	G	C	C
ape	G	C	T	G	T	G	G	G	C	T	G	A	G	T	C	C	T	C	T	T	T	G	C	C
human	G	C	T	G	T	G	G	G	C	T	G	A	G	T	C	C	T	C	T	T	T	G	C	C

Boreoeutherian	A	V	G	W	V	I	F	A
euArc	A	V	G	W	V	I	F	A
primate	A	V	G	W	V	L	F	A
ape	A	V	G	*	V	L	F	A
human	A	V	G	*	V	L	F	A

Figure 11.3 Part of the reconstructed history of the ACYL3 gene (NM_177028), which was lost in both human and chimpanzee. Boreoeutherian = the reconstructed sequence in the Boreoeutherian ancestor; euArc refers to the Euarchontoglires ancestor; primate refers to the primate common ancestor; and here ape refers to the human–chimpanzee common ancestor. The `G` to `A` transition is highlighted in the DNA multiple sequence alignment (top). The consequence, a change from a tryptophan codon (W) to a stop codon, is also illustrated in the alignment with codon translation (bottom).

1.3 Genome reconstruction provides an additional dimension for comparative genomics

All placental mammals living today show a wide range of variation. However, since these species are descended from a common ancestor, they all have inherited specific DNA sequences from the ancestral genome. Therefore, given the genomes of related species, we can use computational analysis to work backwards and determine most of the specific DNA changes that probably occurred, reconstructing the history of the genetic changes for all the individual bases. With the reconstructed history, we will be able to explain the genomic changes on any given lineage, including the human lineage. This will provide an extremely illuminating vertical map, in the sense that we can view the evolutionary changes from past to present directly, decoding the molecular basis for the extraordinary diversity of mammalian forms and capabilities.

Here, we use two examples to show that genome reconstruction can provide an additional dimension for comparative genomics analysis and facilitate discoveries. Figure 11.3 shows a gene called acyltransferase 3 (ACYL3), which was present in archaea, bacteria, and eukaryotes. ACYL3 is still found in the genomes of many mammals, such as rhesus, rat, mouse, and dog, but has been lost in both human and chimpanzee [6]. What happened? Figure 11.3 illustrates the reconstructed history of this gene, which gives us a direct sense of what transpired from past to present. A close look reveals that there was a `G` to `A` transition that happened after the primate common ancestor and before the ape common ancestor. This nonsense mutation changed the tryptophan codon (W) to a stop codon and made this gene non-functional.

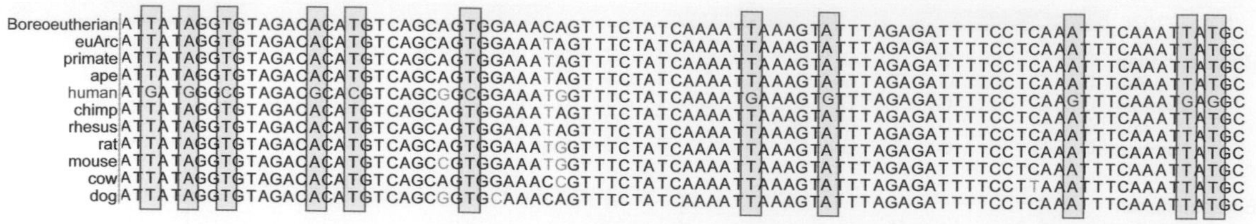

Figure 11.4 Part of the reconstructed history of the Human Accelerated Region 1 (HAR1). Mutations that accumulated in human after diverging from the chimpanzee common ancestor are highlighted.

The second example is a region called Human Accelerated Region 1 (HAR1) [7] with 118 base pairs. Almost all the bases are conserved in mammalian species; furthermore, only two bases differ between chimpanzee and chicken (310 MY divergence time). However, human has surprisingly accumulated 18 substitutions since human and chimpanzee divergence. Figure 11.4 shows the reconstructed history of part of the HAR1 region, highlighting 11 of the 18 substitutions. Scientists believe that this region has experienced accelerated evolution in the human genome due to positive selection. It turns out that HAR1 is part of a novel RNA gene that is expressed specifically during a critical window in embryonic development for a specific set of neurons that guide the development of the layers of the cerebral cortex.

The above examples have demonstrated that if we can create such a reconstructed evolutionary history, we will be able to make many discoveries like this, which will be enormously exciting for human biology. But what kind of computational methods should we use to create such a vertical map that documents all the important genomic changes in mammalian evolution?

1.4 Base-level ancestral reconstruction

In addition to point mutations, which are the most common small-scale genomic changes, various other types of genomic changes can occur. In multiple alignment for sequences from different species, we often see gaps in some of the sequences. What do those gaps mean? Let's examine the following example.

```
  human    ATCAGC------GGCGAT
  chimp    ATCAGC------GGCGAT
macaque    ATCAGCCGGATCGGCGAT
  mouse    ATCAGCCGGATCGGCGAT
    rat    ATCAGCCGGATCGGCGAT
    dog    ATCAGCCGGATCGGCGAT
    cow    ATCAGCCGGATCGGCGAT
```

Actually, the gaps in the alignment correspond to insertion and deletion (indel) events. In the above example, we can infer that the gaps in human and chimpanzee reflect a deletion event that happened before human–chimp common ancestor but after human–macaque common ancestor, which by the principle of parsimony is more likely than any other scenarios. Determining the most plausible indel scenario is the basic idea of inferring indel events from the multiple alignment.

Note that the quality of multiple alignment is critically important for base level reconstruction. The reconstruction methods usually assume that the alignments are evolutionarily correct, i.e. all the bases are placed in the same alignment column as long as they are derived from the same ancestral base, and the boundaries of gaps are placed perfectly consistently with the indel events. Unfortunately, perfect alignment is in practice hard to achieve, especially for genomic regions that have repeatedly undergone various types of genomic changes. The good news is that the majority of the mammalian genomes can be aligned with high confidence. Blanchette *et al.* (2004) [8] showed that given a large genomic region in which there has been no shuffling of bases since the most recent common ancestor, the Boreoeutherian ancestral sequence can be recovered with an accuracy as high as 98% from only 20 optimally chosen modern mammals. Now, how can we reconstruct the entire ancestral genome? The challenge remains: for whole-genome analysis, we must consider large-scale chromosomal changes between different species.

2 Cross-species large-scale genomic changes

2.1 Genome rearrangements

A chromosome is a thread-like macromolecular complex. In eukaryotic cells, chromosomes have a linear form rather than circular. Each chromosome has two arms; the shorter one is called the p arm, while the longer one is the q arm. A chromatid is one of the two identical parts of the chromosome after the synthesis phase. Two chromatids are attached at an area called the centromere. The telomere is the region found at either end of a linear chromosome.

Different kinds of organisms have different numbers of chromosomes. For example, humans have 23 pairs of chromosomes, dogs have 39 pairs, and mice have 20 pairs. A graphic representation of all the chromosomes in a cell of any species is called a karyotype. Karyotype diversity among different species is caused by chromosome rearrangements. Dobzhansky and Sturtevant (1938) [9] reported the observation of inversion events between two *Drosophila* species, thus pioneering the study of chromosome rearrangement. Since then, many studies have concentrated on understanding

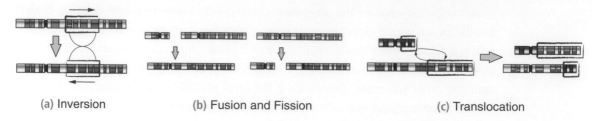

(a) Inversion   (b) Fusion and Fission   (c) Translocation

Figure 11.5 Different types of genomic rearrangements. Each green or red rectangle is a chromosome. In each figure, the large arrow indicates what the chromosomes look like before and after the rearrangement operation.

the differences between genome architectures from an evolutionary perspective. These rearrangements are genomic "earthquakes" [10] that change the chromosomal architecture of an organism. We know that there are a number of different types of rearrangement operations that can be accumulated during chromosomal evolution. In general, these rearrangements are comprised of inversions, translocations, fusions, and fissions.

Figure 11.5 illustrates these four rearrangement operations. In an inversion operation, a genomic segment on one chromosome is reversed and complemented (e.g. AAGTCAT becomes ATGACTT). In a translocation operation, the end part of one chromosome is swapped with the end of another chromosome. In a fusion operation, two chromosomes are joined to form one chromosome; while in a fission operation, a single chromosome is broken into two chromosomes. Among these operations, inversions are the most common events in chromosomal evolution. For translocations, there are two main types, reciprocal (as shown in Figure 11.5c) and Robertsonian. A Robertsonian translocation involves two chromosomes, in which their long arms fuse at the centromere and the remaining two short arms are lost. It has been suggested that Robertsonian translocation also occurred in mammalian genome evolution.

In the general mathematical model of chromosome evolution, a chromosome can be represented as a string of signed numbers (or signed permutation), and a genome as a set of these strings, e.g. 1 2 3 4 5 • 6 7 8, where • separates chromosomes. Numbers could represent any genomic content, e.g. a single base, a gene, or a longer DNA sequence. Numbers have signs, either + or −, which indicate the relative orientation of the genomic content.

Here are some examples of chromosome rearrangements within this mathematical structure. Inversion: 1 2 3 4 5 • 6 7 ⇒ 1 −4 −3 −2 5 • 6 7 (in bioinformatics literature, inversion is also called reversal); translocation: 1 2 3 4 5 • 6 7 ⇒ 1 7 • 6 2 3 4 5; fusion: 1 2 3 4 5 • 6 7 ⇒ 1 2 3 4 5 6 7; fission: 1 2 3 4 5 • 6 7 ⇒ 1 2 • 3 4 5 • 6 7.

Overlapping or nested operations form composite operations. For example, 1 2 3 4 5 6 7 can be transformed to 1 −4 −6 • −5 2 3 7 by two overlapping inversions followed

by a fission: 1 2 3 4 5 6 7 $\Rightarrow$ 1 –4 –3 –2 5 6 7 $\Rightarrow$ 1 –4 –6 –5 2 3 7 $\Rightarrow$ 1 –4 –6 •
–5 2 3 7.

2.2 Synteny blocks

Identifying the genomic content that signed permutations can represent has always been an essential problem in studying genome rearrangements. Nadeau and Taylor (1984) [11] first introduced the term "conserved segment" to name a maximal genomic segment with preserved gene orders that are not disrupted by rearrangements between species. In the past decade, using comparative gene mapping to find orthologous gene loci as the evolutionary markers played an important role in testing algorithms and understanding rearrangement scenarios (the term "orthologous" means that two loci share the same ancestry). However, although this approach works well in small genomes, e.g. virus genomes or mitochondrial genomes, reliable gene annotation and orthology assignment in the entire mammalian genome are technically very difficult, partly because of the great number of duplicated genes existing in mammals. This problem is further complicated by the large proportion of non-coding regions throughout the genome.

Pevzner and Tesler (2003) [12] proposed the GRIMM-Synteny algorithm to partition the genomes into segments which tolerate a certain amount of local micro-rearrangements that are smaller than the size of the segments. These segments are called "synteny blocks," which conceptually is similar to conserved segments. Based on this method, multi-way synteny blocks can be created for multiple species. The GRIMM-Synteny algorithm greatly improved the resolution and precision for whole-genome rearrangement studies.

In recent years, improved whole-genome alignments have allowed us to produce synteny blocks with higher coverage and higher resolution for ancestral genome reconstruction. Ma *et al.* (2006) [13] describe one of these methods. The basic idea can be summarized in Figure 11.6. If two synteny blocks are adjacent in one species and separate in the other, that reflects a breakpoint between these two species. The algorithm processes the whole-genome alignment and partitions the genome every time it encounters a breakpoint in one of the species. In the example in Figure 11.6, if we set the synteny block threshold as 50 kb (i.e. any rearrangements smaller than 50 kb are ignored), this region can be partitioned into 5 synteny blocks.

When we construct synteny blocks, resolution (size threshold) is always a factor to consider (low resolution = large blocks and high resolution = small blocks). If we construct higher-resolution synteny blocks, we can capture more interesting rearrangements, but the smaller ones may not be very reliable due to potentially problematic sequence alignment. If we build lower-resolution synteny blocks, we will have more

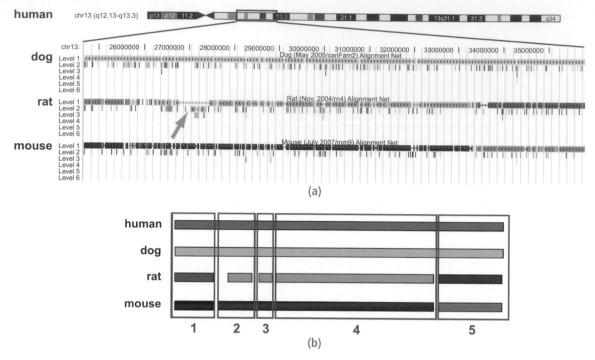

Figure 11.6 Constructing synteny blocks based on whole-genome alignment. (a) A region on human chromosome 13 and its corresponding regions in mouse, rat, and dog (based on their pairwise alignments with human). Different colors refer to different chromosomes in dog, rat, and mouse. This is a snapshot of the UCSC genome browser for this region on human chromosome 13. Each track is a pairwise alignment net between human and a secondary species. In the figure, net identifies putative orthologous genomic segments between two genomes. Level 1 net shows the primary alignment of the region. For example, this human region is roughly orthologous to three regions in different chromosomes in mouse (shown by three colors). Level 2 and beyond show additional nets, which indicate rearrangements (smaller than level 1). For example, the orthologous region on rat chromosome 12 (the green part) has a big net as a level 2 net (indicated by an orange arrow), suggesting a rearrangement. (b) An abstract version of (a), where this genomic region can be partitioned into five synteny blocks.

reliable evolutionary conserved synteny blocks, but we certainly miss a lot of rearrangements that are under the size threshold. In Ma *et al.* (2006) [13], for human, mouse, rat, and dog, 1,338 synteny blocks (size threshold = 50 kb) were constructed, covering about 95% of the human genome.

Once we have the synteny blocks, the next step is to figure out what the ancestral order and orientation of these blocks were in a certain ancestor and what kinds of evolutionary events caused the dramatic shuffling of these blocks in different descendant species.

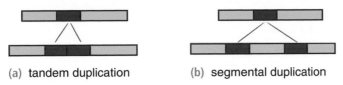

(a) tandem duplication   (b) segmental duplication

Figure 11.7 (a) Tandem duplication, where the two copies are adjacent to each other after the duplication. (b) Segmental duplication, where the target copy is far away from the source copy after the duplication.

2.3 Duplications and other structural changes

Besides the rearrangement operations mentioned above, chromosome architecture can also be changed by other large-scale operations. For example, transposition is a more complicated rearrangement in which a segment of DNA is removed from its original location and then gets inserted into a new location. Duplication is another major source of large-scale genomic change. There are generally two types of duplication events, tandem duplication and segmental duplication (Figure 11.7). In addition, large-scale insertion and deletion can also happen. Even more complex operations are occasionally observed in human disease-associated genome rearrangements [14].

All these operations may happen in nested or overlapping forms during evolution. As a result, genomic architectures between different modern species can be highly distinct. An ancestral genomic segment can be broken into several fragments in an extant genome and widely scattered to different chromosomes and different positions (e.g. Figure 11.1).

 3 ## Reconstructing evolutionary history

3.1 Ancestral karyotype reconstruction

In fact, the problem of ancestral mammalian karyotype reconstruction has been studied for quite a long time. The development of comparative gene mapping and cytogenetic methods have provided biologists with powerful tools in their attempt to solve the puzzle. However, the number of chromosomes in the mammalian common ancestor is still not fixed and is believed that 24 or 25 is currently the best guess. Even though there is no solid evidence of the number of chromosomes in the ancestral eutherian karyotype, some configurations have been widely confirmed, e.g. Hsa14/Hsa15 ("Hsa" refers to a human chromosome.), which means human chromosome 14 and chromosome 15 were in the same ancestral chromosome (in other words, a chromosomal fission happened on the path leading to human).

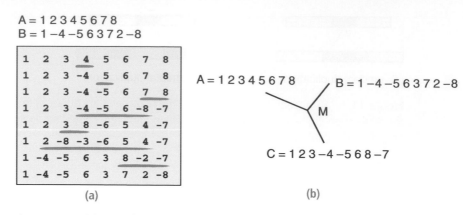

Figure 11.8 (a) One of the most parsimonious solutions of sorting by reversal between A and B. (b) An example of the Median Problem. The median $M = 1\,2\,3\,-4\,-5\,6\,-8\,-7$, with $d(A, M) + d(B, M) + d(C, M) = 8$.

In the past decade, the primary experimental technique used in the study of chromosomal evolution is chromosome painting, in which fluorescently labeled chromosomes from one species are hybridized to chromosomes from another species so that breakpoints can be identified. Although the requirement of optical visibility means that the chromosome painting approach can only recognize rearrangements with long conserved segments and cannot identify intrachromosomal rearrangements, the chromosomal painting approach has the advantage that data are available for more species because we do not need to sequence the genome.

3.2 Rearrangement-based ancestral reconstruction

Indeed, for the past 15 years, genome rearrangement problems have fascinated computational biologists. Computer scientists have also tried to reconstruct the ancestral genome architecture using bioinformatic algorithms in a parsimony framework based on certain distance measurements.

Sankoff pioneered the theoretical study of reversal distance [15] and phylogenetic analysis using gene order data [16]. The analysis of the most parsimonious rearrangement scenarios is the central part of theoretical genome rearrangement study, among which the most well studied is sorting by reversals. Sorting by reversals is the problem of converting one permutation into another using the minimum number of reversal operations. The minimal number of reversals is regarded as reversal distance between two permutations. For example, in Figure 11.8(a), the reversal distance between A and B, abbreviated $d(A, B)$, is 7 because 7 is the minimum number of inversions needed to transform A into B. For these kind of signed permutations, which are practically very important to model mammalian genomes, Hannenhalli and Pevzner (1995) [17] gave the first efficient algorithm to solve the sorting by reversal problem.

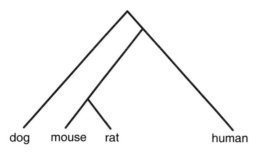

Figure 11.9 The phylogeny of human, mouse, rat, and dog.

However, when we need to use reversal distance to perform phylogenetic analysis (in which we need more than two species), the problem suddenly becomes computationally intractable. A typical problem is the Median Problem: given three signed genomes A, B, and C, as well as the distance measure d, find a median genome, which is a genome M such that $\sum d = d(A, M) + d(B, M) + d(C, M)$ is minimal, as illustrated in Figure 11.8(b). Unfortunately, this problem is computationally intractable. Note that the Median Problem is the simplest problem for the genome reconstruction problem based on reversal distance, in which we have two descendant genomes A and B as well as an outgroup species C. However, there are heuristic approaches available to solve this problem, e.g. multiple genome rearrangements (MGR) [18].

3.3 Adjacency-based ancestral reconstruction

Two synteny blocks are adjacent if they are next to each other on a chromosome. Ma *et al.* (2006) [13] observed that the adjacencies of genomic content in modern species can be used to infer the ancestral adjacencies. The problem can be described as: given a tree, predict the ancestral order and orientation based on adjacencies in modern genomes. That is, consider the end of a synteny block x that does not correspond to a human telomere or centromere. How can we identify the segment that was adjacent to x in the ancestral genome?

If the segment that is currently adjacent in human is identical to the one that is adjacent in dog (but a different segment is adjacent in mouse and rat), the most parsimonious assumption (based on the phylogeny of human, mouse, rat, and dog as shown in Figure 11.9) is that the first and second segments were adjacent in the ancestral genome (and that a disruption occurred in the rodent lineage at this genomic position).

If the same segment is adjacent to the chosen segment in human, mouse, and rat, but not in dog, we need more information to confidently predict the ancestral configuration, since there is a chance that the dog adjacency is ancestral and that the breakage occurred on the short branch from the human–dog ancestor to the human–rodent ancestor. To help resolve these cases, we can add outgroup information, e.g. the opossum sequence. Figure 11.10 shows an example that demonstrates this principle. This snapshot from the UCSC genome browser clearly shows the relative orientations from which the ancestral orientation can be inferred by parsimony. This region can be partitioned into three synteny blocks: 1, 2, and 3. Human, rhesus, mouse, and rat share the order 1 2 3, while dog and opossum have the order 1 –2 3. Based on the parsimony principle discussed above, we can infer that 1 is followed by –2 and 3 is preceded by –2 in the human–dog common ancestor, which creates the ancestor order 1 –2 3. How can we generalize this procedure algorithmically?

The approach is inspired by Fitch's method [19], which was originally used to infer minimum substitutions in a specified tree topology. For that problem, one is given a phylogenetic tree and a letter for every position in each leaf of the tree (corresponding to the contents of orthologous sequence sites). The problem is to infer the ancestral letters (corresponding to internal nodes of the tree), so as to minimize the number of substitutions, i.e. differences between the letters at each end of an edge in the tree.

The algorithm works sequentially, in two stages. For each position, in a bottom-up fashion, it first determines a set M_π of candidate nucleotides at each node π in the tree according to the following rule: if π is a leaf, M_π just contains its nucleotide character; otherwise, if π has children τ and φ, then M_π equals either intersection $M_\tau \cap M_\varphi$ or the union $M_\tau \cup M_\varphi$ depending on whether M_τ and M_φ are disjoint or not. That is,

if M_τ and M_φ do not overlap
    then $M_\pi \leftarrow M_\tau \cup M_\varphi$
    else $M_\pi \leftarrow M_\tau \cap M_\varphi$

Then, in a top-down fashion, it assigns a character b_π from M_π to π according to the following rule: let ρ be the parent of π; if the character b_ρ assigned to ρ belongs to M_π, then, $b_\pi = b_\rho$. Otherwise, set b_π to be any character in M_π. Although character assignment in this second stage may not be unique, any assignment gives an evolutionary history with the minimum number of substitution events.

The rationale behind Fitch's method is as follows. If the character b_π belongs to both children of π, then an optimal strategy for labeling nodes in the subtree rooted at π is to put b at each of π, τ, and φ, and label the subtrees of τ and φ optimally. If there is no such b, then the strategy is to put a character from either M_τ or M_φ at π, pay for one substitution to reach the other child, and optimally label the two subtrees. See Figure 11.11 for an example. The characters at leaves are given. Then we do a post-order tree traversal (i.e. visiting each node in the tree by recursively processing all subtrees and finally processing the root) and create sets in the internal nodes until we reach the root. In this example, the ancestral nucleotide A will give us the minimum number of substitutions, which is 2, for this position.

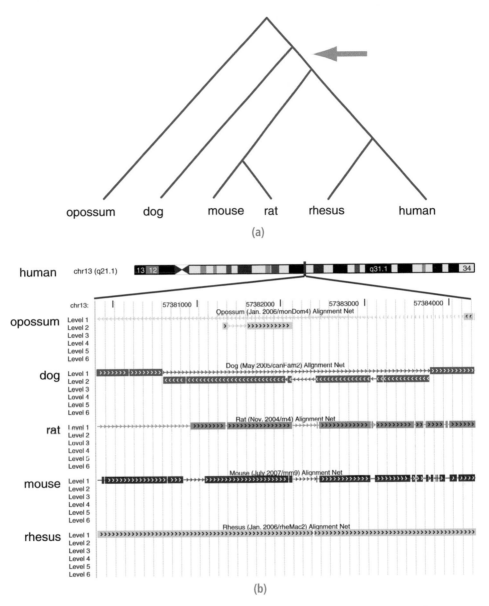

Figure 11.10 (a) is the phylogenetic tree of human, rhesus, mouse, rat, dog, and opossum, where opossum is an outgroup of the placental mammals (all the descendants of the Boreoeutherian common ancestor). (b) is a snapshot of the UCSC genome browser of this region. Each track is a pairwise alignment net between human and a secondary species. In this region, both dog and opossum have level 2 net that reflects an inversion in the alignment with human. Based on the tree in (a), we infer that the inversion happened on the branch leading from the Boreoeutherian common ancestor to the Euarchontoglires common ancestor (the primate-rodent ancestor), as highlighted by the orange arrow. The corresponding human region is hg18.chr13:57,380,591-57,383,765.

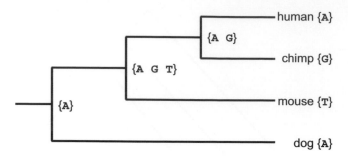

Figure 11.11 An example of Fitch's algorithm.

Let's now formally prove by induction that the Fitch algorithm constructs the most parsimonious solution for the total number of substations. Let $k(\pi)$ denote the minimum number of substitutions in the subtree rooted at π. Let τ and φ be the two children of π. *Basis*: if tree height $h = 1$, then τ and φ are leaves in the phylogeny. If τ and φ are the same, then no substitution is needed; $k(\pi) = 0$. Otherwise, only 1 substitution is needed; $k(\pi) = 1$. *Induction*: if we assume the Fitch algorithm constructs the most parsimonious solution for the subtree height is h, then prove this is the case for height $h + 1$. If the intersection of M_τ and M_φ is not empty, then we can have $k(\pi) = k(\tau) + k(\varphi)$ by assigning any character in the intersection to π. Otherwise, $k(\pi)$ is $k(\tau) + k(\varphi) + 1$, by assigning any character in the union of M_τ and M_φ. This completes the proof.

In our case, we deal with sequences of signed integers, rather than characters of nucleotides or amino acids, and instead of keeping track of letters at a particular sequence position, we track the synteny blocks for each of the immediately adjacent positions. Based on this logic, for a certain ancestor, we can infer what would be the most parsimonious neighbors of each synteny block in the ancestral genome.

We first define predecessor and successor. If modern genome g contains synteny block i, then the predecessor $p_g(i)$ is defined as the signed block that immediately precedes i on the same chromosome relative to the original orientation. In the opposite orientation, $p_g(-i)$ immediately precedes $-i$ in the reverse complement of the same chromosome. We set $p_g(i) = \phi$ if i appears first on a chromosome. The successor $s_g(i)$ of i is defined analogously; we set $s_g(i) = \phi$ if i appears last on a chromosome. For instance, let g have the chromosome (1 −4 −3 5 2). Then in the positive orientation, we have: $p_g(1) = 0$, $p_g(2) = 5$, $p_g(-3) = -4$, $p_g(-4) = 1$, $p_g(5) = -3$, while $s_g(1) = -4$, $s_g(2) = 0$, $s_g(-3) = 5$, $s_g(-4) = -3$, $s_g(5) = 2$. In the opposite orientation, (−2 −5 3 4 −1), we have: $p_g(-1) = 4$, $p_g(-2) = 0$, $p_g(3) = -5$, $p_g(4) = 3$, $p_g(-5) = -2$, while $s_g(-1) = 0$, $s_g(-2) = -5$, $s_g(3) = 4$, $s_g(4) = -1$, $s_g(-5) = 3$.

We consider keeping track of the set of all possible synteny blocks that follow a fixed synteny block in a most parsimonious evolutionary scenario. In the genome that corresponds to node π, block i could be followed by any block that follows i in both τ and φ, without requiring any rearrangements on the branches leading from π to its children. Otherwise, i can be followed by any block that follows i in one of π's children, at the cost of a chromosomal break next to i along the branch leading from π to the other child. This is all closely analogous to the case of substitutions, as sketched above.

Thus, for any genome g, we associate with each block i two sets of signed blocks, denoted $P_g(i)$ and $S_g(i)$, giving potential predecessors and successors of i relative to chromosomes of g. If g is a modern genome, $P_g(i) = \{p_g(i)\}$ and $S_g(i) = \{s_g(i)\}$, for each i. If g does not contain i, then both sets are empty.

The algorithm GET-PREDECESSOR-SUCCESSOR(R) constructs $P_g(i)$ and $S_g(i)$ for each synteny block i of every ancestral genome g in the tree (π is a tree node; τ and φ are π's children in the tree; N is the total number of synteny blocks).

GET-PREDECESSOR-SUCCESSOR (π)

1   **if** π is non-leaf node
2   **then** GET-PREDECESSOR-SUCCESSOR (τ)
3   GET-PREDECESSOR-SUCCESSOR (φ)
4   **for** $i \leftarrow -N$ **to** N ($i \neq 0$)
5   **do if** $P_\tau(i)$ and $P_\varphi(i)$ do not overlap
6   **then** $P_\pi(i) \leftarrow P_\tau(i) \cup P_\varphi(i)$
7   **else** $P_\pi(i) \leftarrow P_\tau(i) \cap P_\varphi(i)$
8   **if** $S_\tau(i)$ and $S_\varphi(i)$ do not overlap
9   **then** $S_\pi(i) \leftarrow S_\tau(i) \cup S_\varphi(i)$
10   **else** $S_\pi(i) \leftarrow S_\tau(i) \cap S_\varphi(i)$

Finally, there is an algorithm to connect the synteny blocks in the ancestor based on possible predecessor/successor relationships into continuous ancestral regions (CARs) which resemble ancestral chromosomes. Using 1,338 synteny blocks constructed from human, mouse, rat, and dog, the karyotype of the Boreoeutherian ancestral genome (shown in Figure 11.12) can be reconstructed with relatively high accuracy [13, 20]. The accuracy can be assessed by comparing with experimental chromosomal painting results and computational simulations.

3.4 Challenges and future directions

The method discussed in the previous section, which was based on adjacencies of synteny blocks, reduced the number of discrepancies between computational and

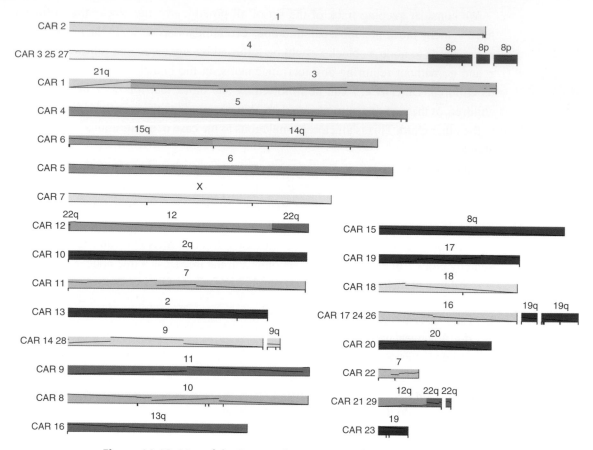

Figure 11.12 Map of the Boreoeutherian ancestral genome. Numbers above bars indicate the corresponding human chromosomes. 1,338 synteny blocks are constructed from whole genome sequences of human, mouse, rat, and dog (size threshold = 50 kb, covering about 95% of the human genome).

experimental large-scale genome reconstruction. The result, in much higher resolution than previous studies, has proven to be reliable [20]. However, such an adjacency-based reconstruction, albeit undoubtedly informative, provides no direct knowledge of the detailed evolutionary operations transforming the ancestor to the present day genomes. Therefore, models that handle sophisticated genomic operations are needed.

With regard to models of evolutionary operations, a key step was the unification of inversion, translocation, fusion, and fission into the general operation of double-cut-and-join (DCJ) [21] (also termed as "2-break operation," see Figure 11.13). Other types of operation were also studied, e.g. transposition and indels. More importantly, duplications cannot be left out of the analysis given their critical role in mammalian evolution. Regarding recovering complex operations on genomes, a recent paper by Ma *et al.* [22] formalized the problem of recovering (by parsimony) the evolutionary history of a set of genomes that are related to an unseen common ancestor genome by

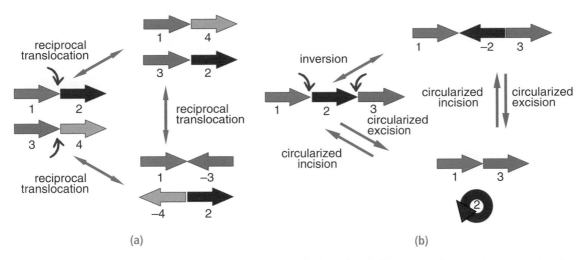

(a)   (b)

Figure 11.13 2-break operations, in which we break the genome in two places, creating four free ends, and then we rejoin the four free ends. (a) Two breakpoints are on the different chromosomes. This models translocation. (b) Two breakpoints are on the same chromosome. This models inversion and indels.

operations of deletion, insertion, duplication, and rearrangement of segments of bases, and by speciation events. The authors show that as the number of bases ("sites") in the genome approaches infinity, the problem of reconstructing the simplest history of operations becomes tractable.

There are a number of computational challenges ahead. For example, so far most algorithms assume that each operation is equally likely to happen in the genome. To be more realistic, each of the different types of operations could have a different cost, and the goal would be to find an evolutionary history with minimal total cost. This method is called weighted parsimony. Models that consider weighted parsimony based on empirical data from practice will be very useful.

In addition, breakpoint reuse, in which the same genomic location is broken more than once during evolution, arises in real data, partly because the synteny block construction method often cannot pinpoint the breakpoint to 1-base resolution. It is also still a challenge to locate more precise breakpoints caused by structural changes, widely believed to contain enriched genomic variation and very interesting biology [23].

 ## 4 Chromosomal aberrations in human disease genomes

Many individual human genomes have been entirely sequenced, including Nobel Laureate James Watson, a Han Chinese, a Korean, Yoruban individuals, etc. These data revealed that, between different normal human individuals, our genomes also show

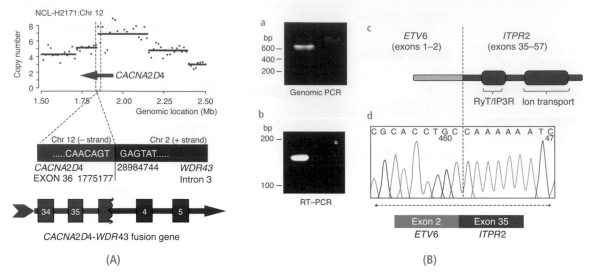

Figure 11.14 Fusion genes in cancer genomes. (A) CACNA2D4-WDR43 fusion gene identified in the NCI-H2171 lung cancer cell line. The 5′ portion of the CACNA2D4 gene is amplified. A rearrangement breaks the gene in exon 36, fusing it into intron 3 of WDR43. The sequence at the breakpoint creates an almost perfect splice-donor site, resulting in a fusion transcript with a shortened exon 36 from CACNA2D4. Figure (A) and caption are from [24]. (B) ETV6-ITPR2 fusion gene in the primary breast cancer PD3668a. [B-a]: Across-rearrangement PCR to confirm the rearrangement in genome. [B-b]: RT-PCR of RNA between ETV6 exon 2 and ITPR exon 35 to confirm the expressed transcript. N, normal; T, tumor. [B-c]: Diagram of the protein domains fused in the ETV6-ITPR2 fusion protein. [B-d]: Sequence from RT-PCR product shown in B-b confirming ETV6 exon 2 fused to ITPR2 exon 35. Figure (B) and caption are from [25].

a large amount of structural variation. One may wonder: how representative is the reference human genome sequenced by the Human Genome Project a decade ago?

We now know that many human diseases are associated with structural genomic changes. New technologies are allowing researchers to map disease-causing structural changes to the genome in much finer resolution. When multiple changes have occurred to the genome and created a genetic state that causes diseases, the algorithms of genome reconstruction discussed above may be useful in better understanding the detailed scenario of these changes, as well as identifying the specific operations that have occurred and the properties of the DNA sequences near their breakpoints.

Cancer is another group of genetic diseases associated with a massive amount of structural genomic changes. Much as germline genomes undergo various chromosomal structural changes over an evolutionary timescale, the genomes of somatic cells also undergo structural changes during cancer progression, including rearrangements, insertions and deletions, and duplications. Recent rapid advancement in high-throughput sequencing technologies have enabled us to use paired-end reads to map novel DNA segment adjacencies caused by different types of rearrangements in individual tumor genomes. A paired-end read consists of two stretches of sequenced DNA with an unsequenced insert of known size between them. Thus, after mapping the paired-end read from a tumor genome to a normal genome, if the distance between those two stretches of DNA changes, then we know there must be a structural genomic change. Interested readers can read [26] for computational approaches to utilize paired-end data.

Figure 11.14(A) shows a CACNA2D4-WDR43 fusion gene in NCI-H2171, a lung cancer cell line [24]. Figure 11.14(B) shows an ETV6-ITPR2 fusion gene generated by a 15-Mb inversion in breast cancer sample PD3668a [25]. Stephens *et al.* (2009) [25] reported rearrangement patterns in 24 breast cancer genomes. With these cancer breakpoint data coming in, the rearrangement-based algorithms may help us better dissect the evolutionary history of individual tumors and understand molecular signatures of different cancers.

DISCUSSION

Our ability to sequence the entire human genome and other mammalian species has given us an unprecedented opportunity to peer into our origins and decode our own genomes. Based on computational analysis of the genomes of modern mammals, it would be extremely exciting to discover the critical genetic changes that led to the remarkable differences among these species. As the genomic data grow exponentially, the idea of ancestral genome reconstruction is an elegant way to organize a large number of related species, creating a vertical map so that we can navigate the genomes and trace the history from past to present. Even when we study genomic variation in the human population and human disease genomes, it is always important to put the genomic data into the evolutionary context to approach these problems. As Theodosius Dobzhansky said: "Nothing in biology makes sense except in the light of evolution."

QUESTIONS

(1) Assume that the synteny block A is followed by B in human, but it is followed by C in chimpanzee, mouse, and dog. What would be the most parsimonious situation for the block that follows A in the human–chimpanzee common ancestor?

(2) Based on Figure 11.12, the map of the Boreoeutherian ancestral genome, identify the interchromosomal breakpoints that occurred on the branch leading to human.

(3) How can we evaluate the performance of the algorithm GET-PREDECESSOR-SUCCESSOR? If you choose a simulation-based approach, what kind of experiment will you design?

REFERENCES

[1] W. Miller, K. Makova, A. Nekrutenko, and R. Hardison. Comparative genomics. *Annu. Rev. Genomics. Hum. Genet.*, 5:15–56, 2004.

[2] E. Margulies, G. Cooper, G. Asimenos, *et al.* Analyses of deep mammalian sequence alignments and constraint predictions for 1% of the human genome. *Genome Res.*, 17(6):760, 2007.

[3] W. Murphy, T. Pringle, T. Crider, M. Springer, and W. Miller. Using genomic data to unravel the root of the placental mammal phylogeny. *Genome Res.*, 17(4):413–421, 2007.

[4] R. Green, J. Krause, S. Ptak, *et al.* Analysis of one million base pairs of Neanderthal DNA. *Nature*, 444:330–336, 2006.

[5] W. Miller, D. Drautz, A. Ratan, *et al.* Sequencing the nuclear genome of the extinct woolly mammoth. *Nature*, 456(7220):387–390, 2008.

[6] J. Zhu, J. Sanborn, M. Diekhans, C. Lowe, T. Pringle, and D. Haussler. Comparative genomics search for losses of long-established genes on the human lineage. *PLoS Comput. Biol.*, 3(12):e247, 2007.

[7] K. Pollard, S. Salama, N. Lambert, *et al.* An RNA gene expressed during cortical development evolved rapidly in humans. *Nature*, 443:167–172, 2006.

[8] M. Blanchette, E. Green, W. Miller, and D. Haussler. Reconstructing large regions of an ancestral mammalian genome in silico. *Genome Res.*, 14(12):2412–2423, 2004.

[9] T. Dobzhansky and A. Sturtevant. Inversions in the chromosomes of *Drosophila pseudoobscura. Genetics*, 23(1):28–64, 1938.

[10] M. Alekseyev and P. Pevzner. Are there rearrangement hotspots in the human genome. *PLoS Comput. Biol.*, 3(11):e209, 2007.

[11] J. Nadeau and B. Taylor. Lengths of chromosomal segments conserved since divergence of man and mouse. *Proc. Natl Acad. Sci. U S A*, 81(3):814–818, 1984.

[12] P. Pevzner and G. Tesler Genome rearrangements in mammalian evolution: Lessons from human and mouse genomes. *Genome Res.*, 13(1):37–45, 2003.

[13] J. Ma, L. Zhang, B. B. Suh, *et al.* Reconstructing contiguous regions of an ancestral genome. *Genome Res.*, 16(12):1557–1565, 2006.

[14] J. Lee, C. Carvalho, and J. Lupski. A DNA replication mechanism for generating nonrecurrent rearrangements associated with genomic disorders. *Cell*, 131(7):1235–1247, 2007.

[15] D. Sankoff. Edit distances for genome comparisons based on non-local operations. In: *Combinatorial Pattern Matching*, pp. 121–135, 1992.

[16] D. Sankoff, G. Leduc, N. Antoine, B. Paquin, B. F. Lang, and R. Cedergren. Gene order comparisons for phylogenetic inference: Evolution of the mitochondrial genome. *Proc. Natl Acad. Sci. U S A*, 89(14):6575–6579, 1992.

[17] S. Hannenhalli and P. A. Pevzner. Transforming cabbage into turnip: Polynomial algorithm for sorting signed permutations by reversals. In: *ACM Symposium on Theory of Computing*, pp. 178–189, 1995.

[18] G. Bourque and P. A. Pevzner. Genome-scale evolution: Reconstructing gene orders in the ancestral species. *Genome Res.*, 12(1):26–36, 2002.

[19] W. M. Fitch. Toward defining the course of evolution: Minimum change for a specific tree topology. *Syst. Zool.*, 20:406–416, 1971.

[20] M. Rocchi, N. Archidiacono, and R. Stanyon. Ancestral genomes reconstruction: An integrated, multi-disciplinary approach is needed. *Genome Res.*, 16(12):1441, 2006.

[21] S. Yancopoulos, O. Attie, and R. Friedberg. Efficient sorting of genomic permutations by translocation, inversion and block interchange. *Bioinformatics*, 21(16):3340–3346, 2005.

[22] J. Ma, A. Ratan, B. J. Raney, B. B. Suh, W. Miller, and D. Haussler. The infinite sites model of genome evolution. *Proc. Natl Acad. Sci. U S A*, 105(38):14254–14261, 2008.

[23] D. Larkin, G. Pape, R. Donthu, L. Auvil, M. Welge, and H. Lewin. Breakpoint regions and homologous synteny blocks in chromosomes have different evolutionary histories. *Genome Res.*, 19(5):770–777, 2009.

[24] P. Campbell, P. Stephens, E. Pleasance, *et al.* Identification of somatically acquired rearrangements in cancer using genome-wide massively parallel paired-end sequencing. *Nat. Genet.*, 40(6):722–729, 2008.

[25] P. J. Stephens, D. J. McBride, M. L. Lin, *et al.* Complex landscapes of somatic rearrangement in human breast cancer genomes. *Nature*, 462:1005–1010, 2009.

[26] P. Medvedev, M. Stanciu, and M. Brudno. Computational methods for discovering structural variation with next-generation sequencing. *Nat. Methods*, 6:13–20, 2009.

PART IV

PHYLOGENY

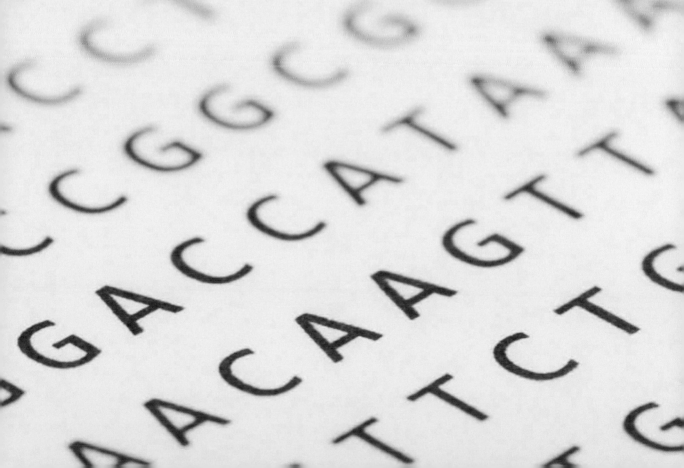

CHAPTER TWELVE

Figs, wasps, gophers, and lice: a computational exploration of coevolution

Ran Libeskind-Hadas

This chapter explores the topic of coevolution: the genetic change in one species in response to the change in another. For example, in some cases, a parasite species might evolve to specialize with its host species. In other cases, the relationship between two species may be mutually beneficial and coevolution may serve to strengthen the benefits of that relationship.

One important way to study the coevolution of species is through a computational technique called *cophylogeny reconstruction*. In this technique, we first obtain the evolutionary (phylogenetic) trees for the two species and then try to map one tree onto the other in the "simplest" (most parsimonious) possible way. We can then use these mappings to determine how likely it is that the two species coevolved.

This chapter begins with descriptions of several pairs of species that are believed to have coevolved: figs and the wasps that polinate them; gophers and the lice that infest them; and a bird species that "tricks" another species to tend to its young. Next, we describe the cophylogeny reconstruction problem, its computational complexity, and a technique for finding good solutions for this problem. Finally, the reader is invited to use this computational method – through a freely accessible software package called Jane – to investigate the relationships between the pairs of species described at the beginning of the chapter.

Bioinformatics for Biologists, ed. P. Pevzner and R. Shamir. Published by Cambridge University Press.
© Cambridge University Press 2011.

1 Introduction

I can understand how a flower and a bee might slowly become, either simultaneously or one after the other, modified and adapted in the most perfect manner to each other, by the continued preservation of individuals presenting mutual and slightly favourable deviations of structure. (Charles Darwin, *The Origin of Species*)

The prescient thought experiment that Darwin describes in *The Origin of Species* is, in fact, borne out in bees and flowers (as documented in the book *The Sex Life of Flowers* [1]). One particularly interesting example is the symbiotic relationship between figs, their tiny flowers, and the miniature wasps that pollinate them.[1]

The story goes something like this. The flowers or "florets" of a fig are in its interior and are protected by the fig's thick membrane. Pollinating a fig is a real challenge! However, each fig species has a species of wasp (usually just one species, but sometimes more) that pollinates it. When a female wasp of the right species finds a fig that she likes, she tunnels into the interior, generally losing her wings in the process. Once inside, she lays her eggs on some of the tiny interior flowers, and, in the process, pollinates the fig. As the host fig develops, the wasp eggs hatch and the larvae feed on the fig tissue. After several weeks, the wasps reach maturity. The wingless males have a short life with only two objectives: they mate with the females and then burrow holes to help the females escape from the fig. The males then die inside the fig and the females fly off in search of their own fig homes to repeat the reproductive cycle. This bizarre story is true [2, 3] and not merely a *fig*ment of our imagination!

Biologists refer to the genetic change of one species in response to the change in another as *coevolution*. In the case of figs and wasps, the coevolution is known as *mutualism* since the two species are mutually dependent on one another for their survival. While there are several hundred varieties of figs (*Ficus*) and fig wasps, many pairs of fig and wasp species have become highly specialized to one another over approximately 60 million years of evolution.

Coevolution is not always mutually beneficial. For example, there are a variety of species of pocket gophers and an equal variety of lice that have specialized to their particular gopher hosts. This form of coevolution, known as *parasitism*, is a sort of evolutionary war: the gophers have evolved to defend themselves from the parasitic lice and the lice have evolved along with them to defeat their hosts' defenses [4].

A truly bizarre form of parasitism arises between finches from the family Estrildidae and another family commonly known as indigobirds [5, 6]. Each species of indigobird has evidently specialized to exploit a specific finch host species. The parasitic indigobirds very slyly lay their eggs in the nests of the host finches. The indigobird eggs look

---

[1] Wasps are not bees, but they are in the same order called Hymenoptera.

virtually identical to the corresponding host finch eggs and the juvenile indigobirds have markings and begging behaviors that are nearly identical to those of their finch nestmates. In this way, the parasitic indigobirds trick the host finches into caring for their eggs and feeding their young!

Finally, an urgent and compelling case of parasitism is the evolution of HIV. Studies of the evolutionary history of HIV indicate that it has close relatives including SIV (simian immunodeficiency virus) that infects non-human primates and FIV (feline strains) that infects cats. Interestingly, SIV and FIV do not appear to have deleterious effects on their hosts. By understanding the relationships between these different parasite viruses and their human, non-human primate, and feline hosts, researchers hope to develop better treatments and, ultimately, vaccines against HIV [7].

Indeed, there are countless cases of coevolution that have been studied, both of mutually beneficial and parasitic types. How do biologists determine whether two taxa coevolved and, if there is evidence that they did, what did that coevolution look like? This is known as the *cophylogeny problem* and is the topic of this chapter.

 ## 2 The cophylogeny problem

While we will soon examine figs and wasps, gophers and lice, and finches and indigo-birds, let's begin with a simpler case of contrived taxa that we'll call Groodies and Cooties. (Google "Purves Groody" to learn about Groodies.)

Imagine that biologists have observed that Cooties are parasites of their Groody hosts and have constructed evolutionary histories, or *phylogenetic trees*, for Groodies and similarly for Cooties as shown in Figure 12.1.[2] The Groody tree is shown in black on the left and the Cootie tree is shown in blue on the right. From now on, we'll refer to one of the trees as the *host tree* (the Groody tree, in our example) and the other the *parasite tree* (the Cootie tree in this case).

The *nodes* in a tree represent hypothesized ancestral species. The end nodes, or "tips," of each tree represent the currently living, or *extant*, species. In Figure 12.1, we've given these names Groody 1 through 4 and Cootie 1 through 4. All the other nodes in the trees represent hypothesized species, named X, Y, Z in the Groody tree and x, y, z in the Cootie tree. More precisely, those nodes represent speciation events when the hypothesized ancestral species divided into two new species. Therefore, an *edge* in the tree can be thought of as the lifetime of the species with the node at the end

---

[2] The construction of phylogenetic trees is itself a fascinating and important field in computational biology, but here we'll assume that the phylogenetic trees have already been constructed using one of several known techniques.

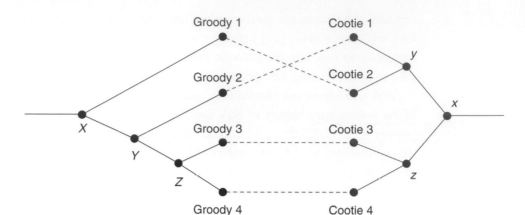

Figure 12.1 A tanglegram for Groodies and Cooties.

of that edge indicating the speciation event. Finally, the associations between the tips of the Groody and Cootie trees are indicated by dotted lines. A figure like this showing two phylogenetic trees and the associations between their tips is called a *tanglegram*.

You might expect that coevolution should imply that the Groody and Cootie trees are exactly identical. However, such perfect congruence almost never happens even for species that have coevolved. Figure 12.2(a) and (b) show two possible ways in which the species might have coevolved. In each case, the Cootie tree in blue is superimposed on the Groody tree in black. Each of these is called a *reconstruction* since it attempts to reconstruct the histories of the two species.

In the reconstruction in Figure 12.2(a), we see that Cootie speciation event z occurs at the same time as Groody event Z. This is called a *cospeciation* event and corresponds to two lineages speciating contemporaneously. For example, consider a species of louse living on a species of gopher. Imagine that the gopher species becomes geographically distributed with one population living in a warmer climate and another in a colder climate. Eventually, the gopher species splits into two new species, one with short hair and one with thick long hair adapted for the colder climate. The parasitic louse species may also split to specialize to the two new species of gophers – one new louse species may adapt to the short-haired gophers and the other to the thick long-haired gophers. In general, if two species coevolved, we would expect to see a significant number of cospeciation events between their two phylogenetic trees.

Notice that in Figure 12.2(a), events x and y in the Cootie tree occurred in the "prehistory" of the Groody species, that is, before the first inferred Groody speciation event. Speciation events in the Cootie tree that are not contemporaneous with speciation events in the host tree are called *duplications*. Duplications suggest that the Cootie speciation was independent of the Groody speciation, which does not contribute to

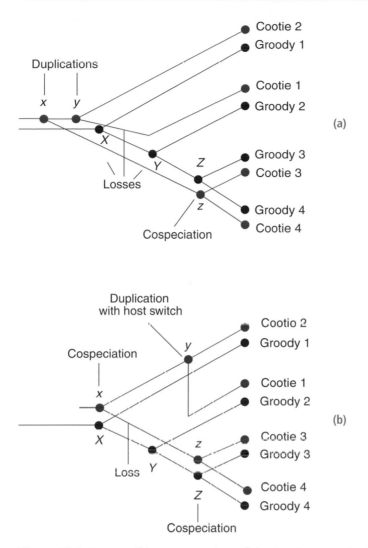

Figure 12.2 Two possible reconstructions of the Cootie tree on the Groody tree.

evidence of coevolution of the two species. Finally, the edge from y to Cootie 1 passes through X and Y as does the edge from x to z. These are called *loss* events. Loss events may be due to a failure of the Cootie lineage to speciate, or there may have been a speciation but one of the lineages became extinct.

The reconstruction in Figure 12.2(b) suggests another possible way in which the two species may have coevolved with two cospeciation events (x maps to X and z maps to Z), a loss event at Y, and a duplication event where y occurs independently of a speciation event in the Groody tree. Another interesting thing happens here: one of the two descendant lineages from y switches to a different part of the Groody tree.

This is called a *host switch*, or *horizontal transfer* event; such events are thought to be quite common in evolution. For example, it is known that one strain of HIV host switched from chimpanzees to humans sometime around the end of the nineteenth century [7].

There are many other possible reconstructions of these two phylogenetic trees and biologists would like to know which reconstructions, if any, are most plausible under the assumption that the two species coevolved. One approach is to estimate the relative likelihood of each of the four types of events (cospeciation, duplication, host switch, and loss) assuming coevolution has occurred and assign each such event a numerical "cost" so that likely events have low cost and unlikely ones have a higher cost. For example, cospeciation is a very likely event under the assumption that our two species coevolved, so the cost of this event might be 0 whereas duplication is a much less likely event and would therefore have some positive cost.

Now our objective becomes that of finding a reconstruction of minimum total cost under the given cost scheme. This is called the *cophylogeny reconstruction problem*. If there exists a reconstruction of very low cost, this gives strong evidence of coevolution. For example, imagine that cospeciation is assigned a cost of 0 and each duplication, host switch, and loss is assigned cost 1. Then, in the reconstruction in Figure 12.2(a), the total cost is 5 (2 duplications plus 3 losses), whereas in the reconstruction in Figure 12.2(b) the total cost is 3 (1 duplication, 1 loss, and 1 host switch). You might be wondering if there is a better reconstruction for the Groody and Cootie trees. The answer is yes, there is a reconstruction of cost 2 and you might want to pause here to find it. (Note that event x in the Cootie tree could be associated with something after X in the Groody tree. Moreover, the edge leading into x is not considered to be involved in loss events because we have no putative ancestor for x.)

Imagine that we enumerated every possible reconstruction of the Groody and Cootie trees and, for each one, we computed its total cost. We then selected the reconstruction of minimum total cost. In our example, that cost is 2. How do we know whether that cost of 2 suggests coevolution? Certainly, if the cost had been 0, we'd probably feel pretty confident that there was coevolution here because that would mean that the two trees were identical. However, is a cost of 2 suggestive of that as well?

One way to find out is to use a basic idea in statistical hypothesis testing. Specifically, we can formulate the *null hypothesis* that the two phylogenies and the associations of their tips were random. Under this hypothesis, we'd like to measure the probability that there was a reconstruction of cost 2 or less. We can do so by writing a computer program that generates random pairs of trees and associations between their tips.[3]

---

[3] There is some controversy on the issue of what should be randomized in such tests. Generally, the host tree is not modified but the parasite tree is randomized. Another school of thought is that neither tree should be changed but only the associations between the tips should be randomized.

Next, we find the reconstruction of least cost and record that value. We repeat this computational experiment some large number of times, say 100 times. Imagine that we did this and discovered that for 96% of these random pairs, the cost of a minimum reconstruction was 3 or higher and in only 4% were the minimum costs 2 or less. In this case, we would say that the *p-value* is 0.04 because the probability of doing at least as well as 2, assuming that the trees were just random, is 0.04. If the p-value is low (typically less than or equal to 0.05), then we can reject the null hypothesis that the pairs of trees were simply random.

3 Finding minimum cost reconstructions

Our statistical hypothesis testing depends on our ability to solve the cophylogeny reconstruction problem. Moreover, once biologists are confident that a pair of species coevolved, they would like to see what minimum cost reconstructions look like to get a sense of some plausible ways in which the species coevolved.

Unfortunately, there are far too many different possible reconstructions for a pair of phylogenetic trees for us to enumerate them all. The number of possible reconstructions for two trees, each with n tips, can be shown to be an exponential function of n. Just to get a sense of how bad that is, imagine that there were "only" 2^n possible reconstructions for a pair of host and parasite trees with n tips each. (The actual number of reconstructions can be significantly larger than this!) If we have a pair of trees with 100 tips each (small relative to some of the trees that biologists would like to evaluate), we have 2^{100} reconstructions to evaluate. Even if we had a supercomputer capable of examining a billion reconstructions per second, it would take over 40 *trillion* years to examine them all! Considering that the sun will burn out in about nine *billion* years, this is very very bad news.

"Let's just wait a few years for faster computers; they should be able to do the job!" you might be thinking to yourself. Let's explore that for a moment. Under the very optimistic assumption that computers get twice as fast every year, waiting 20 years would result in computers that are about one million times faster than they are now. With such a fast computer we could solve the problem for trees with 100 tips in a mere 40 *million* years! In the off chance that this seems like a significant improvement, consider that if we increased the number of tips in the trees from 100 to 120, we'd be back to taking 40 *trillion* years to solve the problem, even with our super-fast futuristic computer. Considering that biologists have developed cophylogeny data sets in which the trees have over 200 tips, it appears that we're in serious trouble if we try to solve the problem this way. The moral of this story is that computational methods that consider an exponential number of possibilities are useless for even relatively small phylogenetic trees.

For some computational problems, there are clever ways of finding the desired optimal solution without brute-force examination of every possible option. For example, you've probably used a program like Mapquest or Google Maps and asked for driving directions from one location to another. Those programs can find the shortest path between two locations without actually looking at every one of the large number of different paths. Computer scientists have found very clever algorithms that are absolutely guaranteed to find you a shortest path and the computation time is lightning fast.

It would be nice if this was possible for the cophylogeny reconstruction problem. Unfortunately, this appears to be very unlikely. The cophylogeny reconstruction problem was recently shown to be *NP-hard*, which essentially means that a fast algorithm for solving the cophylogeny reconstruction problem probably doesn't exist [8].

So what is to be done about the cophylogeny reconstruction problem? If the NP-hardness of the problem meant that there was absolutely no hope, then evolutionary biologists would be very disappointed and this chapter would be over. Fortunately, computational biologists have developed several strategies for solving the cophylogeny problem *reasonably well*. One approach is to try to use clever computational techniques to avoid examining certain reconstructions that can't be optimal. Professor Michael Charleston, at the University of Sydney in Australia, has developed a technique called *jungles* [9] that does exactly this. This approach still takes exponential time in many cases so it can only be used with relatively small trees. The technique has been implemented in a software tool called TreeMap [10].

Another approach is to use *heuristics*. A heuristic is a computational method that doesn't guarantee an optimal solution but foregoes optimality for efficiency. For example, Professors Daniel Merkle and Martin Middendorf at the University of Leipzig in Germany developed a very fast heuristic [11] used in a package called *Tarzan* [12]. (First there were jungles and then there was Tarzan.) Tarzan is known to find solutions that are not necessarily optimal and sometimes even finds solutions that don't quite make sense biologically (e.g. reconstructions that are impossible because they require a speciation event x to occur before another speciation event y but also for y to occur before x, creating an irreconcilable inconsistency). Nonetheless, Tarzan often finds very good solutions and can handle very large phylogenetic trees.

We have recently developed a different kind of heuristic for cophylogeny reconstruction that uses a paradigm, called *genetic algorithms*, that computer science has borrowed from biology. The irony here is that we are trying to use computational methods to solve a biological problem but the computational method was one that computer scientists learned from biology! Unlike jungles, but like Tarzan, our approach does not guarantee optimal solutions. However, our approach is guaranteed to always produce

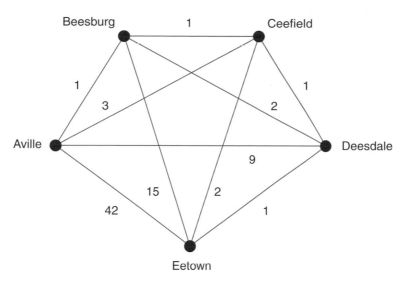

Figure 12.3 Cities and flight costs.

good and biologically reasonable solutions in a reasonable amount of time. Continuing the jungles and Tarzan theme, our software is called *Jane*. In section 5, we explain how Jane works. Then, you'll have a chance to try it out for the fig/wasp, gopher/louse, and finch/indigobird relationships. In the meantime, you can download Jane from http://www.cs.hmc.edu/~hadas/jane.

 ## Genetic algorithms

In this section we'll examine genetic algorithms. In the next, we'll see how Jane uses genetic algorithms to solve the cophylogeny problem. Finally, we'll use Jane to explore some real data in coevolution.

To explain the concept of a genetic algorithm – the key idea behind the Jane software – we now take a short aside to discuss a famous computational problem called the *Traveling Salesperson Problem*. The problem goes like this. Imagine that you are a salesperson who needs to travel to a set of cities to show your products to potential customers. The good news is that there is a direct flight between *every pair of cities* and, for each pair, you are given the cost of flying between those two cities. Your objective is to start in your home city, visit each city *exactly once*, and return back home. For example, consider the set of cities and flights shown in Figure 12.3 and imagine that your start city is Aville.

A tempting approach to solving this problem is to use an approach like this: starting at our home city, Aville, fly on the cheapest flight. That's the flight of cost 1 to Beesburg.

From Beesburg, we could fly on the least expensive flight to a city that we have not yet visited, in this case Ceefield. From Ceefield we would then fly on the cheapest flight to a city that we have not yet visited. (Remember, the problem stipulates that you only fly to a city once, presumably because you're busy and you don't want to fly to any city more than once – even if it might be cheaper to do so.) So now, we fly from Ceefield to Deesdale and from there to Eetown. Uh oh! Now, the constraint that we don't fly to a city twice means that we are forced to fly from Eetown to Aville at a cost of 42. The total cost of this "tour" of the cities is $1 + 1 + 1 + 1 + 42 = 46$. This approach is called a "greedy algorithm" because at each step it tries to do what looks best at the moment, without considering the long-term implications of that decision. This greedy algorithm didn't do so well here. For example, a much better solution that goes from Aville to Beesburg to Deesdale to Eetown to Ceefield to Aville has a total cost of $1 + 2 + 1 + 2 + 3 = 9$. In general, greedy algorithms are fast, but often fail to find optimal or even particularly good solutions.

It turns out that finding the optimal tour for the Traveling Salesperson Problem is very difficult. Of course, we could simply enumerate every one of the possible different tours, evaluate the cost of each one, and then find the one of least cost. However, there are a huge number (exponential or worse!) of different tours and this approach is not viable for even a moderate number of cities. Like the cophylogeny reconstruction problem, the problem is in the category of NP-hard problems – problems for which there is strong evidence that no fast algorithms exist. So, we are in the same predicament for the Traveling Salesperson Problem as for cophylogeny reconstruction.

Now for the clever idea that computer scientists borrowed from biology. Let's call the cities in Figure 12.3 by their first letters: A, B, C, D, and E. We can represent a tour by sequence of those letters in some order, beginning with A and with each letter appearing exactly once. For example, the tour Aville to Beesburg to Deesdale to Eetown to Ceefield and back to Aville would be represented as the sequence $ABDEC$. Notice that we don't include the A at the end because it is implied that we will return to A at the end.

Now, let's imagine a collection of some number of orderings such as $ABDEC$, $ADBCE$, $AECDB$, and $AEBDC$. Let's think of each such ordering as an "organism" and the collection of these orderings as a "population." Pursuing this biological metaphor further, we can evaluate the "fitness" of each organism/ordering by simply computing the cost of flying between the cities in that given order.

Now let's push this idea one step further. We start with a population of organisms/orderings. We evaluate the fitness of each organism/ordering. Now, some fraction of the most fit organisms "mate," resulting in new "child" orderings where each child has some attributes from each of its "parents." We now construct a new population of such children for the next generation. Hopefully, the next generation will be more fit – that is, it will, on average, have less expensive tours. We repeat this process for some number of generations, keeping track of the most fit organism (least cost tour) that we have found and report this tour at the end.

"That's a cute idea," we hear you say, "but what's all this about mating traveling salesperson orderings?" That's a good question – we're glad you asked! There are many possible ways we could define the process by which two parent orderings give rise to a child ordering. For the sake of example, we'll describe a very simple (and not very sophisticated) method; better methods have been proposed and used in practice.

Imagine that we select two parent orderings from our current population to reproduce (we assume that any two orderings can mate): $ABDEC$ and $ACDEB$. We choose some point at which to split the first parent's sequence in two, for example as $ABD|EC$. The offspring ordering receives ABD from this parent. The remaining two cities to visit are E and C. In order to get some of the second parent's "genome" in this offspring, we put E and C in the order in which they appear in the second parent. In our example, the second parent is $ACDEB$ and C appears before E, so the offspring is $ABDCE$.

Let's do one more example. We could have also chosen $ACDEB$ as the parent to split, and split it at $AC|DEB$, for example. Now we take the AC from this parent. In the other parent, $ABDEC$, the remaining cities DEB appear in the order BDE, so the offspring would be $ACBDE$.

In summary, a genetic algorithm is a computational technique that is effectively a simulation of evolution with natural selection. The technique allows us to find good solutions to hard computational problems by imagining candidate solutions to be metaphorical organisms and collections of such organisms to be populations. The population will generally not include every possible "organism" because there are usually far too many! Instead, the population comprises a relatively small sample of organisms and this population evolves over time until we (hopefully!) obtain very fit organisms (that is, very good solutions) to our problem.

Just as evolution makes no promises that it results in optimally fit organisms, this technique cannot guarantee that the solutions that it finds will be optimal. However, carefully crafted genetic algorithms have been shown to find very good solutions to some very hard problems. Now, let's see how these ideas are used in Jane.

5 How Jane works

Earlier, we noted that the cophylogeny reconstruction problem is computationally very hard; the only known approaches for solving this problem would take nearly an eternity. On the other hand, here's some good news: if we happen to know the order in which

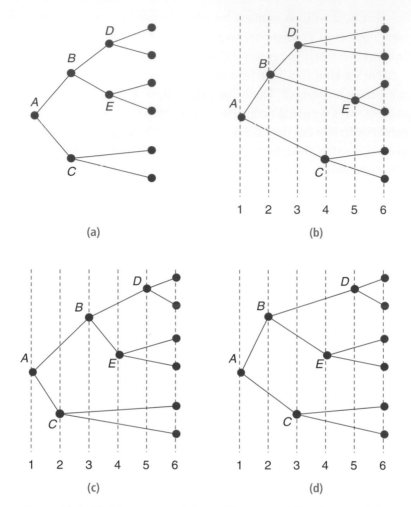

Figure 12.4 (a) A host tree and three different possible orderings of the speciation events shown in (b), (c), and (d).

speciation events occurred in the host phylogeny, the problem turns out to be solvable very quickly!

What do we mean by the order of the speciation events? Consider the host phylogeny shown in Figure 12.4(a). Obviously, speciation event A occurred before speciation events B and C. Similarly, speciation event B occurred before speciation events D and E. However, which speciation event occurred first: B or C? Similarly, did D occur before E, or vice versa? There are many possible orderings for these events and three of them are shown in Figure 12.4(b), (c), and (d). Recall that we assume that all of the tips of the tree occur at the same time – that is, at current time.

Surprisingly, if we happen to know the ordering of the speciation events in the host tree, even if we know nothing about the ordering of the events in the parasite tree, then we can find a least-cost solution in next-to-no-time using a clever computational technique called *dynamic programming* [8]. While we won't go into that technique here, it is one of the mostly widely used methods in computational biology. For example, sequence alignment, RNA folding, and various other computational biology problems can be solved using this technique. In the case of cophylogeny reconstruction, we can solve the problem in about one second (on a typical laptop computer) when the host and parasite trees have 100 tips each. That's fast!

"Wait a second!" we hear you exclaim. "Why does the ordering of the speciation events in the host tree matter at all?" Take a look again at Figures 12.4(c) and (d). In these figures let (A, C) denote the edge from node A to node C and let (B, E) denote the edge from node B to node E. Notice that in the ordering shown in (c), speciation event C occurs before speciation event B. Thus, a parasite that duplicates on edge (A, C) cannot host switch to edge (B, E) because (A, C) ends before (B, E) begins. On the other hand, in the ordering shown in (d), such a switch is possible because speciation event c occurs after speciation event B so edges (A, C) and (B, E) overlap in time. It might be that the best solution (the one that minimizes the total cost of the cospeciation, duplication, host switch, and loss events) requires a switch from (A, C) to (B, E), in which case the ordering in (c) might not be as "good" as the ordering in (d).

There's just one problem. How do we know the order in which the speciation events occurred in the host tree? If we're very lucky, we might have this information from the fossil record, but generally we will have little or no reliable information on the orderings of these events. Perhaps we could just try out all possible orderings of the host tree events and see which one permits us to find the best reconstruction of the parasite tree on the host tree? Unfortunately, there are way too many different orderings of the host (an exponential number, to be specific!), so that's totally impractical.

This is essentially the same problem that we had in the Traveling Salesperson Problem; there were too many possible orderings of the cities to explore them all. So, we used a genetic algorithm that kept a population that was a relatively small sample of the totality of all possible orderings and we artificially "evolved" better solutions.

The Jane software package does exactly this for the cophylogeny reconstruction problem. It starts with a population comprising some relatively small population of random orderings of the speciation events in the host tree as illustrated in Figure 12.5(a). For each such ordering of events in the host tree, we use our very fast dynamic programming algorithm to find the best solution for reconstructing the parasite tree on the host tree with this particular ordering of events. The cost of the best solution can be thought of as the fitness for that ordering. Figure 12.5(b) shows the orderings scored

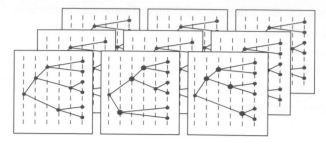

(a) The genetic algorithm maintains a population of "organisms," each of which is a different ordering of the events in the host tree.

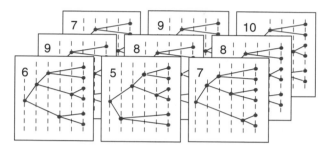

(b) A very fast dynamic programming algorithm is used to find the best reconstruction of the parasite tree onto each of the orderings of the host tree. The cost of that reconstruction is used as the fitness of that ordering. Example fitness scores are shown in the upper left corner of each ordering.

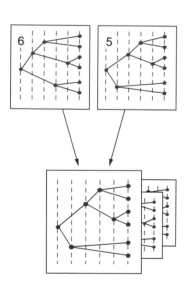

(c) Two orderings are chosen at random, but biased in favor of orderings with lower cost (better fitness). These orderings are then "mated" to construct a new offspring ordering that maintains some properties of its parent orderings. This offspring ordering is placed into the population for the next generation.

(d) The parents are placed back into their mating population and the mating process is repeated until a new population of orderings of the desired size is constructed. We now go back to step (a) using this new generation as the mating population.

Figure 12.5 The steps of the genetic algorithm used by Jane.

by their fitness. Keep in mind that in this context, a lower-cost solution is more fit than a higher-cost solution.

Next, we repeatedly choose pairs of orderings to "mate." While a pair of orderings is chosen at random, our random choice is biased to prefer more fit (lower-cost) orderings to less fit (higher-cost) ones. That is, we tend to prefer orderings of the speciation events in the host tree that permit us to find better solutions. We mate that pair of orderings in some way, resulting in a new ordering that preserves some attributes from each of its two parent orderings.[4] The offspring is a new ordering of the host tree events that has some attributes from each of its two parent orderings. Our hope is that this new ordering of the speciation events in the host tree might be one for which there exists an even better solution. This is illustrated in Figure 12.5(c).

We repeat this process of constructing new offspring orderings until we've built a population of new orderings of some desired size. This is our next generation as illustrated in Figure 12.5(d). We now start all over again with this new population serving as the mating population. This process is iterated for a user-specified number of generations. At the end, we report the best solutions that were found during this evolutionary process.

6   See Jane run

Now that we have an understanding of the computational challenge posed by the cophylogeny reconstruction problem, and the approach taken by Jane, let's try using Jane on some real cophylogeny data for figs and wasps and for gophers and lice. If you haven't done so already, download Jane from the website http://www.cs.hmc. edu/~hadas/jane. After you download it you can simply click on the the icon for that file and Jane will start up on your computer. From the Jane page, there is also a link that contains several example trees for you to download. One file is for figs and wasps, one is for pocket gophers and chewing lice, and one is for finches and indigobirds. You may also wish to read the Jane tutorial on the website, but the following is a self-contained demonstration of Jane in action.

Now click on Jane to start the program. You'll see the Jane window shown in Figure 12.6. In the "File" menu at the top of the Jane window, select "Open Trees" and find the Ficus-Ceratosolen.tree file that you downloaded from the Jane site. These are trees for figs and wasps that pollinate them. When the file loads, you'll see that the Jane window reports that the trees have 16 tips each. Notice that there are sliders in the Jane window that let you choose the "Number of Generations" (the number of generations of the genetic algorithm) and the "Population Size" (the number of tree orderings in each population maintained by the genetic algorithm). The defaults for both of these values are 30, which is fine for now. Click "Go" to start Jane running.

---

[4] We won't go into the details of the mating of orderings here, but if you're interested, you can find a detailed description online at [13].

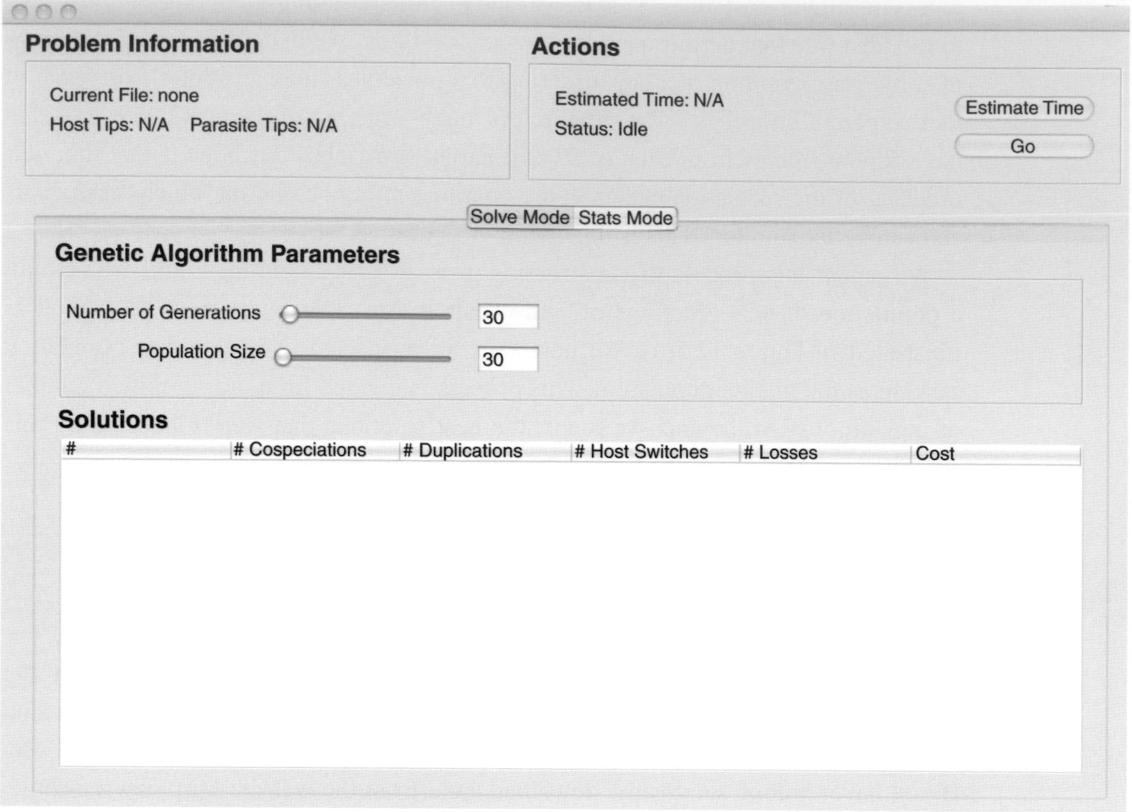

Figure 12.6 The Jane window.

Within a second or so, Jane will complete the genetic algorithm and will display a list of solutions in the "Solutions" window. (Since there is some randomness employed in the genetic algorithm, you won't necessarily get exactly the same solutions that are shown here, nor will you necessarily get the same solutions each time you run Jane.) Jane presents you with a list of best solutions that it found along with their costs. By default, Jane assumes that cospeciations have cost 0, duplications and host switches have cost 1, and losses have cost 2. While these values have been used in many studies, biologists often try to infer appropriate relative values of these costs from other biological data. The values of these parameters can be changed in the "Settings" menu in Jane.

Coming back to our example, you can see that these solutions had 9 cospeciations, 12 duplications, 6 host switches, and 1 loss for a total cost of $9 \times 0 + 12 \times 1 + 6 \times 1 + 1 \times 2 = 20$. These are valid solutions, but since Jane uses a heuristic, there is no guarantee that they are optimal solutions.

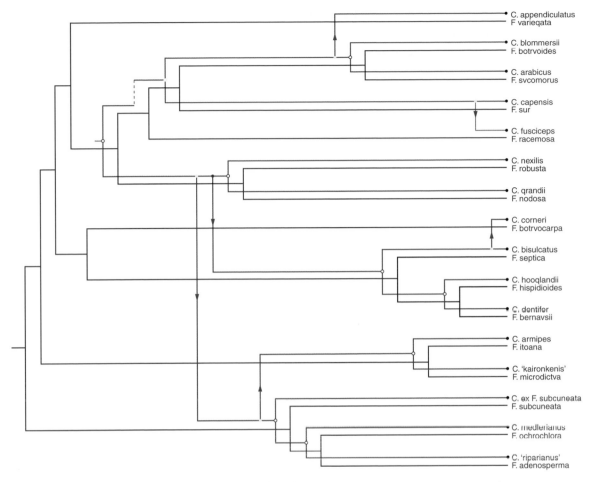

Figure 12.7 A sample solution found by Jane.

Now, click on a solution to see what it looks like. You will see a new window with a solution that might look something like the one shown in Figure 12.7. The black tree is the host tree and the blue tree is the parasite tree. The hollow dots indicate cospeciation events while the solid red dots indicate duplication events. Some duplication events are accompanied by host switches as can be seen by the edges with arrows on them. Finally, loss events are indicated by dashed lines. To learn more about the meaning of the colors of the nodes, please read the tutorial on the Jane website. (You might notice that there appear to be only 6 duplications rather than 12. In this cost model, each duplication actually counts as two duplications – one for each of the two child species that result from the duplication event.) Try this out for the gopher_louse.tree file that you downloaded.

Next, let's take a look at the finch and indigobird data set in the file Vidua.tree. The trees here are larger than the others that you've experimented with previously; the host tree has 33 tips and the parasite tree has 21 tips (some host species have no parasites). Open this file in Jane and, this time, choose the "Number of Generations" used in the genetic algorithm to be small – let's try 3 generations. Similarly, let's use a small population size in the genetic algorithm – let's make it 4. Click on "Go" and Jane will run its genetic algorithm for 3 generations with 4 orderings per generation. You'll see some solutions reported in the "Solutions" window – these are the best solutions that resulted from our artificial evolution of solutions in this case. Note the cost of these solutions.

As biologists, we know that natural selection works slowly and more effectively in large populations. So, let's now increase the "Number of Generations" to a larger value – say 20 – and let's increase the size of the population in each generation to something larger as well, perhaps 100. Now, click "Go" again. The old solutions will still be listed here, but below them will be the new solutions found from this longer and larger evolutionary simulation. Take a look at the cost of these solutions! You should see that much better solutions were found in this second run.

Now, you can perform a statistical experiment to get a sense of whether or not the cost of the best solution found by Jane is suggestive of coevolution. More precisely, you can test the null hypothesis that the best solution found for the observed data – that is, the least-cost mapping of the given parasite tree onto the host tree given the observed mapping between the tips of the parasite tree and the tips of the host tree – is no better than we would find for random trees and tip mappings. If that's true, then the case for coevolution for these species is weak. If it's false, we are likely to accept that coevolution was at work here.

To try this out for yourself, click on the "Stats Mode" tab in the middle of the Jane window. By clicking "Go," Jane will find the best solution it can for the observed data and compare it with the best solution it can find for 50 random samples, each of which is the same pair of trees but with a completely random mapping between the tips of the host and parasite trees. The histogram at the bottom right shows the costs of the 50 samples: our original tip mapping is indicated in the histogram in red and the 50 random mappings are indicated by blue bars. If the majority of the random samples have higher cost than the original mapping, it is likely that the low cost for the observed tip mapping is not due to randomness. In particular, if 5% or fewer of the random solutions are better than the observed, this is considered strong evidence against the null hypothesis. Notice that you can change the sample size from 50 to any value that you like. Try it!

You can also test an alternative null hypothesis that the solution for the observed data is no better than random when the parasite tree *and* the tip mapping are randomized.

To do so, click on the "Random Parasite Tree" button in the "Statistical Parameters" panel and then press "Go" again. Now, try these computational experiments all over again with the other data sets. You will discover that, indeed, the case for coevolution is very compelling in each case.

DISCUSSION

This chapter has explored aspects of the field of cophylogeny – the study of the evolutionary associations of species. Since we can't travel backwards in time to study these relationships *in vivo*, we do the next best thing and study them *in silico* – that is, using computational methods. We've explored one computational approach for cophylogeny reconstruction and the *Jane* software that uses this approach.

Using computational tools, biologists are developing a better understanding of how parasites such as HIV and malaria have coevolved with their primate hosts which may ultimately lead to new approaches to combatting these diseases. Professor Michael Charleston, one of the leading researchers in the field of cophylogeny writes: "The global melt-down of ecological diversity is leading to greater chances of unrelated organisms interacting, leading in turn to greater potential of new pathogens crossing the species barrier into the human population. Understanding the way in which such cross species transmissions occur is of fundamental importance and it is through phylogenetic tools such as cophylogenetic maps which will shed the light we need."[14]

In addition to this pragmatic need, cophylogeny allows us to explore some of the beautiful and surprising ways that nature works, as Darwin himself imagined over 150 years ago.

QUESTIONS

(1) The Jane website (http://www.cs.hmc.edu/~hadas/jane) contains a number of sample host and parasite trees, including several that were discussed in this chapter. If you haven't done so already, download the "Ficus and Ceratsolen" file (called Ficus-Ceratosolen.tree) for the fig/wasp mutualism. Open this file in Jane and you will see in the upper-left corner of the Jane panel that these trees both have 16 tips.

(a) Use Jane to find solutions for this pair of trees. You may use the default settings of 30 generations and a population size of 30. Jane will present a number of different solutions found. Click on a solution to view it. Then, click on another solution to view it. Finally, click on a third solution. You will now have three solution windows open. These solutions will differ in some places but will agree in others. Describe where these solutions differ.

(b) Next, enter "Stats Mode" and click the "Go" button. Take a look at the histogram produced. The dashed red line shows the cost of the best solution found for the original data and the blue bars indicate the best solutions found for 50 random samples. What do these results suggest?

(2) Using the Ficus–Ceratosolen data set, make a note of the number of cospecation, duplication, host switch, and losses in the solutions found by Jane. (If you are still in "Stats Mode," you will need to go back to "Solve Mode" to do this.) Jane allows biologists to set the relative costs of each of these four event types. This is done by clicking on the "Settings" menu and selecting "Set Costs." (You will be asked if you would like to clear the solution table. Click "Yes".) Now, change the cost of a loss (sorting) event from 2 to 1, click "Go" to re-solve the problem, and note the number of each of the four event types used in the best solutions found. Explain why the solutions to the first case differ from the second case.

(3) Do a web search for "cophylogeny" and/or "host parasite" to find at least one more example of a host-parasite system. Briefly describe this system and the results found by the authors.

REFERENCES

[1] B. Meeuse and S. Morris. *The Sex Life of Flowers*. Facts on File, 1984.

[2] figweb. http://www.figweb.org/.

[3] G. D. Weiblen and G. W. Bush. Polination in fig pollinators and parasites. *Molec. Ecol.*, 11:1573–1578, 2002.

[4] M. S. Hafner and S. A. Nadler. Phylogenetic trees support the coevolution of parasites and their hosts. *Nature*, 332:258–259, 1988.

[5] J. DaCosta and M. Sorenson. http://www.indigobirds.com.

[6] M. D. Sorenson, C. N. Balakrishnan, and R. B. Payne. Clade-limited colonization in brood parasitic finches (*Vidua* spp.). *System. Biol.*, 53:140–153, 2004.

[7] Understanding evolution: HIV's not-so-ancient history. http://evolution.berkeley.edu/evolibrary/news/081101_hivorigins.

[8] R. Libeskind-Hadas and M. Charleston. On the computational complexity of the reticulate cophylogeny reconstruction problem. *J. Comput. Biol.*, 16(1):105–117, 2009.

[9] M. Charleston. Jungles: A new solution to the hostparasite phylogeny reconciliation problem. *Math. Biosci.*, 149:191–223, 1998.

[10] Michael Charleston. TreeMap. http://www.it.usyd.edu.au/ mcharles/software/treemap/ treemap.html.

[11] D. Merkle and M. Middendorf. Reconstruction of the cophylogenetic history of related phylogenetic trees with divergence timing information. *Theor. Biosci.*, 123(4):277–299, 2005.

[12] D. Merkle and M. Middendorf. Tarzan. http://pacosy.informatik.uni-leipzig.de/pv/ Software/Tarzan/PV-Tarzan.engl.html.

[13] C. Conow, D. Fielder, Y. Ovadia, and R. Libeskind-Hadas. Jane: A new tool for cophylogeny reconstruction problem. *Algorith. Mol. Biol.*, 5(16), 2010. http://www.almob.org/content/5/ 1/16.

[14] M. Charleston. Principles of cophylogeny maps. In M. Lässig and A. Valleriani (eds) *Biological Evolution and Statistical Physics.* Springer-Verlag, 2002.

CHAPTER THIRTEEN

Big cat phylogenies, consensus trees, and computational thinking

Seung-Jin Sul and Tiffani L. Williams

Phylogenetics seeks to deduce the pattern of relatedness between organisms by using a phylogeny or evolutionary tree. For a given set of organisms or taxa, there may be many evolutionary trees depicting how these organisms evolved from a common ancestor. As a result, consensus trees are a popular approach for summarizing the shared evolutionary relationships in a group of trees. We examine these consensus techniques by studying how the pantherine lineage of cats (clouded leopard, jaguar, leopard, lion, snow leopard, and tiger) evolved, which is hotly debated. While there are many phylogenetic resources that describe consensus trees, there is very little information regarding the underlying computational techniques (such as sorting numbers, hashing functions, and traversing trees) for building them written for biologists. The pantherine cats provide us with a small, relevant example for exploring these techniques. Our hope is that life scientists enjoy peeking under the computational hood of consensus tree construction and share their positive experiences with others in their community.

Bioinformatics for Biologists, ed. P. Pevzner and R. Shamir. Published by Cambridge University Press.
© Cambridge University Press 2011.

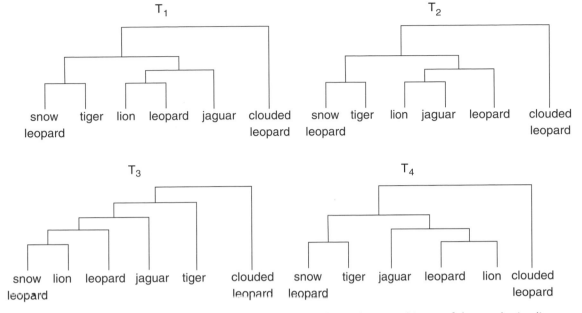

Figure 13.1 Four phylogenies representing the evolutionary history of the pantherine lineage. Trees T_1, T_2, T_3, and T_4 were published by Johnson *et al.* in 1996 [6], Johnson *et al.* in 2006 [7], Wei *et al.* in 2009 [8], and Davis *et al.* in 2010 [3], respectively. Each tree was reconstructed using different biological data. For all trees, the clouded leopard is the most distantly related taxon and serves as the outgroup to root each tree.

1 Introduction

For millennia, scholars have attempted to understand the diversity of life, scrutinizing the behavioral and anatomical form of organisms (or taxa) in search of the links between them. These links (or evolutionary relationships) among a set of organisms form a phylogeny, which served as the only illustration for Charles Darwin's landmark publication *The Origin of Species*. Phylogenetic trees most commonly depict lines of evolutionary descent and show historical relationships, not similarities [1]. That is, evolutionary trees communicate the evolutionary relationships among elements, such as genes or species, that connect a sample of taxa. Figure 13.1 shows several phylogenies that hypothesize how the pantherine lineage of cats (clouded leopard, jaguar, leopard, lion, snow leopard, and tiger) evolved. The evolution of these big cats is hotly debated [2, 3]. Being one of the most threatened of all carnivore groups, we must understand all that we can about these great cats. The true phylogeny for a group of taxa such as the pantherine cats can only be known in rare circumstances (for example, where the pattern of evolutionary branching is created in the laboratory and observed directly as it occurs [4]). Since fully resolved and uncontroversial phylogenies are rare, the generation, testing, and updating of evolutionary hypotheses is an active and highly debated area of research [5].

In this chapter, we examine how to summarize the different hypotheses reflected in a group of phylogenetic trees into a single, evolutionary history (or consensus tree). We use the phylogenies of the pantherine lineage of cats as the basis for understanding evolutionary trees and constructing their consensus. The appealing feature of consensus trees is that life scientists can study a single tree with the most robust branching patterns of how the taxa evolved from a common ancestor. While there is some debate over the use of consensus trees [9], they remain critical for phylogenetics.

Many references exist to describe the numerous types of consensus tree approaches [9–11]. Unfortunately, little information is provided to help life scientists understand the computational ideas behind the algorithms. The consensus tree problem encompasses several fundamental computational concepts, such as sorting branching patterns, hashing functions, and traversing trees. Computational thinking [12] is a new way of solving problems that leverages fundamental concepts in computer science. Furthermore, computational thinking is very relevant for life scientists. In a recent report [13], the Committee on Frontiers at the Interface of Computing and Biology for the National Research Council concluded that computing and biology have converged and that "Twenty-first century biology will be an information science, and it will use computing and information technology as a language and a medium in which to manage the discrete, nonsymmetric, largely nonreducible, unique nature of biological systems and observations." We hope that by providing a window into the underlying algorithms behind building consensus trees, life scientists will appreciate the computational ideas involved in solving biological problems and share their experiences with their interdisciplinary colleagues.

 ## 2   Evolutionary trees and the big cats

The pantherine lineage diverged from the remainder of modern Felidae less than 11 million years ago. The pantherine cats consist of the five big cats of the genus *Panthera*: *P. leo* (lion), *P. tigris* (tiger), *P. onca* (jaguar), *P. pardus* (leopard), and *P. uncia* (snow leopard), as well as the closely related *Neofelis* species (clouded leopards), which diverged from *Panthera* approximately six million years ago. These cats have received a great deal of scientific and popular attention because of their charisma, important ecological roles, and conservation status due to habitat destruction and over-hunting. Dissimilar patterns of diversification, evolutionary history, and distribution

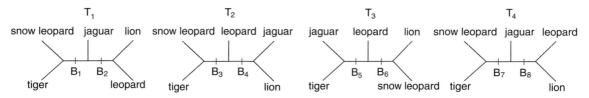

Figure 13.2 Unrooted phylogenies of the *Panthera* genus based on the trees in Figure 13.1.

make these species useful for characterizing genetic processes. Furthermore, extensive descriptive information is available on their natural histories, morphology, behavior, reproduction, evolutionary history, and population genetic structure, which provides a rich basis for interpreting genetic data.

Despite their highly threatened status, the evolutionary history of these cats has been largely obscured. The difficulty in resolving their phylogenetic relationships is a result of (i) a poor fossil record, (ii) recent and rapid radiation during the Pliocene, (iii) individual speciation events occurring within less than one million years, and (iv) probable introgression between lineages following their divergence [3]. Multiple groups have attempted to reconstruct the phylogeny of these cats using morphological as well as biochemical and molecular characters. However, there is great disparity between these phylogenetic studies.

2.1 Evolutionary hypotheses for the pantherine lineage

Davis *et al.* [3] show 14 phylogenetic trees (including the tree that they reconstructed) from different studies of these cats. Figure 13.1 shows 4 of the 14 pantherine trees in the Davis *et al.* work. Trees T_1, T_2, and T_4 produce the hypothesis that the *Panthera* genus is composed of two main clades consisting of (i) snow leopard and tiger, and (ii) jaguar, leopard, and lion. Furthermore, in trees T_1 and T_4, lion and leopard are sister taxa with jaguar sister to these species. Tree T_3 shows a completely different evolutionary picture, in which snow leopard and lion are sister taxa. Based on numerous phylogenetic studies, clouded leopard is assumed to be the most distantly related species and serves as the outgroup taxon in order to root the phylogenetic tree. However, the relationships among the five big cats of the *Panthera* genus are still under debate given the numerous incongruent findings by scientists. Thus, unrooted trees are used to focus attention on the big cats in the *Panthera* genus as shown in Figure 13.2.

The resulting consensus trees for the *Panthera* genus are shown in Figure 13.3. While there are a variety of approaches for building consensus trees, we concentrate on majority and strict consensus trees, which are the most commonly used approaches. Majority consensus trees consist of those branching patterns that exist in a majority of the trees. Strict consensus trees contain evolutionary relationships that appear in all of

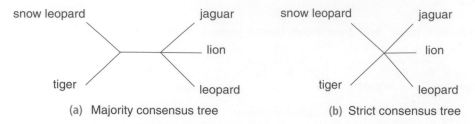

Figure 13.3 Majority and strict consensus trees of the *Panthera* genus of big cats based on unrooted trees shown in Figure 13.2.

the trees. For example, one branching pattern that appears in the majority tree is the relationship that shows snow leopard and tiger as sister taxa, which appears in three of the four trees in Figure 13.2. Instead of looking at all four pantherine trees, one simply examines the consensus trees to understand the evolutionary relationships among the taxa.

Finally, we note that while we show topological conflict among phylogenetic studies performed by different research groups, there can also be topological conflict within the same phylogenetic study. Such conflicts are often resolved using consensus trees as well.

2.2   Methodology for reconstructing pantherine phylogenetic trees

Below, we summarize how the four trees shown in Figure 13.1 were reconstructed. Although each of the studies below were conducted on the pantherine lineage of cats, no one phylogenetic study was performed in exactly the same manner.

2.2.1   Tree T_1: Johnson, Dratch, Martenson, and O'Brien

Tree T_1 is based on RFLP (Restriction Fragment Length Polymorphisms) of complete mitochondrial DNA (mtDNA) genomes using 28 restriction endonucleases [6]. Johnson, Dratch, Martenson, and O'Brien believed that mtDNA has several traits which make it useful for phylogenetic analysis, including nearly complete maternal, clonal inheritance, a general lack of recombination, and a relatively rapid rate of evolution, and that RFLP analysis has the advantage of rapidly sampling the entire mitochondrial genome. In their study, estimated sizes of fragments were summed for general concordance with domestic cat mitochondrial DNA, which has a length of 17 kb, disregarding putative nuclear mitochondrial (numt) DNA fragments. Percentage inter-species variation was estimated using FRAG_NEW. Phylogenetic relationships among individuals within each set of RFLP data were constructed from the distance data by the minimum-evolution method estimated by the Neighbor Joining algorithm [14] implemented in PHYLIP [15], and from the character data using the Dollo parsimony model implemented in PAUP* [16], followed by the bootstrapping option with 100 resampling. For comparison, trees were also reconstructed by maximum parsimony using PAUP*.

2.2.2 Tree T_2: Johnson, Eizirik, Pecon-Slattery, Murphy, Antunes, Teeling, and O'Brien

Johnson *et al.* [7] found tree T_2 using the largest molecular database to date, consisting of X- and Y-linked DNA, autosomal DNA, and mitochondrial DNA sequences, which consisted of 19 autosomal, 5 X, 4 Y, 6 mtDNA genes (23,920 bp) sampled across 37 living felid species plus 7 outgroup species representing each feliform carnivoran family. They present a phylogenetic analysis for nuclear genes (nDNA). First, the eight Felidae lineages are strongly supported by bootstrap analyses and Bayesian posterior probabilities (BPP) for the nDNA data and most of the other separate gene partitions. Second, the four species previously unassigned to any lineage have been placed, and the hierarchy and timing of divergences among the eight lineages are clarified. Third, the phylogenetic relationships among the non-felid species of hyenas, mongoose, civets, and linsang corroborate previous inferences with strong support.

2.2.3 Tree T_3: Wei, Wu, and Jiang

Tree T_3 was found by Wei, Wu, and Jiang [8] based on 7 mtDNA genes (3,816 bp). They constructed the tree based on the concatenated 7 mtDNA genes from 10 species with the data set obtained from GenBank. Maximum likelihood using PAUP* and Bayesian inference using MrBayes [17] were used for the reconstruction of the phylogenetic tree. Their result indicated that snow leopard and tiger are sister taxa, which is incongruent with previous findings.

2.2.4 Tree T_4: Davis, Li, and Murphy

Most recently Davis, Li, and Murphy [3] published tree T_4 using intronic sequences contained within single-copy genes on the felid Y chromosome which was combined with previously published data from Johnson *et al.* [7], and newly generated sequences for four mitochondrial and four autosomal genes, highlighting areas of phylogenetic incongruence. More specifically, they sequenced the 12S, CYTB, ND2, and ND4 gene segments using in-house DNAs with reagent and thermal cycler protocols. Their 47.6 kb combined data set was analyzed as a supermatrix with respect to individual partitions using maximum likelihood and Bayesian phylogenetic inference, in conjunction with Bayesian estimation of species trees (BEST) [18, 19] which accounts for heterogeneous gene histories. They emphasized that the Y chromosome has a very low level of homoplasy in the form of convergent, parallel, or reversal substitutions and renders the vast majority of substitutions phylogenetically informative. Their analysis fully supported the lion and leopard as sister taxa with the jaguar being sister to these species. In Figure 13.1, Tree T_1 by Johnson *et al.* and tree T_4 by Davis *et al.* are identical trees but reconstructed over different phylogenetic data.

2.3 Implications of consensus trees on the phylogeny of the big cats

The majority consensus tree in Figure 13.3(a) shows that the four phylogenetic studies considered in this chapter agree that there are two distinct clades of the big cats. Lions, leopards, and jaguars share a specific set of common characteristics that distinguish them from the second clade consisting of tiger and snow leopard. Moreover, this majority consensus tree agrees with studies by Hemmer that examined morphological, ethological, and physiological features [20]. The analysis of excretory chemical signals by Bininda-Emonds *et al.* [21] also supports these two distinct clades. Davis et al. [3] state that published molecular studies that failed to fully support this two clade distinction (lion–leopard–jaguar and tiger–snow leopard) probably relied heavily on mtDNA sequences that had not been vetted as true cytoplasmic mitochondria (cymt) amplifications, suffered from species misidentification, or lacked sufficient phylogenetic signal. The strict consensus tree in Figure 13.3(b) shows a star tree topology and gives us no information regarding the evolution of the big cats. Even if 99.9% of the trees agree on a clade, it would not appear in the strict consensus tree. Hence, majority trees are preferred over their strict counterparts.

3　Consensus trees and bipartitions

As shown in Figure 13.2, there is incongruence among the trees across different phylogenetic studies of the *Panthera* genus. While we are able to build a consensus tree by hand for this small data set, much larger trees are also of interest to the phylogenetic community. For example, Janecka *et al.* [22] analyzed 8,000 trees on 16 Euarchontoglires using MrBayes [17]. Hence, we need computational approaches for building consensus trees – especially as the size of phylogenetic studies continues to increase. The key to computational approaches for constructing majority and strict consensus trees is identifying the shared evolutionary relationships (or bipartitions) among a group of trees.

Table 13.1 The bipartitions and their bitstring representations for the trees in Figure 13.2. The bistrings are based on the taxa being in the following order: snow leopard, tiger, jaguar, lion, and leopard, where snow leopard represents the first bit, tiger the second bit, etc. TID and BID represent tree and bipartition indexes, respectively.

TID	BID	Bipartition	Bitstring
T_1	B_1	{snow leopard, tiger | jaguar, lion, leopard}	11000
	B_2	{snow leopard, tiger, jaguar | lion, leopard}	11100
T_2	B_3	{snow leopard, tiger | leopard, jaguar, lion}	11000
	B_4	{snow leopard, tiger, leopard | jaguar, lion}	11001
T_3	B_5	{snow leopard, lion | leopard, jaguar, tiger}	10010
	B_6	{snow leopard, lion, leopard | jaguar, tiger}	10011
T_4	B_7	{snow leopard, tiger | jaguar, leopard, lion}	11000
	B_8	{snow leopard, tiger, jaguar | lion, leopard}	11100

3.1 Phylogenetic trees and their bipartitions

Let $\mathcal{T}$ represent the set of trees of interest that we want to summarize into a single consensus tree. For example, in Figure 13.2, $\mathcal{T} = \{T_1, T_2, T_3, T_4\}$. The branches (or bipartitions) of interest in the trees are denoted by vertical bars. In tree T_1, there are two bipartitions labeled B_1 and B_2. If we remove the bipartition B_1, then the tree will be split into two pieces. One part of the tree will have snow leopard and tiger. The other side will contain jaguar, lion, and leopard. We will represent this bipartition B_1 as {snow leopard, tiger | jaguar, lion, leopard}, where the vertical bar separates the taxa from each other. Bipartition B_2 represents the bipartitions {snow leopard, tiger, jaguar | lion, leopard}. For any bipartition, how taxa are ordered on a particular side of the tree has no impact on its meaning. That is, {tiger, snow leopard, jaguar | leopard, lion} is another valid representation of bipartition B_2.

Table 13.1 provides a listing of the bipartitions for each of the four trees. Each tree has two bipartitions. Every evolutionary tree is uniquely and completely defined by its set of bipartitions. That is, bipartitions B_5 and B_6 can only define the relationships in tree T_3. It is not possible for two different trees to have the same bipartitions. If two trees share the same bipartitions, then they are equivalent. So, based on Table 13.1, trees T_1 and T_4 are identical, although in Figure 13.2 they are drawn differently in terms of the placement of the lion and leopard taxa names.

Finally, we note that the bipartitions in Figure 13.2 are non-trivial bipartitions. Trivial bipartitions are bipartitions that every tree is guaranteed to have. These are branches that connect to a taxon such as {snow leopard | tiger, jaguar, lion, leopard}, {jaguar |

snow leopard, tiger, lion, leopard}, etc. Every tree must have n of these bipartitions, where n is the number of taxa. In order to build a consensus tree, every input tree must be over the same taxa set, which results in every tree having the same set of trivial bipartitions. Thus, we do not consider trivial bipartitions in our explanation of algorithms for building consensus trees.

3.2   Representing bipartitions as bitstrings

A convenient way to represent a bipartition is as a bitstring. Each taxon will be represented by a bit, which means that the bitstring length will be equal to the number of taxa in our trees. Taxa that are on the same side of the tree receive the same bit value of either a "0" or a "1." To use a bitstring notation, we need to establish the ordering of the taxa. Any ordering will do as long as the taxa names are not duplicated. We choose the following taxa ordering: snow leopard, tiger, jaguar, lion, and leopard. So, snow leopard will represent the first leftmost bit, tiger the second leftmost bit, jaguar the third leftmost bit, etc. In Figure 13.2, bipartition B_2, which is {snow leopard, tiger, jaguar | lion, leopard}, would be represented by the bitstring 11100. Here, taxa on the same side of a bipartition as taxon snow leopard receive a "1." For every bipartition shown in Figure 13.2, Table 13.1 also shows its shorter bitstring representation. PAUP* [16], a general-purpose software package for phylogenetics, uses the symbols "." and "*" (instead of "0" and "1") to represent bipartitions when outputting them to the user.

4   Constructing consensus trees

The consensus tree algorithm consists of the following three steps: (i) collecting bipartitions from a set of trees, (ii) selecting consensus bipartitions, and (iii) constructing the consensus tree. Steps 1 and 3 are the same regardless of whether a majority or strict consensus tree is the desired result. For step 2, if a majority tree is desired, then the consensus bipartitions are those that appear in over half of the trees. For strict trees, consensus bipartitions appear in all of the trees. In the subsections that follow, our examples will be based on building a majority consensus tree. The examples can be adapted easily to accommodate building strict consensus trees.

4.1   Step 1: collecting bipartitions from a set of trees

Our first step in building a majority consensus tree is collecting all of the bipartitions from the phylogenetic trees of interest. For our big cats example, it is not difficult to list the bipartitions in the trees by hand. However, for larger trees, we would like a

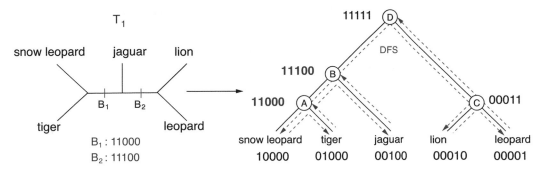

Figure 13.4 Using depth-first traversal to collect the bipartitions from tree T_1.

computational procedure to make the task easier. Consider Figure 13.4. The left side of the figure shows tree T_1 and the two bitstrings that represents its bipartitions. The right side of the figure shows how to obtain those bitstrings.

First, we root tree T_1 arbitrarily, which in this example is at bipartition B_2. A rooted tree allows us to use a depth-first traversal of the tree to obtain the bipartitions systematically. Second, we initialize each taxa with a 5-bit bitstring to represent the trivial bipartitions. Starting at node D, we visit each left-hand side node ($D \rightarrow B \rightarrow A$). Upon reaching node A, we gather the bitstrings of its children (snow leopard and tiger bitstrings) and OR them together. Computing the OR between the two child bipartitions requires visiting each of the five columns of these two bitstrings. To compute the OR operation, if one of the children's bits in column j is a "1," then a "1" bit is produced for column j in the bitstring representation of the parent. The result of the OR operation at node A produces a bitstring of 11000, which reflects that snow leopard and tiger are on one side of the tree and jaguar, lion, and leopard are on the other side of the tree. Moreover, bitstring 11000 is also identified as bipartition B_1 in tree T_1.

After visiting node A, we return to node B since we know node A's bitstring. The result of the OR operation on the bitstrings of node A and the jaguar bitstring results in a bitstring of 11100 for node B. Next, we return to node D to get its bitstring, but we do not yet know the bitstring of node C. Once the bitstring of node C is known (which is 00011), then we can compute the bitstring for the root node D, which is 11111. Given that this is a star bitstring, we do not collect it explicitly, but we do take advantage of its presence in our consensus tree building routine described in Section 4.3. The root node's bitstring will always consist of 1s since there is no division of the taxa on a particular side of the tree. Notice that the bipartition for node C is the exact complement of the bitstring for node A. Both of these bitstrings represent the bipartition {snow leopard, tiger | jaguar, lion, leopard}. As a result, both of these bipartitions are not needed, and node C's bitstring is thrown out since we assume that

Table 13.2 Processing the bitstrings from Table 13.1. The first (leftmost) column puts the bitstrings in order based on the trees they originated from. The first column also shows the value of the conversion from a bitstring (binary number) to a decimal value. The second (middle) column puts the bitstrings in sorted ascending order based on their decimal value, and the final (rightmost) column removes the redundant bitstrings and shows the frequency that each unique bitstring or bipartition appeared in the trees.

Unsorted		Sorted		Sorted and filtered	
Bitstring	Value	Bitstring	Value	Bitstring	Frequency
B_1: 11000	24	B_5: 10010	18	10010	1
B_2: 11100	28	B_6: 10011	19	10011	1
B_3: 11000	24	B_1: 11000	24	11000	3
B_4: 11001	25	B_3: 11000	24	11001	1
B_5: 10010	18	B_7: 11000	24	11100	2
B_6: 10011	19	B_4: 11001	25		
B_7: 11000	24	B_2: 11100	28		
B_8: 11100	28	B_8: 11100	28		

any taxa on the same side of snow leopard will be represented by a "1" bit. Node C assumes the opposite.

The above depth-first traversal procedure is applied to each tree to obtain all of the bipartitions across the trees. For this example, there are eight total bipartitions.

4.2 Step 2: selecting consensus bipartitions

4.2.1 Our first selection algorithm: sorting bitstrings

Once we have collected all of the bipartitions, then we are in a good position to select the majority bipartitions, which we will later use to build the majority consensus tree. Table 13.2 shows the results of this stage of the algorithm in the leftmost column. We use our shorthand bitstring notation to represent the bipartitions. Every bitstring is a binary number that can be represented by a decimal value. The rightmost bit has a decimal value of 2^0 or 1, the second rightmost bit has a value of 2^1 or 2, etc. For example, the bitstring 11000 for bipartition B_1 is $1 \cdot 2^4 + 1 \cdot 2^3 + 0 \cdot 2^2 + 0 \cdot 2^1 + 0 \cdot 2^0$ or a decimal value of 24.

Next, we sort the collected bipartitions according to their decimal representations. The second column of Table 13.2 shows the result. Given the sorted bitstrings, it is easier to find the frequencies of the bipartitions. First, we start a new empty list to store unique bipartitions. Then, we scan our sorted list, starting at our first sorted bipartition. We copy this bipartition to our list of unique bipartitions and set the frequency count of this bipartition to 1. We visit the next bipartition in the sorted list. If it is the same bipartition that we just visited, then we increment its frequency counter in the unique bipartition list by 1. If it is not the same, then we have found a new unique bipartition, and copy it to the unique bipartition list, and we initialize its frequency count to 1. We repeat the above process until all bipartitions in our sorted list have been processed.

The final column of Table 13.2 shows the result of filtering the unique bipartitions and the resulting frequency counts. There are four unique bipartitions out of the eight processed. The only majority bipartition is 11000 (or {snow leopard, tiger | jaguar, lion, leopard}), which occurs three times in the input trees. From our list, we can also see that the bipartition {snow leopard, tiger, jaguar | lion, leopard} represented by bitstring 11100 appeared twice, which was not enough for it to be a majority bipartition. We'll discuss how to use the majority bitstrings to build a majority tree in Section 4.3.

4.2.2 Our second selection algorithm: using hash tables

Now that we have a technique for finding the majority bipartitions within a set of trees, can we do better? Our first approach collected the bipartitions from each of the trees, sorted them, and ended with a filtering process to collect the unique bipartitions and their frequency. In Table 13.1, the first column is the input to constructing a majority consensus tree. The final column is the desired output in terms of producing a frequency table of the unique bipartitions. Is it possible to get rid of the sorting step (the second column) so that we can perform the computation faster?

In our second attempt at constructing majority consensus trees, we will use a technique known as hashing in order to get rid of the sorting step in our first selection algorithm. A few algorithms [23, 24] have been developed that leverage the power of hash functions to construct consensus trees. A hash function examines the input data (hash keys) and produces an output hash value (or code). For us, the input data are the list of bipartitions. The output data are the list of unique bipartitions. The advantage of hashing is that each time we put our data through the hash function we know exactly where to find it in the table. In our first selection algorithm, once we put the bitstrings in the table, we had to perform a number of steps to organize the list later so that it would be useful. With hash tables, our hashing function will keep our data organized and quickly accessible.

Figure 13.5 shows an example of how to use hash tables to organize the bipartitions of our big cat trees. We have a hash table with 13 slots labeled from 0 to 12. The arrows show where each bitstring will be placed in the hash table. For example, the bitstring for bipartition B_1 will be placed in location 11 of the hash table. Bipartition B_8 is placed in location 2. It appears that the bipartitions are placed randomly in the hash

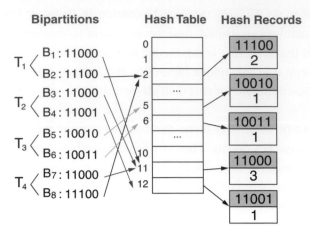

Figure 13.5 An illustration depicting how the bipartitions from the four big cat phylogenies are stored in a hash table. Each location in the hash table stores the bitstring representation of a bipartition and its frequency among the four phylogenetic trees.

table. However, if placement in the hash table was purely random, then bipartitions with the same bitstring would not be placed in the same location making it difficult to update our frequency counts.

Each bitstring in Figure 13.5 is given to a hash function h defined as

$$h(b) = x \bmod m, \tag{13.1}$$

where x is the decimal value of a bitstring b and m is the size of the hash table. In our example, m is 13. The output of the function h provides the location in the hash table to store the bipartition. The notation mod is shorthand for the modulo function. Given two numbers, a (the dividend) and b (the divisor), a modulo b (abbreviated as $a \bmod b$) is the remainder on division of a by b. For instance, 24 mod 13 would evaluate to 11, while 28 mod 13 would evaluate to 2.

Each tree's bipartition bitstrings are fed to a hashing function h and the output determines the location where the bitstring will reside in the hash table. Each time we insert a bitstring into the hash table, we determine whether the hash table location is empty. If location $h(b)$ is empty in the hash table, then we insert the bitstring and initialize the frequency to 1. Otherwise, the bipartition bitstring is already there and we simply update the frequency count by 1. The beauty of hashing resides in its ability to find a bitstring with one retrieval operation. For example, if the bitstring is 11001, $h(11001)$ returns the hash table location 25 mod 13 or 12. Accessing location 12 of the hash table directly gets the number of times the bitstring 11001 appeared among the phylogenetic trees, which was once.

While hash functions are elegant, there is one caveat to using them. There is a possibility for two different bitstrings to reside in the same location in the hash table. Such a condition is called a *collision*. Different bitstrings colliding to the same location in the hash table is analogous to different people having the same credit card number. Collisions not only slow down the algorithm, but could lead to erroneous results. Ideally, we would like a perfect hash function which maps different inputs to different outputs. Thus, much research has been conducted on how to construct good hashing functions that attempt to simulate the behavior of a perfect hashing function.

Both Amenta *et al.* [23] and Sul *et al.* [24] employ more sophisticated hashing techniques such as universal hashing functions to reduce the probability of different bipartition bitstrings colliding in the hash table. In our examples, the decimal value of the bitstring $b_4 b_3 b_2 b_1 b_0$ is evaluated as

$$b_4 \cdot 2^4 + b_3 \cdot 2^3 + b_2 \cdot 2^2 + b_1 \cdot 2^1 + b_0 \cdot 2^0. \tag{13.2}$$

For example, the bitstring 11001, where $b_4 = 1, b_3 = 1, b_2 = 0, b_1 = 0$, and $b_0 = 1$ evaluates to 25. Under universal hashing functions, a random number, r_i, is used instead of 2^i. As a result, the decimal value for a bitstring $b_4 b_3 b_2 b_1 b_0$ becomes

$$b_4 \cdot r^4 + b_3 \cdot r^3 + b_2 \cdot r^2 + b_1 \cdot r^1 + b_0 \cdot r^0. \tag{13.3}$$

If $r_4 = 197, r_3 = 17, r_2 = 49, r_1 = 997$, and $r_0 = 5$, then the bitstring 11001 evaluates to 219.

Under universal hashing, a different set of random numbers is generated each time the algorithm is used. Since the hashing function is being changed each time with a different set of random numbers, the bitstrings will evaluate to different values. As a result, the probability of two different bitstrings hashing (or more appropriately colliding) at the same location will be very low. Imagine the chance of identity theft if you received a new credit card number each time you made a purchase. While inconvenient for credit card use, a new set of random numbers is quite convenient when using universal hashing functions to organize bipartitions in a hash table in a collision-free manner to construct consensus trees.

4.3 Step 3: constructing consensus trees from consensus bipartitions

Initially, the majority consensus tree is a star tree of n taxa. In Figure 13.6, the left-most tree is a star of five taxa since there are no bipartitions that separate the taxa

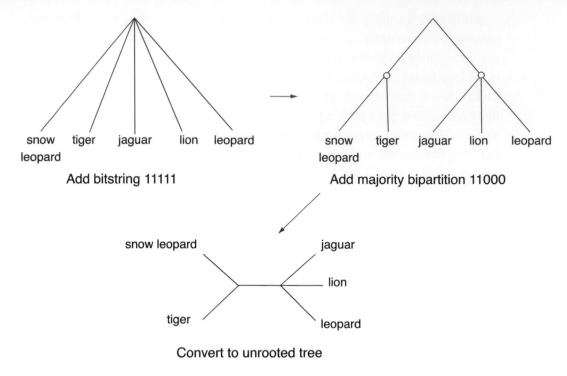

Figure 13.6 Creating the majority consensus tree for the phylogenies shown in Figure 13.2. There is only one majority bipartition {snow leopard, tiger | jaguar, leopard, lion}, or bitstring 11000.

on different sides of the tree. This star tree is represented by the bitstring 11111. Bipartitions are added to refine the majority tree based on the number of 1s in its bitstring representation. (The number of 0s could have been used as well.) The greater the number of 1s in the bitstring representation, the greater the number of taxa that are grouped together by this bipartition. For each of the majority bitstrings, we count the number of 1s it contains. Bitstrings are then sorted in descending order, which means that bipartitions that group the most taxa appear first. The bipartition that groups the fewest taxa appears last in the sorted list of "1" bit counts. For each bipartition, a new internal node in the consensus tree is created. Hence, the bipartition is scanned to put the taxa into two groups: taxa with "0" bits compose one group and those with "1" bits compose the other group. The taxa indicated by the "1" bits become children of the new internal node. The above process repeats until all bipartitions in the sorted list are added to the consensus tree.

In Figure 13.5, for example, bitstring 11000 appears in three trees among four input trees which means it is a majority bipartition. Figure 13.6 shows the steps to construct

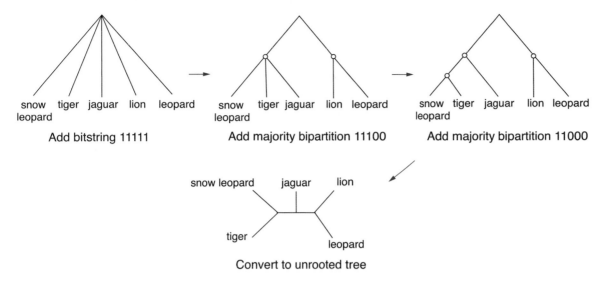

Figure 13.7 Another illustration of creating a consensus tree. Here, we assume the majority bipartitions are represented by the bitstrings 11100 and 11000.

a majority consensus tree using this bipartition. Starting from a star tree constructed from the bitstring 11111, the majority bipartition 11000 determines that the taxa snow leopard and tiger should be in the same group. Two internal nodes are inserted into the starting star tree and the edges are updated. Since we have only one non-trivial majority bipartition in our example, the construction of the majority tree is finished. The resulting tree is converted into an unrooted tree, which is also the majority tree shown in Figure 13.3. Rooting the tree is done in order to construct the consensus tree, but it has no biological meaning. A separate process is performed in order to root the tree for biological significance. For example, for the *Panthera* genus, the clouded leopard is used as an outgroup taxon in order to root the tree. As previously mentioned, this is a separate process from building consensus trees.

Suppose we have more than one majority bipartition. Figure 13.7 provides an example of two majority bipartitions (11000 and 11100) making up the majority consensus tree. Again, the bipartitions are sorted in descending order by the number of 1s. Thus 11100 is first selected for processing which shows that the snow leopard, tiger, and jaguar taxa reside in the same group. Next, 11000 is used to further resolve the intermediate tree. In other words, the {snow leopard, tiger, jaguar} clade can be resolved so that snow leopard and tiger exist in a same group. Finally, as described in the previous example, the root tree is converted to an unrooted, majority consensus tree.

DISCUSSION

In this chapter, we explored several fundamental computational techniques (sorting bitstrings, hashing functions, traversing trees) to build consensus trees using phylogenies constructed from the pantherine lineage of cats. The *Panthera* genus consists of the lion, tiger, jaguar, leopard, and snow leopard. There is much dispute concerning the true phylogeny of these big cats. Given that there is no universally accepted tree at this time, we used several published trees depicting different hypotheses of evolution. Afterward, we used those trees to explore how to build a consensus tree to summarize the various hypotheses of how these big cats evolved.

While many phylogenetic resources give a definition of how to construct a consensus tree, few resources actually give the reader insight into the computational techniques for solving the problem. While a few published algorithms describe how to build majority consensus trees [23, 24], they are not suitable for someone not well versed in computer science. In this chapter, we give scientists a taste of the beauty of computational ideas as they relate to phylogenetics. Although constructing majority consensus trees is a simple problem to explain, it has a wealth of hidden jewels that form the foundation of many computational algorithms such as sorting numbers, hashing bitstrings, and traversing trees.

Overall, we hope that our investigation of consensus tree computation inspires life scientists to learn about other computational ideas in bioinformatics. Furthermore, we encourage scientists well versed in computational ideas to seek opportunities to share their experiences in a language that interdisciplinary scientists can appreciate and share with their colleagues.

QUESTIONS

(1) Why are consensus trees important in studies of the pantherine lineage of cats?
(2) Why is it difficult to reconstruct the evolutionary history of the big cats?
(3) Why is computational thinking important for biologists?
(4) Besides constructing consensus trees, what other computational problems in biology can take advantage of hashing functions?

REFERENCES

[1] D. A. Baum, S. D. Smith, and S. S. S. Donovan. EVOLUTION: The tree-thinking challenge. *Science*, 310(5750):979–980, 2005.

[2] P. Christiansen. Phylogeny of the great cats (Felidae: Pantherinae), and the influence of fossil taxa and missing characters. *Cladistics*, 24(6):977–992, 2008.

[3] B. W. Davis, G. Li, and W. J. Murphy. Supermatrix and species tree methods resolve phylogenetic relationships within the big cats, Panthera (Carnivora: Felidae). *Molec. Phylogen. Evol.*, 56(1):64–76, 2010.

[4] D. M. Hillis, J. Bull, M. White, M. Badgett, and I. K. Molinoux. Experimental phylogenetics: Generation of a known phylogeny. *Science*, 255:589–592, 1992.

[5] T. R. Gregory. Understanding evolutionary trees. *Evo. Edu. Outreach*, 1, 2008.

[6] W. Johnson, P. Dratch, J. Martenson, and S. O'Brien. Resolution of recent radiations within three evolutionary lineages of Felidae using mitochondrial restriction fragment length polymorphism variation. *J. Mammal. Evol.*, 3: 97–120, 1996.

[7] W. E. Johnson, E. Eizirik, J. Pecon-Slattery, *et al.* The late Miocene radiation of modern Felidae: A genetic assessment. *Science*, 311(5757):73–77, 2006.

[8] L. Wei, X. Wu, and Z. Jiang. The complete mitochondrial genome structure of snow leopard *Panthera uncia. Molec. Biol. Rep.*, 36:871–878, 2009.

[9] D. Bryant. A classification of consensus methods for phylogenetics. *DIMACS Ser. Discr. Math. Theor. Comput. Sci.*, 61:163–184, 2003.

[10] J. Felsenstein. *Inferring Phylogenies.* Sinauer Associates, Sunderland, MA, 2005.

[11] R. D. M. Page and E. C. Holmes. *Molecular Evolution: A Phylogenetic Approach.* Wiley-Blackwell, Hoboken, NJ, 1998.

[12] J. M. Wing. Computational thinking. *Commun. ACM*, 49(3):33–35, 2006.

[13] Committee on Frontiers at the Interface of Computing and Biology. *Catalyzing Inquiry at the Interface of Computing and Biology.* National Academy Press, Washington, DC, 2005.

[14] N. Saitou and M. Nei. The neighbor-joining method: A new method for reconstructiong phylogenetic trees. *Molec. Biol. Evol.*, 4:406–425, 1987.

[15] J. Felsenstein. Phylogenetic inference package (PHYLIP), version 3.2. *Cladistics*, 5: 164–166, 1989.

[16] D. L. Swofford. PAUP*: Phylogenetic analysis using parsimony (and other methods). Available: http://paup.csit.fsu.edu/.

[17] F. Ronquist and J. P. Huelsenbeck. MrBayes 3: Bayesian phylogenetic inference under mixed models. *Bioinformatics*, 19(12):1572–1574, 2003.

[18] L. Liu and D. K. Pearl. Species trees from gene trees: Reconstructing Bayesian posterior distributions of a species phylogeny using estimated gene tree distributions. *Syst. Biol.*, 56(3):504–514, 2007.

[19] L. Liu, D. K. Pearl, R. T. Brumfield, and S. V. Edwards. Estimating species trees using multiple-allele DNA sequence data. *Evolution*, 62(8):2080–2091, 2008.

[20] H. Hemmer. Die evolution der pantherkatzen: Modell zur überprüfung der brauchbarkeit der hennigschen prinzipien der phylogenetischen systematik für wirbeltierpaläontologische studien. *Paläontolog. Zeitschr.*, 55:109–116, 1981.

[21] O. R. P. Bininda-Emonds, D. M. Decker-Flum, and J. L. Gittleman. The utility of chemical signals as phylogenetic characters: An example from the Felidae. *Biol. J. Linn. Soc.*, 72(1):1–15, 2001.

[22] J. E. Janecka, W. Miller, T. H. Pringle, *et al.* Molecular and genomic data identify the closest living relative of primates. *Science*, 318:792–794, 2007.

[23] N. Amenta, F. Clarke, and K. S. John. A linear-time majority tree algorithm. *Workshop on Algorithms in Bioinformatics*, 2168:216–227, 2003.

[24] S.-J. Sul and T. L. Williams. An experimental analysis of consensus tree algorithms for large-scale tree collections. In: *Proc. 5th International Symposium on Bioinformatics Research and Applications*. Springer-Verlag, Berlin, Heidelberg, 2009, 100–111.

CHAPTER FOURTEEN

Phylogenetic estimation: optimization problems, heuristics, and performance analysis

Tandy Warnow

Phylogenetic trees, also known as evolutionary trees, are fundamental to many problems in biological and biomedical research, including protein structure and function estimation, drug design, estimating the origins of mankind, etc. However, the estimation of a phylogeny is enormously challenging from a computational standpoint, often involving months or more of computer time in order to produce estimates of evolutionary histories. Even these month-long analyses are not guaranteed to produce accurate estimates of evolution, for a variety of reasons. In addition to the errors in phylogeny estimation produced by limited amounts of data, there is the added – and critically important – fact that all the best phylogeny estimation methods are based upon heuristics for optimization problems that are difficult to solve. Consequently, large data sets are often "solved" only approximately. In this chapter, we discuss the issues involved in phylogeny estimation, as well as the technical term from computer science, "NP-hard."

Bioinformatics for Biologists, ed. P. Pevzner and R. Shamir. Published by Cambridge University Press.
© Cambridge University Press 2011.

1   Introduction

One of the most exciting research topics in biology is the investigation of how life evolved on earth, ranging from questions concerned with very early evolution (e.g. What did the earliest organisms look like? Are fungi or plants closer to each other than either is to animals?) to more recent evolution (e.g. What is the relationship between humans, chimps, and gorillas? Where did human life begin? How did human populations migrate around the world?). However, interest in evolutionary histories is not restricted to species trees, as biologists are also interested in how protein families have evolved, and the evolution of function within protein families. All these questions are addressed through the use of computational methods that estimate evolutionary trees, most typically on molecular sequence alignments, but also sometimes on morphological characters. The good news is that in the last few decades, increasingly accurate and powerful methods have been developed for these analyses, and genome sequencing projects have generated more and more sequence data; consequently, phylogenetic analyses of very large data sets (with hundreds or thousands of sequences) are not unusual. As a result, while there are still substantial debates about much of the Tree of Life, many questions are now reasonably well resolved. For example, scientists now believe that humans are more closely related to chimps than to gorillas, the human species began in Africa, birds are derived from dinosaurs, and whales are more closely related to hippopotamus than to other species.

All these phylogenetic analyses are the result of a combination of fieldwork, wetlab work, and computational methods. In this chapter we discuss the computational problems and methods that are used for these computational analyses. In the course of this chapter, we will consider questions such as: *What does it mean for a method to solve a computational problem? How can we determine if a method is able to solve its problem?* As we shall see, some computational problems have been formally shown to be "hard" to solve (the formal term is "NP-hard"), and computational problems of interest to biologists are often NP-hard. Furthermore, when a problem is NP-hard, the ability to solve it correctly generally requires techniques that can be unacceptably inefficient. Therefore, NP-hard problems will require computationally expensive methods for exact solutions, and conversely, efficient methods are likely to give suboptimal solutions in some cases.

This chapter will illustrate these issues through problems that arise in the context of estimating evolutionary trees. As we will see, certain computational problems posed in this context can be solved exactly by methods whose running times are bounded by polynomials in the input size (i.e. a function like n^3, where the input has size n). Whether this is considered efficient or not will depend upon how big n can get and the degree of the polynomial, so that quadratic time is often acceptable, but running times that grow like n^4 or worse, despite being "polynomial," are not considered all that "efficient." On the other hand, some problems seem not to admit any exact algorithms that are guaranteed to run in polynomial time. For these problems, exact solutions may require a technique such as exhaustive search, which will have exponential running times (i.e. functions like 2^n, where the input has size n) on some inputs. Since techniques like exhaustive search are computationally intensive on many large data sets, the most commonly used methods are not guaranteed to solve their problems exactly. Understanding the difference between methods that have accuracy guarantees and those that have no guarantees is important – without this understanding, interpretation of a computational analysis for an NP-hard problem can be difficult. Therefore, in particular, interpreting trees produced by the most popular methods of phylogenetic analysis is difficult, since these are almost entirely attempts to solve NP-hard problems.

 ## 2 Computational problems

We begin by discussing some very simple computational problems which will help illustrate concepts such as "algorithm," "heuristic," "polynomial time," and "NP-hard."

Imagine you have a kid brother, and you need to arrange a birthday party to which all his friends will be invited. The problem is that some of the friends don't get along with each other, and if you invite kids who don't get along, they'll fight and that will spoil the party. Fortunately, you know exactly which pairs of children don't get along. Since your brother wants *all* his friends to be invited, you propose having a few parties, but dividing up the friends so that everyone who's invited to a party likes everyone else at the party. Your brother likes the plan, so that's what you do.

Of course, since planning a party takes time and energy (plus money), you are hoping to do this with as few parties as possible. You already know two of his friends don't get along, so you can't do it with one party. Can you do it with two parties, you wonder?

Suppose your brother's friends are Sally, Alice, Henry, Tommy, and Jimmy, but Sally and Alice don't get along, Henry and Sally don't get along, Henry and Tommy don't get along, and Alice and Jimmy don't get along. Can you invite them to two parties?

Here you have the brilliant observation that you can figure this out using logic. Suppose Sally is invited to the first party. Since you have to invite everyone, but Sally doesn't get along with Alice and Henry, it follows that Alice and Henry have to be invited to the second party. And since Henry doesn't get along with Tommy, Tommy has to be in the first party. Similarly, since Alice and Jimmy don't get along, Jimmy

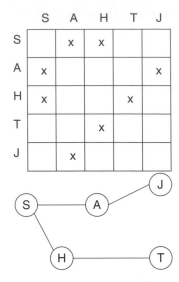

Figure 14.1 A matrix and a graphical representation of which people don't get along with each other. A refers to Alice, S refers to Sally, H refers to Henry, J refers to Jimmy, and T refers to Tommy.

has to be in the first party. So, your solution is: Sally, Tommy, and Jimmy get invited to the first party, and Henry and Alice are invited to the second party. This works, since Sally, Tommy, and Jimmy all get along, and Henry and Alice get along. You tell your brother, and he's happy. The parties will be planned, and all is well.

Note that figuring this out was easy, and didn't take very much time. How much time did it take? One way of analyzing this is to count "operations," where looking at your information counts as one operation, assigning someone to a party counts as an operation, etc. To be formal about this, you have to describe how you represent your information. Suppose you store this information about which friends get along in a square matrix, with a row and column for each of your brother's friends. You put an X in a square if the pair of kids don't get along. Thus, for the instance we described above, the matrix would be as in Figure 14.1.

Now, to solve this problem, you can put the first friend in one party, and then go through the row for that person, putting everyone who's got an X for that row in the second party. After that, you go to someone you just put into the second party, and go through his/her row, putting everyone who doesn't get along with him/her in the first party, and so forth.

It is clear that this algorithm works correctly – but what is the running time? Every time you process a row of the matrix, you use as many operations as there are people in the set (remember, every examination of your input information counts as an operation). Also, you have to repeat this processing of rows as many times as there are people (well, one less time). Suppose there are n people (friends of your brother,

I mean!). Then this discussion shows that this algorithm uses roughly n^2 time (there are n^2 entries in the matrix, after all). Since this is a rough estimate of the time, we write this as $O(n^2)$ time, to hide the extras here and there. What $O(n^2)$ time means is that the number of operations used by the algorithm is bounded from above by Cn^2, where C is some positive constant. This bound holds no matter what the value of n is (that is, the constant C doesn't depend upon n), and holds for any possible input with n people. (By the way, this is pronounced "big-oh of n squared.")

Running times like these are *polynomial* because they are bounded from above by polynomials, and so we call this a *polynomial time algorithm*. If the degree of the polynomial is small (say, at most two), this means the amount of time it takes to use this algorithm won't be very large, even for pretty large values for n. By contrast, exponential functions grow quickly; their initial values may be small, but quite soon the numbers are quite large. Large degree polynomials still grow quickly, but not quite as quickly as functions that grow exponentially. What this means is that for any polynomial and any exponential function, there will be some value for n after which point the exponential function is larger than the polynomial. This is why the distinction is important.

We return to the computational problem and our proposed method. In general, this problem is formulated as a problem about a *graph*, where a graph has vertices (also called nodes) and edges between certain pairs of vertices. Here, the people would each be represented by a vertex in the graph, and if two people don't get along, then the vertices representing them would be connected by an edge, as we did for the graph in Figure 14.1. In this framework, we are looking for a partition of the vertices into two sets, A and B, so that no two vertices within A (or within B) are connected by an edge. Such a partition may not exist, of course, but when it does, the partition gives a solution to the problem of dividing the friends into two sets: the ones who go to one party (corresponding to the vertices in A) and the ones who to go the other party (corresponding to the vertices in B). The usual way of describing this problem is that we would like to color the vertices of the graph with two colors, say red and blue, so that no edge connects two red vertices or two blue vertices. If such a coloring can be produced, then the vertices colored red would constitute the set A, and the vertices colored blue would constitute the set B. A coloring with this property is called a "2-coloring" of the vertices, and the problem we figured out how to solve is the "2-colorability problem."

2.1 The 2-colorability problem

Input: Graph G with vertex set V and edge set E.

Output: A coloring of the vertices in V with red and blue, so that no edge connects vertices of the same color, if it exists, and otherwise the statement "Fail."

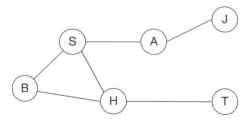

Figure 14.2 Graph representing the incompatibilities when you add Bobby to the problem.

To summarize the discussion above, what you figured out is that we can solve the 2-coloring problem in $O(n^2)$ time, where V contains n vertices.

However, let's return to the problem of coming up with parties for your brother. You draw the graph representing the information you have, and the graph has five vertices, one for each of your brother's friends. You name these vertices S for Sally, J for Jimmy, A for Alice, T for Tommy, and H for Henry. There is an edge between vertices A and S, since Alice and Sally don't get along. There is also an edge between vertices H and S, between H and T, and between A and J. This graph is given in Figure 14.1.

You then color the vertices of the graph with red and blue, and get Jimmy, Sally, and Tommy colored red, and Alice and Henry colored blue. This coloring means that Jimmy, Sally, and Tommy go to one party, and Alice and Henry go to the other. Thus, you can invite all the friends with just two parties.

So, you are happy. You have figured out how to have everyone invited to a party, and you can do it in two parties. All is well. And on top of that, you are proud of yourself for coming up with a nice algorithm to solve the problem.

But your brother, being a bit of a difficult kid (as all kid brothers can be, I suspect), interrupts you at dinner to say "I forgot I have to invite Bobby." You groan. Why? Because Bobby is kind of difficult himself, and doesn't get along with many people. Your brother insists, however, so you add Bobby. Bobby doesn't get along with Sally and Henry, but he does get along with the others. Can you still do it in two parties? You redraw the graph by adding a vertex (B) for Bobby, and including edges between B and S, and between B and H (Figure 14.2). But when you redo your algorithm, you discover a problem. You try to 2-color this graph: B gets colored red, then S must be colored blue, and so what can H be colored? The problem is that vertex H is adjacent to both B and S, and so cannot be colored either blue or red. (Notice that this analysis doesn't depend upon what color you gave the first vertex; so if you start by coloring B blue, you still end up with a problem.) In other words, there is no way to have two parties with Bobby in the picture. You tell your brother, and he cries a bit, but then you come up with the plan: use three parties, and let Bobby be in the third party.[1]

And now you are happy again, but only for a short time. Your brother remembers he has to invite some other friends. Ten more friends, in fact. And now you have 16 people, and you'd like to figure out the minimum number of parties you need to invite everyone. You know you can't manage with only two parties (the first six people needed three parties), but now you'd like to figure out if you can do it with *only* three. How are you going to solve this?

Unfortunately, figuring out how to do it in three parties is by no means straightforward. You can start as before, putting Sally in the first party, but then you are stuck. Sally doesn't get along with Alice or Henry, but which parties should Alice and Henry go to? The same party, or different ones? Any decision you make now may be wrong. This is distinctly different from the situation you faced when you only had two parties to deal with; there, all decisions were obviously correct. And so with 16 people to put into three parties, it gets complicated. Very complicated. You are very frustrated. You try a few different attempts, but don't come up with a way of putting them all into three parties ... and you are about to give up. But then, you realize that you may have missed a solution, and you had better just try all the possible ways of doing this. So you try to enumerate all the possible solutions, and you check them, one by one. Each one you check takes only a minute to write down and check (you are very good at this!), and so you are sure you can be done very quickly. The only problem is that there are many possible solutions. That is, each person can be put in any one of three parties, and so there are $3^{16} = 43,046,721$ possible ways of putting them into parties. And at one minute per assignment, this is 717,445 hours, which is 29,893 days, or almost 82 years. Let's see. You are 21 now, and that means that if you don't sleep at all, you'll be 103 when you are done. That will take too long (and your kid brother isn't that patient).

This kind of method is called "exhaustive search," because it is defined by a search strategy that explicitly examines every possible solution in the search for an optimal solution. Exhaustive search techniques are provably correct, but they are infeasible for many inputs. (Even using computers, such techniques quickly hit their limits in running time, so that analyses using exhaustive search can take years on small inputs, and millennia on some only moderately large inputs.)

So you can't do it this way.

How will you do this?

At this point, you say to your brother, "Sorry, kiddo, but I can't figure this out. I don't know if we can do it in three parties. I think we can't, but I am not sure. Do you

---

[1] You can always move some people, such as Tommy and Alice, into the third party, if you are feeling sorry for Bobby. That is, there may not be a unique solution to this problem!

care very much if we do it with the smallest number of parties? Maybe we should try something else, like not inviting *everyone*?"

Your brother is a bit concerned, but he's willing to consider the new approach. He asks you to try to invite as many people as you can, but just to one party. And you try to figure that out. It seems like an easier problem.

Once again, you think about this as a graph problem. The same graph will work: the people are the vertices, and edges mean they don't get along. And since you want a group of people who all get along, and you want that group to be as large as possible, you are looking for what is called a "maximum independent set": a subset of the vertices in which no two vertices are connected by an edge, and such that the subset is as big as possible.

2.2 Maximum independent set

Input: A graph G with vertex set V and edge set E.

Output: A subset V_0 of the vertex set V so that V_0 is an independent set (no two vertices in V_0 are connected by an edge) and has maximum size among all such subsets.

How would you try to solve this problem?

You start hopefully, thinking since Sally gets along with lots of people the best solution will probably include her (besides, you like Sally and you hope she'll be at the party so you can get to know her better). You take out the two people (Henry and Alice) she doesn't like, and you look at the rest. Now, if you include Tommy, you can't include the people Tommy doesn't get along with, and unfortunately there are some people in the group that Tommy doesn't like. But this basic problem is true for everyone in the set: no one is an obvious addition. So you just hope for the best, and add Tommy, and throw out the ones he doesn't like, and see what happens. Hoping for the best, you put together a group of people where all the people get along. Unfortunately, you don't know if it's the largest group. So you try again. This time, you begin with Sally, but this time you don't include Tommy ... and you get a slightly smaller group. So you try again, including Tommy and Alice, but making some other decisions differently, and each decision gives you a different group. You do this many times, and eventually get tired. You see that you have a group of 8 people (out of 16, not so great, perhaps). You ask your brother if this is okay.

He says: "Is this the best you could do?"

And honestly, you don't know. Maybe a better solution could be found. You try to figure out if you can find an optimal solution, and you wonder about using some "exhaustive search" technique. You'd have to look at all possible subsets of people, and then check each subset to see if everyone got along. How many subsets are there of n

people? For each person, you can either include them in the subset or not. Thus, each subset is defined by the sequence of n choices you make (include or don't include), one for each person. Since there are two possible choices, there are 2^n possible subsets of n people. For 16 people, there are 2^{16} subsets, but one of these is the empty set (has no one in it), and so you only have to look at $2^{16} - 1$ subsets. How big is that number? Unfortunately, it's big: 65,535. Not as big as the previous number, but still big enough. And if each subset took one minute to process, it would take 1,092 hours, or 45 days. Not nearly as bad as the previous problem, but still too long.

So you say to yourself, I can't use an exhaustive search technique. Let me think about doing this differently, where I don't have a guarantee of getting an optimal solution, but maybe it will work. I'll find a set of people who get along, and then try to modify it. I'll look at someone not in the group, and see what happens if I add that person to the group. If they don't get along with some people in the group, I'll throw out the ones they don't get along with. That will make the number of people in the group go down, and maybe my set will then be smaller. But if I remove these people, I might be able to add some others to the group who get along with everyone in the group, so it might be better. And, in any event, it will make it possible to keep exploring possible sets. Maybe I'll do better this way.

And so you try this. And after a while, you find a set of nine people you can invite (before you only had eight, so this is an improvement). But you don't find a bigger set. And you say to your brother – "Hey, we can invite nine of your friends. How's that?" He's not happy and asks you "Can you do better"? You aren't sure. You just aren't sure. How can you be sure? But you are tired of looking for a larger set, and you are pretty fed up. By now, you aren't sure you want to do this party for him *at all*. (As an aside, many heuristics have been developed for this maximum independent set problem, for example, [1].)

So he accepts the plan. You have a party for nine people, and you give up being his social organizer for the future. You still love your kid brother, but you won't be trying to arrange his parties in the future!

NP-hardness, and lessons learned

You are not alone in having a very hard time with finding effective techniques for solving these problems. These problems are really hard. So hard, in fact, that computer scientists have studied them for decades, and some computer scientists believe that it is *not possible* to solve these problems exactly and efficiently. I'll explain what this means.

Remember how you came up with an algorithm to determine if you could manage to invite everyone with two parties? That is, you showed how to solve the 2-colorability problem for n vertices in $O(n^2)$ time. On the other hand, trying to figure out if you could invite everyone with just three parties was hard, and you couldn't find an algorithm that solved that problem without resorting to *exhaustive search*. And your exhaustive search technique used more than 3^n operations, because there were 3^n ways of assigning n people to three parties. The difference in growth between these two functions – n^2 and 3^n – is dramatic (just look at the difference in value when $n = 20$, and when $n = 100$). That is, n^2 is polynomial in n, and 3^n is exponential in n. Functions that are exponential in their parameter grow much more quickly than functions that are polynomial in their parameter. Therefore, while both functions may have reasonably small values for small n, the exponential function will be much larger than the polynomial function at some point, and then stay larger. And, worse, the running time of the algorithm, if it is described by an exponential function, will be too large for all but pretty small values of n.

The fact that the running time of the exact algorithm you developed for the "three-party problem" (otherwise known as the "3-colorability problem") is exponential is not at all surprising, because this problem has been proven to be an "NP-hard" problem (this is bad news!). Similarly, the maximum independent set problem is also NP-hard. It was just your bad luck that you tried to solve two NP-hard problems!

NP-hardness has a technical definition [2], which we'll not go into here. The main consequence of saying that a problem is NP-hard, though, is that to date, no one has ever been able to find an algorithm that can solve an NP-hard problem and that runs in polynomial time. So, you were in very good company. Your inability to come up with a technique to solve this problem correctly, and which runs in polynomial time, is shared with many very famous and smart mathematicians and computer scientists.

What does a computer scientist do when confronted with an NP-hard problem? Often, they develop *heuristics* for these problems, by which we mean methods that try to find good solutions that may not be exactly correct. In the context of the 3-colorability problem, they might try to develop a method that is sometimes able to find 3-colorings, but may fail on occasion to find a 3-coloring even when the graph can be 3-colored. In the context of the maximum independent set problem, they might try to find a heuristic to produce an independent set, and they'd hope that the set they produce is the largest possible ... but on some inputs, it wouldn't be the largest possible. If they are lucky, the heuristic will be fast, but often it won't be. In fact, if you think back to your attempt to solve the maximum independent set problem, your approach tried to modify the current independent set by adding and subtracting people. How long would that heuristic take? The way you did it, you stopped when you got tired. But you could have put in some kind of stopping rule, such as stopping when the size of the biggest independent set hasn't increased in the last 100 sets you examined. How long would it take before that stopping rule would apply? It's not always easy to predict this, and in general, running times of heuristics like these are hard to analyze.

So, when given an NP-hard problem, you have several options. One is to try to solve it exactly, which typically will mean an approach that essentially involves a technique that includes some exhaustive search method. These techniques are computationally intensive, and limited to smallish data sets (even if you use a computer). Or, you can design a heuristic which is not guaranteed to solve the problem correctly. These heuristics have often produced very good results, sometimes even the correct result!, on many inputs. The problem with heuristics is that you generally aren't able to be sure that your result is optimal, and you also can't predict the running time.

How does this relate to phylogeny estimation?

4   Phylogeny estimation

The phrase "phylogeny estimation" refers to the action of producing a hypothesis of the evolutionary tree (also called a "phylogeny" or "phylogenetic tree") for a given set of taxa. Thus, this is also called "phylogenetic tree estimation" or "evolutionary tree construction."

The relationship of the material in Section 3 to phylogeny estimation is that almost every computational approach in phylogeny estimation is based upon an NP-hard problem. That is, the computational methods that biologists typically use for estimating evolutionary trees are methods that try to solve an optimization problem that is NP-hard. Here, we will talk about one of these problems, maximum parsimony.

4.1   Maximum parsimony

Maximum parsimony is a very natural optimization problem for phylogeny estimation; here we describe it in the context of estimating evolutionary trees ("phylogenies") from DNA sequences which all have the same length. However, you could use techniques for maximum parsimony on some other kind of biological "character" data, such as morphological features, RNA sequences, amino acid sequences, etc.

Suppose you have DNA sequences, all of the same length (and without any gaps), such as the following.

The maximum parsimony problem asks you to find a tree, with leaves labeled by the sequences in the input and with the internal nodes labeled by additional sequences, all of the same length as the input sequences, which minimizes the total number of

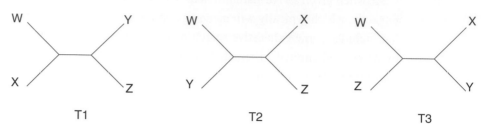

W = ACATTAGGGAGG
X = ACATAAGGGAGG
Y = CCATGAGGGAGG
Z = CCATCGGGAAGG

Figure 14.3 The three unrooted fully resolved trees on leaf set {*W, X, Y, Z*}.

substitutions on the tree. Thus, to compute the "cost" of the tree (given the sequences at every node), you would count up the number of substitutions implied by each edge. (To define the number of substitutions on an edge, you just compare the sequences at the endpoints of the edge, and note the number of positions in which they have different values. Thus, an edge with endpoints $AACCTA$ and $AACTTG$ would have two substitutions, since the endpoints are different in positions 4 and 6.) The tree with the minimum possible total would be returned by maximum parsimony.

4.1.1   Maximum parsimony

Input: Set S of strings (e.g. nucleotide sequences) of the same length k.

Output: Tree T with leaves identified with the different elements of S, and with other strings of length k labeling the internal nodes, so that the total number of substitutions is minimized.

When a tree is given for the set S, and the objective is to find the best sequence labels for each node, we have the "Fixed-tree Maximum Parsimony problem."

Let's try to solve this problem on this input. We'll do this by exhaustive search, examining every possible tree, and trying to find the sequences at the internal nodes that give the minimum total cost.

The first thing to notice about this problem is that how you root the tree doesn't matter, since the number of changes on each edge doesn't depend upon the rooting. Therefore, you only need to look at unrooted trees. The next thing to notice is that the optimal score would be obtained by a tree that is fully resolved: each non-leaf vertex in the tree has three edges coming out of it. Therefore, since there are only four sequences in the input, you only need to look at three different trees (Figure 14.3).

The first, $T1$, has W and X siblings, and Y and Z siblings. We denote this tree by $(WX|YZ)$. The second tree, $T2$, is denoted by $(WY|XZ)$, and the third tree, $T3$, is denoted by $(WZ|XY)$. Now, we look at how to label the internal nodes optimally for each tree.

Consider the first tree, $T1$. Let us call the internal nodes a_1 and b_1, with a_1 adjacent to W and X, and b_1 adjacent to Y and Z. How shall we assign sequences to a_1 and b_1? Note that minimizing the total number of substitutions on the tree is the same as minimizing the total number of times each site changes on the tree. Hence, we calculate the optimal sequences for the internal nodes by considering the sites (columns), one by one. The first thing to notice is that whenever a site is constant on all the taxa (that is, all the taxa have exactly the same nucleotide for that site), then we will label all internal nodes with that state as well for that site. This is optimal, these sites won't change at all on the tree, and will therefore contribute 0 to the total tree cost. This observation takes care of most of the sites in the tree.

Now, let's consider the remaining sites. The first site has W and X having the nucleotide A and Y and Z having nucleotide C. It's very easy to see that this site must change at least once on the tree, and that if we set a_1's state to A and b_1's state to C, we will achieve that minimum.

The second through fourth sites are all constant, so we set a_1 and b_1 to be the constant state for those sites. The fifth site is interesting: every leaf has a different state. Therefore, the minimum possible number of times this site will change on this tree is three, and we can achieve that by labeling a_1 and b_1 by the same state. We pick A for the two internal nodes, but we could have achieved the same value using C, T, or G – as long as they both have the same state.

The sixth site is also interesting: three leaves have the same state (A), and the fourth leaf has a different state. We label a_1 and b_1 with A. Note that under this labeling, the site changes once on the tree, and that this is the minimum possible (since two states appear for this site).

The seventh and eighth sites are also constant. The ninth site is like the sixth – three leaves have the same state (G), so we label the internal nodes with G.

The tenth through twelfth sites are constant.

Hence, we produce the sequences a_1 = ACATAAGGGAGG and b_1 = CCATAAGGGAGG. Thus, a_1 and b_1 differ in exactly one position only, a_1 and X are identical as sequences, and b_1 is different from every other sequence.

The six sequences labeling the nodes of this tree are given in Table 14.1.

To count how many changes there are on this tree, we can just look at each edge in the tree, in turn. There are five edges: $e_1 = (W, a_1)$, $e_2 = (X, a_1)$, $e_3 = (a_1, b_1)$, $e_4 = (b_1, Y)$, and $e_5 = (b_1, Z)$. The cost of the tree will be the sum of the edge costs, i.e. $cost(e_1) + cost(e_2) + cost(e_3) + cost(e_4) + cost(e_5)$. Note that $cost(e_2) = 0$ since X

Table 14.1   Sequences labeling the nodes of tree $T1$.

W	=	ACATTAGGGAGG
X	=	ACATAAGGGAGG
Y	=	CCATGAGGGAGG
Z	=	CCATCGGGAAGG
a_1	=	ACATAAGGGAGG
b_1	=	CCATAAGGGAGG

Table 14.2   Edge $e_1 = (W, a_1)$ in tree $T1$; note $cost(e_1) = 1$.

W	=	ACATTAGGGAGG
a_1	=	ACATAAGGGAGG

Table 14.3   Edge $e_2 = (X, a_1)$ in tree $T1$; note $cost(e_2) = 0$.

a_1	=	ACATAAGGGAGG
X	=	ACATAAGGGAGG

and a_1 are identical sequences. We calculate the cost of each edge, one by one; see Tables 14.2–14.6. Based upon our edge cost calculations, we see that the total cost of this tree is 6.

We now compute the cost of tree $T2$; this tree has W and Y adjacent, and X and Z adjacent. Let's call the internal nodes a_2 and b_2, with a_2 adjacent to W and Y, and b_2 adjacent to X and Z. To set the sequences a_2 and b_2 we go through each site, one by one, using the same techniques as we used for the tree $T1$. The same analysis we did for $T1$ can be applied to sites 2 through 12, but the first site requires more discussion.

Note that on site 1, W and X have A, while Y and Z have C. The best we can do for this tree is to label both a_2 and b_2 with A (or both with C), and for this label we would have the site changing twice on this tree – it is not possible to have the site change only once! Therefore, we can assign identical labels for a_2 and b_2, with $a_2 = b_2 = $ ACATAAGGGAGG. Note that $a_2 = b_2 = X$. See Table 14.7 for the set of six sequences labeling the tree $T2$.

Table 14.4 Edge
$e_3 = (a_1, b_1)$ in tree $T1$;
note $cost(e_3) = 1$.

a_1	=	ACATAAGGGAGG
b_1	=	CCATAAGGGAGG

Table 14.5 Edge
$e_4 = (b_1, Y)$ in tree $T1$;
note $cost(e_4) = 1$.

b_1	=	CCATAAGGGAGG
Y	=	CCATGAGGGAGG

Table 14.6 Edge
$e_5 = (b_1, Z)$ in tree $T1$;
note $cost(e_5) = 3$.

b_1	=	CCATAAGGGAGG
Z	=	CCATCGGGAAGG

The total cost of this tree, with this labeling, can be computed either by adding up the changes on each site, or by adding up the changes on each edge. We demonstrate this calculation by computing this on an edge-by-edge basis. Recall that a_2 is adjacent to W and Y and b_2 is adjacent to X and Z, and the five edges in the tree are therefore (W, a_2), (Y, a_2), (a_2, b_2), (b_2, X), and (b_2, Z). Since $a_2 = b_2 = X$, there are no changes on edges (a_2, b_2) or (b_2, X), and so the only edges on which there are any changes are (W, a_2), (Y, a_2), and (b_2, Z). By examining Table 14.7 we see that edge (W, a_2) has cost 1, edge (Y, a_2) has cost 2, and edge (b_2, Z) has cost 4, giving the total cost of 7.

Finally, if we look at $T3$, we can do the same analysis, and produce the optimal sequences for its internal nodes. This tree will also have a total cost of 7. (This is left to the reader as an exercise!)

Thus, the best solution to maximum parsimony on this four-sequence input is $T1$, and it has total cost 6. Note that we computed this by hand. The technique is: for each tree, we determined the sequences at each internal node site-by-site, using the pattern at the leaves. Once the sequences at the internal nodes were computed, we then calculated the cost of the tree by computing the cost of each edge, and adding

Table 14.7 The six sequences for the tree T2.

X	=	ACATAAGGGAGG
W	=	ACATTAGGGAGG
Y	=	CCATGAGGGAGG
Z	=	CCATCGGGAAGG
a_2	=	ACATAAGGGAGG
b_2	=	ACATAAGGGAGG

them up. A running time analysis for this special case of four-leaf trees shows that this approach takes $O(k)$ time, where k is the number of sites (columns) in input sequences.

This is good, but can we apply this technique to larger data sets?

Suppose we had a five-taxon input to maximum parsimony. We could look at all the unrooted fully resolved trees on five leaves, and try to find the optimal sequences for the internal nodes. How much time would this take? The first thing to note is that while there were only three trees on four leaves, there are 15 trees on five leaves (go ahead and write them out!). So this will take more time. But what about scoring each tree, i.e. finding the optimal sequences for the internal nodes? This, it turns out, can still be done in polynomial time. How this is done is beyond the scope of this chapter, but it works! And rest assured, it is not too difficult to learn. The algorithm for finding the optimal sequences for the internal nodes of a given tree uses a special algorithmic technique, called Dynamic Programming, to solve the problem exactly. The running time for computing these optimal sequences is $O(nk)$, where there are n leaves and k sites. That's a pretty efficient algorithm – it's "linear-time" in the input size (the matrix itself uses $O(nk)$ space). This is important enough that we will highlight it as a theorem:

Theorem 1. *Let $s_1, s_2, \ldots, s_n$ be DNA sequences with k sites. Let T be a tree on leaf set $\{s_1, s_2, \ldots, s_n\}$. Then we can compute the optimal sequences for the internal nodes of the tree T so as to minimize the total cost of the tree (its parsimony score) in $O(nk)$ time. In other words, we can solve Maximum Parsimony on a fixed tree in $O(nk)$ time.*

See [3] for more information about this algorithm.

Using this algorithm to compute the cost of a tree allows us to consider an exhaustive search technique, whereby we examine every tree for the input sequences, score the tree (that is, compute the optimal sequences giving the smallest total cost), and return the tree that has the best cost. How much time does this take? The running time is the product of the number of trees and the cost of computing the score of each

Table 14.8 The number of unrooted fully resolved trees on n leaves.

Number of leaves	Number of trees
4	3
5	15
6	105
7	945
8	10,395
9	135,135
10	2,027,025
20	2.2×10^{20}

tree. Can we express the number of fully resolved, unrooted trees on n leaves with a formula? Yes! Unfortunately, it is a large number – the number of these trees is $(2n - 5) \times (2n - 7) \times \ldots 3$, and this is a big number even for relatively small values of n (see Table 14.8). Thus, the number of trees on 10 leaves is already more than 2,000,000. So attempts to solve maximum parsimony by hand are limited to very small numbers of taxa. With a good computer, exact analyses can be performed on data sets with about 20 or (sometimes) 30 taxa. However, analyses of larger data sets cannot be done exactly; even today's supercomputers cannot enable exhaustive search analyses of data sets of the size that biologists want to analyze!

To summarize this discussion, since solving maximum parsimony on a single n leaf tree takes $O(nk)$ time, when the input sequences are all of length k, and there are $(2n - 5)!! = (2n - 5)x(2n - 7)x \ldots x3$ trees, the exhaustive search technique will take the product of these two numbers. In other words:

Theorem 2. *The exhaustive search technique for solving Maximum Parsimony uses* $O((2n - 5)!!nk)$ *time, where* $(2n - 5)!! = (2n - 5) \times (2n - 7) \times \ldots \times 3$.

However, since biologists try to solve maximum parsimony on much larger data sets, with hundreds of sequences (and sometimes thousands) [4], what do they do? Here is where our earlier discussion becomes relevant. Unfortunately, like maximum independent set and 3-colorability, maximum parsimony is one of those NP-hard problems. And this too is important, so we make it a theorem:

Theorem 3. *The Maximum Parsimony problem is NP-hard (from [5]).*

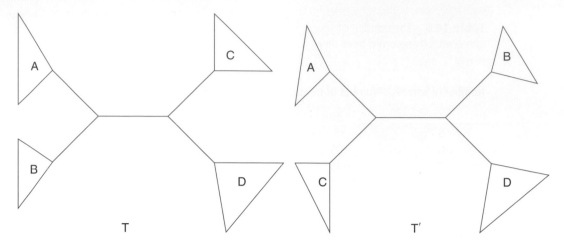

Figure 14.4 Trees *T* and *T'* are related by one NNI move.

And so, while exact algorithms based upon exhaustive search (or branch-and-bound) can be used to solve maximum parsimony, these are limited to small data sets (with up to at most 30 sequences). Beyond such data set sizes, heuristics are used for "solutions" to maximum parsimony.

4.1.2 Heuristics for maximum parsimony

We will now discuss different heuristics for maximum parsimony. Remember that it is an "easy" problem to compute the "cost" of a tree (i.e. to compute the optimal sequences for the internal nodes, so as to have the minimum cost), in that it can be calculated in linear time. We will use that fact throughout this section. Thus, when we say we "score the current tree," or "compute the cost of the current tree," we mean that we will apply the polynomial time algorithm to the current tree with leaves labeled by sequences, in order to score the tree.

The simplest heuristics for maximum parsimony use a "Greedy Algorithm" to find a better tree. These greedy algorithms perform a search through "treespace", and always move to a new tree when the score improves, and never move to the new tree if the score gets worse. One such move is the NNI (nearest neighbor interchange) move, which swaps subtrees that are separated by a single internal edge (Figure 14.4).

It is known that all pairs of trees are connected by some sequence of NNI moves, and so it is possible to explore all possible trees on the input sequence set, using NNI moves. A heuristic search, based upon NNI moves, would have this basic structure:

Step 1: Start by computing an initial tree for the input sequences, and compute its cost. The initial tree can be computed in many ways, including by using a random tree, or by adding sequences sequentially to a tree, each time placing the newly added sequence optimally into the tree so as to minimize the total cost.

Step 2: Modify the current tree by using an NNI move, and score the new tree. If the score improves, replace the current tree by the new tree, and begin again at the start of Step 2. If the score is not better, then explore other NNI moves. If all NNI moves fail to improve the score, then *exit*, and return the current tree as the best tree.

By its structure, this method will only stop when all the trees that are one NNI move from the current tree have a worse score. Thus, when the heuristic stops and returns a tree, that tree will be a "local optimum," meaning that none of its NNI neighbors have a better score. It's very important to realize that trees that are local optima are not necessarily global optima, in that they can have very poor scores compared to the global optima. Also, this definition depends on the definition of "neighbor," and that this in turn depends upon the specific "move" that is used to explore treespace. The algorithm we described above, however, is based on the NNI move, which only has $2(n-3)$ neighbors.[2]

Because all heuristics for maximum parsimony can get stuck in local optima, the best heuristics include techniques to "get out of local optima." Typically, these heuristics accept a move even if it produces a poorer score, with a probability that depends upon the difference in the tree score. By design, these methods could continue indefinitely – getting into local (and perhaps global) optima, using randomness to exit the local/global optima, and repeating the process. To stop this process, the algorithm designer adds a "stopping rule," which ensures that the heuristic will eventually exit and return a tree. Simple stopping rules, based upon some fixed number of iterations or number of hours, can be used. More frequently, however, the stopping rule is based upon the heuristic search not having found any improvement in the score over some number of iterations.

Note that by design, unless the stopping rule is based upon the total number of hours or number of iterations, it is not all that easy (and is sometimes impossible) to predict when heuristics like these will stop. That is, whereas before we were able to talk about running times, and could give upper bounds on the running time of different algorithms, running times of heuristics of this sort are described anecdotally, through empirical studies, on real or simulated data sets.

The combination of effective search techniques, with randomness to exit local optima, has produced the most accurate methods – in the sense that they produce the best scores (smallest total parsimony scores). However, even the best methods can

---

[2] To see this, note that every NNI move is performed around a single internal edge in the tree, that there are two NNI moves around any specific edge, and that there are $n-3$ internal edges in a tree on n leaves.

still take a very long time on some large data sets. Furthermore, there can be many trees with the same optimal score found during a search, and biologists are typically interested in seeing as many of the optimal trees. For these reasons, some phylogenetic analyses have very long running times, using months or years of analysis.

DISCUSSION AND RECOMMENDED READING

Phylogenetic estimation involves solving NP-hard problems, which are by their nature very hard to solve exactly. As a result, when performing a phylogenetic estimation on a large data set, biologists use heuristics to find phylogenetic trees that have good scores, but which may not have the optimal scores for their input data sets. In particular, the best methods for maximum parsimony (one of the major approaches for phylogeny estimation, and an NP-hard problem) are not guaranteed to produce the true optimal solutions, even when run for a very long time. Because of the importance of phylogenetic estimation, biologists are willing to dedicate many weeks (sometimes months or years) of computational effort in order to obtain highly accurate phylogenetic trees. This means that new heuristics are still being developed, in order to make it possible for highly accurate results to be obtained on the large data set analyses that are to come.

This chapter focused on the maximum parsimony method of phylogeny estimation, but there are other methods of phylogeny estimation that are very popular. For further reading into this important research area, see [6–12].

QUESTIONS

(1)   What does it mean to say that a computational problem is NP-hard?
(2)   How do biologists compute evolutionary trees?
(3)   Why is computing evolutionary trees difficult?

REFERENCES

[1] A. Grosso, M. Locatelli, and W. Pullan. Simple ingredients leading to very efficient heuristics for the maximum clique problem. *J. Heuristics*, 14(6):587–612, 2008.

[2] M. R. Garey and D. S. Johnson. *Computers and Intractability: A Guide to the Theory of NP-Completeness.* W.H. Freeman, San Francisco, CA, 1979.

[3] W. Fitch. Toward defining the course of evolution: Minimum change for a specified tree topology. *System. Biol.,* 20:406–416, 1971.

[4] U. Roshan, B. M. E. Moret, T. L. Williams, and T. Warnow. Rec-I-DCM3: A fast algorithmic technique for reconstructing large phylogenetic trees. In: *Proc. IEEE Computer Society Bioinformatics Conference (CSB 2004),* Stanford University, 2004.

[5] L. R. Foulds and R. L. Graham. The Steiner problem in phylogeny is NP-complete. *Adv. Appl. Math.,* 3:43–49, 1982.

[6] J. Felsenstein. *Inferring Phylogenies.* Sinauer Associates, Sunderland, MA, 2004.

[7] J. Kim and T. Warnow. Tutorial on phylogenetic tree estimation, 1999. Presented at the ISMB 1999 conference, available online at http://kim.bio.upenn.edu/jkim/media/ISMBtutorial.pdf.

[8] D. Grauer and W.-H. Li. *Fundamentals of Molecular Evolution.* Sinauer Publishers, Sunderland, MA, 2000.

[9] C. R. Linder and T. Warnow. Overview of phylogeny reconstruction. In S. Aluru (ed.) *Handbook of Computational Biology.* Chapman & Hall, CRC Computer and Information Science Series, 2005.

[10] M. Nei, S. Kumar, and S. Kumar. *Molecular Evolution and Phylogenetics.* Oxford University Press, Oxford, 2003.

[11] R. Page and E. Holmes. *Molecular Evolution: A Phylogenetic Approach.* Blackwell Publishers, Oxford, 1998.

[12] D. L. Swofford, G. J. Olsen, P. J. Waddell, and D. M. Hillis. Phylogenetic inference. In D. M. Hillis, C. Moritz, and B. K. Mable (eds) *Molecular Systematics.* Sinauer Associates, Sunderland, MA, 1996.

PART V

REGULATORY NETWORKS

CHAPTER FIFTEEN

Biological networks uncover evolution, disease, and gene functions

Nataša Pržulj

Networks have been used to model many real-world phenomena, including biological systems. The recent explosion in biological network data has spurred research in analysis and modeling of these data sets. The expectation is that network data will be as useful as the sequence data in uncovering new biology. The definition of a *network* (also called a *graph*) is very simple: it is a set of objects, called *nodes*, along with pairwise relationships that link the nodes, called *links* or *edges*. Biological networks come in many different flavors, depending on the type of biological phenomenon that they model. They can model protein structure: in these networks, called *protein structure networks*, or *residue interaction graphs (RIGs)*, nodes represent amino acid residues and edges exist between residues that are close in the protein crystal structure, usually within 5 Å (Figure 15.1). Also, they can model *protein–protein interactions (PPIs)*: in these networks, proteins are modeled as nodes and edges exist between pairs of nodes corresponding to proteins that can physically bind to each other (Figure 15.2a). Hence, PPI and RIG networks are naturally undirected, meaning that edge AB is the same as edge BA. When all proteins in a cell are considered, these networks are quite large, containing thousands of proteins and tens of thousands of interactions, even for model organisms. An illustration of the PPI network of baker's yeast, *Saccharomyces cerevisiae*, is presented in Figure 15.2b. Networks can model many other biological phenomena, including transcriptional regulation, functional associations between genes (e.g. synthetic lethality), metabolism, and neuronal synaptic connections.

Bioinformatics for Biologists, ed. P. Pevzner and R. Shamir. Published by Cambridge University Press.
© Cambridge University Press 2011.

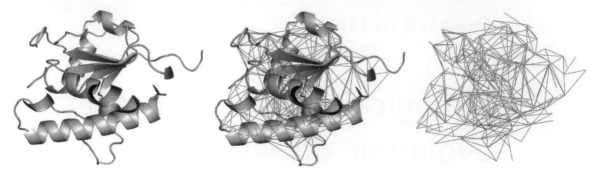

Figure 15.1 An illustration showing a residue interaction graph.

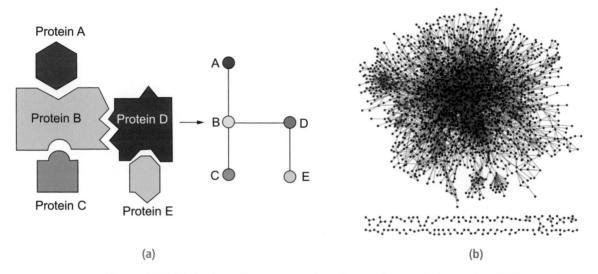

Figure 15.2 (a) A schematic representation of a protein–protein interaction (PPI) network. (b) Baker's yeast protein–protein interaction (PPI) network downloaded from Database of Interacting Proteins (DIP).

In this chapter, we give an introduction to network analysis and modeling methods that are commonly applied to biological networks. We mainly focus on protein–protein interaction (PPI) networks as a biological network example, but the same methods can be applied to other biological networks. The chapter is organized as follows. In Section 1, we describe the main techniques that yielded large amounts of PPI and related biological network data. Then in Section 2, we talk about the main computational concepts related to network representation and comparison. In Section 3, we describe some of the main network models and illustrate their use to solve real biological problems. In Section 4, we show how biological function, involvement in disease and homology can be extracted from analyzing network data sets. Finally, in Section 5, we give an overview of the major approaches for network alignment.

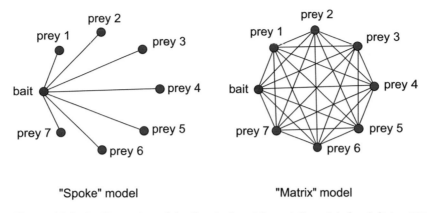

Figure 15.3 An illustration of the "spoke" and "matrix" models for defining PPIs in pull-down experiments.

 ## Interaction network data sets

Experimental techniques have been producing large amounts of network data describing gene and protein interactions. The main techniques include yeast two-hybrid (Y2H) assays (e.g. [1]), affinity purification coupled with mass spectrometry (e.g. [2]), and synthetic-lethal and suppressor networks (e.g. [3]). They have produced partial networks for many model organisms (e.g. [1–3]) and humans (e.g. [4]), as well as for microbes (e.g. [5]), viruses,[1] and human–viral interactions [6]. Since these networks are very large and complex (e.g. see Figure 15.2b), it is not possible to understand them without computational analyses and models.

Our current data sets are noisy due to limitations in experimental techniques. Also, they are largely incomplete, since the experimental techniques are only capable of extracting samples of interactions that exist in the cell. Furthermore, they contain sampling and data collection biases introduced by humans (e.g. see [7]). For example, more data have been collected in parts of the networks relevant for human disease due to increased interest and availability of funding. Another example is the "spoke" versus the "matrix" model that are used to represent interactions obtained from pull-down experiments. In the "spoke" model, interactions are assumed between the tagged "bait" protein and all of the protein interaction target ("prey") proteins, while in the "matrix" model, additional interactions are assumed between all preys as well (Figure 15.3). Both of these models simplify the biological reality by making a broad assumption that is sometime false and thus add noise. Due to such sampling and data collection

---

[1] http://mint.bio.uniroma2.it/virusmint/.

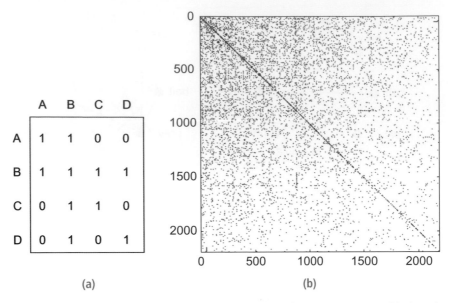

Figure 15.4 (a) The adjacency matrix of the network from Figure 15.2a. (b) The adjacency matrix of the PPI network from Figure 15.2b, illustrating its sparsity.

biases, PPI networks are currently quite sparse with some parts being more dense than others (e.g. parts relevant for human disease).

There are two main standards for representing network data. The first one is called an *edge list*, or an *adjacency list* – it is simply a list of edges in the network. For example, the edge list of the network presented in Figure 15.2a is:

{A, B}
{B, C}
{B, D}
{D, E}

Recall that we are dealing with undirected networks, so for example, edge {A,B} is the same as edge {B,A}. The other standard way of representing a network is an *adjacency matrix*. In an adjacency matrix, rows and columns represent nodes, and the matrix entries are 1s and 0s, with a 1 in location (i, j) corresponding to the presence of an edge connecting node i to node j, and a 0 in location (i, j) corresponding to the absence of such an edge. For example, the adjacency matrix representation of the network in Figure 15.2a is presented in Figure 15.4a. As illustrated in this figure, adjacency matrices of networks with no directions on edges are symmetric, meaning that entry (i, j) is equal to the entry (j, i) in the matrix; this is because edges are undirected. We illustrate the sparsity of the PPI network data by visually displaying

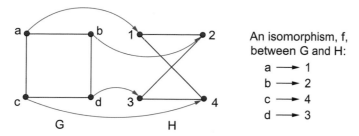

Figure 15.5 Isomorphic graphs G and H with an isomorphism function, f, that maps nodes of G to nodes of H. H is a re-drawing of G, since bijective function f satisfies: ab is an edge of G and $f(a)f(b) = 12$ is an edge of H, bd is an edge of G and $f(b)f(d) = 23$ is an edge of H, dc is an edge of G and $f(d)f(c) = 34$ is an edge of H, and ca is an edge of G and $f(c)f(a) = 41$ is an edge of H.

the adjacency matrix of the yeast PPI network from Figure 15.2b; in its adjacency matrix, presented in Figure 15.4b, the 1s (representing interactions) are displayed as colored dots, while 0s (non-interactions) are not colored. Adjacency list and matrix representations of the data are usually used as input into network analysis software tools (e.g. GraphCrunch [8], Citoscape[2]).

Despite the noise and incompleteness of the interaction networks, these data sets still present a rich source of biological information that computational biologists have begun to analyze. Analyzing these data, comparing them, and finding well-fitting network models to them is non-trivial not only due to the low quality of currently available biological network data, but also due to the provable computational intractability of many graph theoretic problems. Since comparing large networks is computationally hard, approximate or *heuristic* solutions to the problem have been sought. We address this topic in the next section.

 ## Network comparisons

Finding similarities and differences between data sets or between data and models is essential for any data analysis. Hence, if we are dealing with network data, we need to be able to compare large networks. However, comparing large networks is computationally intensive for the following reason. The basis of network comparison lies in finding a *graph isomorphism* between two networks, which can be thought of as re-drawing a graph in a different way [9]. An illustration of an isomorphism is presented in Figure 15.5.

[2] http://www.cytoscape.org.

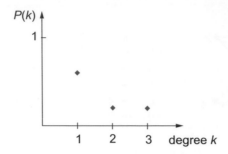

Figure 15.6 The degree distribution of the network from Figure 15.2a.

A *subgraph* of graph G is a graph whose nodes and edges belong to G. For two networks G and H taken as input into a computer program, determining whether G contains a subgraph isomorphic to H is computationally infeasible (the technical term is *NP-complete*, see [9] for details). Furthermore, even if subgraph isomorphism were computationally feasible, it would still be inappropriate to look for exact matches of biological networks due to biological variation. Hence, we want our network comparison methods intentionally to be more flexible, or approximate. Easily computable approximate measures of network topology that are commonly used for comparing large networks are referred to as *network properties*.

Network properties can historically be roughly divided into two main groups: global properties and local properties. Macroscopic statistical *global properties* of large networks are conceptually and computationally easy, and thus they have been extensively studied in biological networks. The most widely used global network properties are the *degree distribution, clustering coefficient, clustering spectra, network diameter*, and various forms of network *centralities* [10]. A global property of a data network and of a model network are computed, and if they are similar, then we say that the model network fit the data with respect to that property. The above-mentioned global properties are defined as follows.

The *degree* of a node is the number of edges touching the node. Hence, in the network presented in Figure 15.2a, nodes A, C, and E have degree 1, node D has degree 2, and node B has degree 3. The *degree distribution* of a network is the distribution of degrees of all nodes in the network. Equivalently, it is the probability that a randomly selected node of a network has degree k (this probability is commonly denoted by $P(k)$). An illustration of the degree distribution of the network from Figure 15.2a is presented in Figure 15.6. Many biological networks have skewed, asymmetric degree distributions with a tail that follows a "power-law" given by the following formula: $P(k) \sim k^{-\gamma}$, for some fixed $\gamma > 0$. All such networks have been termed "scale-free" [10]. This power-law means that the largest percentage of nodes in a scale-free network

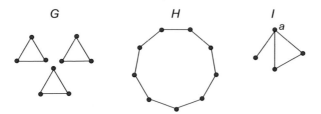

Figure 15.7 *G* and *H* are networks of the same size and the same degree distribution whose structure is very different. The clustering coefficient of network *G* is 1, the clustering coefficient of network *H* is 0, while the clustering coefficient of network *I* is between 0 and 1.

has degree 1, a much smaller percentage of nodes has degree 2, and so forth, but that there exist a small number of highly linked nodes called "hubs."

The *clustering coefficient* of a node is defined as follows. *Neighbors* of node v are nodes that share an edge with v. We look at the neighbors of the node in question, v, and we count how many edges exist between these neighbors as a percentage of the maximum possible number of edges between the neighbors. For example, each node of network G in Figure 15.7 has two neighbors, the neighbors are connected by an edge, and the maximum possible number of edges linking two nodes is 1; thus, the clustering coefficient of each node in network G is $1/1 = 1$. Similarly, we can compute that the clustering coefficient of each node in network H in Figure 15.7 is 0, since there are no edges between the neighbors of any node in H. An example of a clustering coefficient that is strictly between 0 and 1 is that of node a in graph I in the same figure: the clustering coefficient of a is $1/3$, since a has 3 neighbors and only one edge exists between them while the maximum possible number of edges between the 3 neighbors is 3. The clustering coefficient of a network is defined simply as the average of clustering coefficients of all of its nodes. Clearly, it is always between 0 and 1. The clustering coefficient of network G in Figure 15.7 is 1, the clustering coefficient of network H in the same figure is 0, and the clustering coefficient of network I in the same figure is $7/12$ (exercise: verify that the clustering coefficient of network I is equal to $7/12$). Hence, G and H are very different with respect to their clustering coefficients, even though they are of the same size and have the same degree distribution. The *clustering spectrum* of a network is defined as the distribution of average clustering coefficients of degree k nodes over all degrees k in the network.

The *diameter* of a network describes how "far spread" the network is in the following sense. We consider all possible pairs of nodes and for each pair find the shortest path between them; the maximum length over all those paths is the network diameter. We can also take the average of shortest path lengths between all pairs of nodes in a network to obtain the network's *average diameter*.

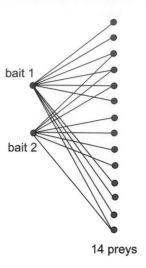

bait 1

bait 2

14 preys

Figure 15.8 An illustration of a bias introduced to the network structure by sampling a much smaller number of baits than preys in pull-down experiments. The baits are forced to be hubs and the preys are of low degree.

Note, however, that networks with exactly the same value for one network property can have very different structures. In the example in Figure 15.7, network G consisting of 3 triangles and H network consisting of one 9-node ring (cycle) are of the same size (i.e. they have the same number of nodes and edges) and have the same degree distribution (each node has degree 2), but their network structure is clearly very different. The same holds for other global network properties [11]. Furthermore, since molecular networks are currently largely incomplete, global network properties of such incomplete networks do not tell us much about the structure of the entire real networks. Instead, they describe the network structure produced by the sampling techniques used to obtain these networks (e.g. [7]). For example, in bait–prey experiments for PPI detection, if the number of baits is much smaller then the number of preys, then all of the baits will be detected as hubs, and all of the preys will be of low degree, as illustrated in Figure 15.8. Thus, global statistics on incomplete real networks may be biased and even misleading with respect to the currently unknown complete networks. Conversely, as mentioned above, certain local neighborhoods of molecular networks are well studied, usually the regions of a network relevant for human disease. Therefore, *local* statistics applied to the well-studied areas of a network are more appropriate.

Local network properties include network motifs and graphlets (e.g. [11–13]). Analogous to sequence motifs, *network motifs* have been defined as subgraphs that recur in a network at frequencies much higher than those found in randomized networks [12]. Recall that a *subgraph* (or a *partial subgraph*) of a network G is a network whose nodes and edges belong to G. An *induced subgraph* of G is a subgraph that contains

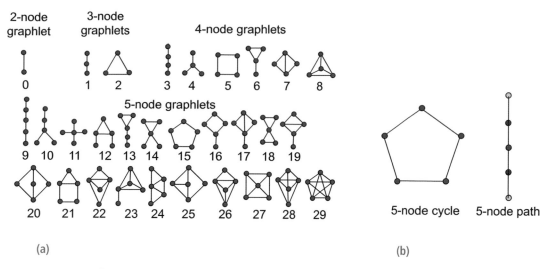

Figure 15.9 (a) All 2-, 3-, 4-, and 5-node graphlets. (b) A 5-node cycle and a 5-node path; all nodes in the cycle are the same, but the nodes on the path are topologically different.

all edges of G connecting the chosen subset of nodes. For example, a 3-node partial subgraph of a triangle can be a 3-node path (a 3-node path is denoted by 1 in Figure 15.9a), but a triangle has only one induced subgraph on 3 nodes, which is a triangle. Note that when we are finding network motifs, it is not clear what subgraphs are more frequent than expected at *random*, since it is not clear what should be expected at random [14]. Nevertheless, motifs have been very useful for finding functional building blocks of transcriptional regulation networks, as well as for differentiating between different types of real networks. Also, being partial subgraphs, they are appropriate for studying biological networks, since not all interactions in real biological networks need to concurrently occur in a cell, while they are all present in the network representations that we study.

Approaches for studying network structure have been proposed that are based on the frequencies of occurrences of *all* small induced subgraphs in a network (not only overrepresented ones), called *graphlets* (Figure 15.9a) [11, 13, 15]. These approaches are free from the biases that motif-based approaches have, namely biases introduced by selection of a random graph model (defined below) for the data that is necessary to define network motifs (graph models are described below), as well as by the choice of partial rather than induced subgraphs for studying network structure. That is, graphlets do not need to be overrepresented in a data network and this, along with being induced, distinguishes them from network motifs. Note that whenever the structure of a graph (or a graph family) is studied, we care about induced rather than partial subgraphs. If we simply find the frequency of each of the graphlets in a network and compare such frequency distributions, we can measure structural similarity between networks [11]. We can further refine this similarity measure by noticing that in some graphlets, the nodes are distinct from each other. For example, in a ring (cycle) of five nodes, every node looks the same as every other, but in a chain (path) of five nodes, there are two end nodes, two near-end nodes, and one middle node (Figure 15.9b). This idea of finding symmetry groups within graphlets can be mathematically formalized [13]. Network analysis and the modeling software package called GraphCrunch[3] provides graphlet-based network comparisons [8].

When we are comparing two networks, one of their network properties can indicate that the networks are similar, while another can indicate that they are different. Recall that networks G and H in Figure 15.7 have identical degree distributions, but different clustering coefficients. There exist approaches that try to reconcile between such contradictions in the agreement of different network properties (see [16] for details).

3 Network models

In this section, first we describe the most commonly used network models and then we discuss how they can be used to learn new biology from biological network data.

There exist many different network (or random graph) models that we could compare the data against, for example, to find network motifs [14]. The earliest such model is the *Erdos–Renyi* random graph model. An Erdos–Renyi random graph on n nodes is constructed so that edges are added between pairs of nodes with the same probability p. Many of the properties of Erdos–Renyi random graphs are mathematically well understood. Therefore, they form a standard model to compare the data against, even though they are not expected to fit the data well. Since Erdos–Renyi graphs, unlike biological networks, have "bell-shaped" degree distributions and low clustering coefficients, other network models for real-world networks have been sought. One such model is the *generalized random graphs* model. In these graphs, the edges are randomly chosen as in Erdos–Renyi random graphs, but the degree distribution is constrained to match the degree distribution of the data (for their construction, see [10]). Another commonly used network model is that of *small-world* networks. In these networks, nodes are placed on a ring and connected to their ith neighbors on the ring for all i smaller than some given number k, but there is also a small number of random links across the ring (as illustrated in Figure 15.10b). Hence, small-world networks have small diameters (meaning that their diameter is an order of magnitude smaller than the number of their nodes) and large clustering coefficients [10]. The scale-free network model has

[3] http://bio-nets.doc.ic.ac.uk/graphcrunch/ and http://bio-nets.doc.ic.ac.uk/graphcrunch2/.

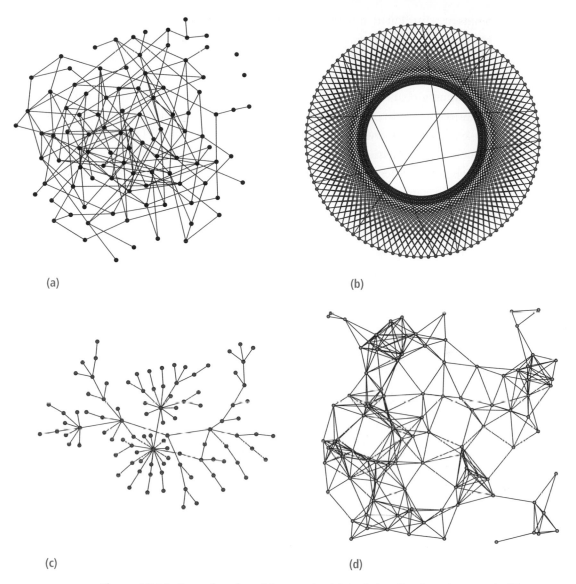

Figure 15.10 Examples of model networks. (a) An Erdos–Renyi random graph. (b) A small-world network. (c) A scale-free network. (d) A geometric random graph.

already been mentioned above; *scale-free* networks include an additional condition that the degree distribution follows a power-law [10]. Another relevant graph class is that of *geometric graphs* defined as follows. If we have a collection of points dispersed in space, we pick some constant distance ϵ and say that two points are "related" if they are within ϵ of each other. The relationship can be represented as a graph, where each point in space is a node and two nodes are connected if they are within distance ϵ. If the points are distributed at random, then it is a *geometric random graph*. Illustrations of networks of about the same size, but that belong to these different network models are presented in Figure 15.10; even without computing any network properties for them, we can just look at them and conclude that their structure is very different. Studies examining global network properties of early PPI networks tried to model them with scale-free networks. Later, the above described graphlet-based measures of local network structure demonstrated that newer and more complete PPI network data are better modeled by geometric graphs [11, 13]. It is important to be aware of different network models, since different biological networks (e.g. metabolic networks, transcriptional regulation networks, neuronal wiring networks) might be best modeled by different network models.

The degree distributions of many biological networks approximately follow a power-law. Hence, many variants of scale-free network growth models have been proposed. For PPI networks, such models are based on biologically motivated gene *duplication and mutation* network growth principles (e.g. [17]): networks grow by duplication of nodes (genes), and as a node gets duplicated, it inherits most of the interactions of the parent node, but gains some new interactions. Similarly, gene duplication and mutation-based geometric network growth models have been proposed [18]. These models are based on the following observations. All biological entities, including genes and proteins as gene products, exist in some multidimensional biochemical space. Genomes evolve through a series of gene duplication and mutation events, which are naturally modeled in the above-mentioned biochemical space: a duplicated gene starts at the same point in biochemical space as its parent, and then natural selection acts either to eliminate one, or cause them to separate in the biochemical space. This means that the child inherits some of the neighbors of its parent while possibly gaining novel connections as well. The farther the "child" is moved away from its "parent," the more different its biochemical properties.

How can we use network models to learn more about biology? Even though modeling of biological networks is still in its infancy, network models have already been used for such purposes. As mentioned above, network models are crucial for network motif identification and network motifs are believed to be functional building blocks of molecular networks. Another example of the use of network models is finding cost-effective strategies for completing interaction maps, which is an active research topic (e.g. see [19]). A scale-free network model has been used to propose a strategy for time- and cost-optimal interactome detection [20]. Using the property that scale-free networks contain hubs, this strategy proposes an "optimal walk" through the PPI network using pull-down experiments, so that we preferentially choose hub nodes as baits, since that way we would detect most of the interactions with the smallest number of expensive pull-down experiments. However, the danger of using an inadequate network model for such a purpose (for instance, if real PPI networks do not have hubs) is that we might waste time and resources. Furthermore, we might end up with a wrong identification of the "complete" interactome maps, since the model might tell us never to examine certain parts of the interactome.

Network models have also been used successfully for other biological applications. In addition to the above-mentioned use of network models for fast data collection, another reason for modeling biological networks is the development of fast heuristic methods for data analysis. One property of every heuristic approach is that it performs poorly on some data. Thus, heuristics are designed, with the help of models, to work well for a particular application domain, for example, for PPI networks. Geometric graph models have been used for this purpose. In particular, they were used to design efficient strategies for graphlet count estimation [21] in PPI networks. Another application is de-noising of PPI network data for which geometric graphs have been used, as follows [22]. A method that directly tests whether PPI networks have a geometric structure was used to assess the confidence levels of PPIs obtained by experimental studies, as well as to predict new PPIs, thus guiding future biological experiments [22]. Specifically, it was used to assign confidence scores to physical human PPIs from the BioGRID database. Also, it was used to predict novel PPIs, a statistically significant fraction of which corresponded to protein pairs involved in the same biological process or having the same cellular localization. This is encouraging, since such protein pairs are more likely to interact in the cell. Moreover, a statistically significant portion of the predicted PPIs was validated in the HPRD database and the newer release of BioGRID.

4   Using network topology to discover biological function

Analogous to extracting biological knowledge by analyzing genetic sequences, biological networks are a new, rich source of biological information from which we started learning about biology. Finding the relationship between network topology and biological function is a step in this direction. Network-based prediction of protein function and the role of networks in disease have been studied [23, 24].

The simplest property of a node in a network is its degree. Hence, early approaches studied correlations between high protein connectivity (i.e. high degree) in a PPI network and its essentiality in baker's yeast [25]. Even though early data sets showed such correlations, this simple technique failed on newer PPI network data [26]. Similar conflicting results have been reported for correlations between protein connectivity and evolutionary rates (e.g. [27]). Similarly, correlations between connectivity and protein function were examined [28].

Other methods for linking network structure to biological function were based on the premise that proteins that are closer in the PPI network are more likely to have similar function (e.g. [29]). Attempts to utilize somewhat more sophisticated graph theoretic methods for this purpose have been examined, including cut-based and network flow-based approaches (e.g. [30]) (informally, a *cut* is a division of a network into disconnected parts, while a *network flow* can be thought of as a flow of fluids in pipes). Also, various *clustering* methods (that usually look for densely interconnected subnetworks) have been applied to PPI networks and functional homogeneity of proteins in the clusters has been used for protein function prediction (e.g. [2, 28, 31]).

Human PPI networks have been analyzed in the search for topological properties of disease-related proteins. The hope is to get insights into diseases that would lead to better drug design. It has been concluded that disease-related proteins have high connectivity, are closer together, and are centrally positioned within the PPI network [24]. However, a controversy arises again, since, as discussed above, disease-causing proteins may exhibit these properties in a network simply because they have been better studied than non-disease proteins.

Graphlets have been used to generalize the node degree into a topologically stronger measure that captures the structural details of individual nodes in a network. This measure has been used to relate the network structure around a node to protein function and involvement in disease (e.g. [15]). The generalization is achieved as follows. Recall that the *degree* of a node is the number of edges it touches. An edge is the only graphlet with two nodes (graphlet 0 in Figure 15.9a). Thus, analogous to the node degree, we can define a *graphlet degree* of node v with respect to each graphlet i in Figure 15.9a, in the sense that the i-degree of v counts *how many graphlets of type i touch node v* [13]. That is, we count not only how many edges a node touches (this is the node degree), but also how many triangles it touches, how many squares it touches, etc. Hence, the node degree is simply the 0-degree. Also, it matters where a node touches a graphlet that is not "symmetric"; for example, an edge is symmetric, but in a 3-node path, the end nodes look the same, but the middle node is different (see [13, 15] for details). Hence, we need to count how many 3-node paths a node touches at an end and also how many 3-node paths it touches at the middle. By counting this for all graphlets, we get the *graphlet degree vector (GDV)* or *GD-signature* of a node. An example of computing a GD-signature is presented in Figure 15.11.

Since the degree of a protein in a PPI network is a weak predictor of its biological function, the question is whether the GD-signature captures the link between network topology and biological function better than the degree. Indeed, it has been shown that GD-signatures correspond to similarity in biological function and involvement in disease that could not have been discovered from node degrees and the function predictions have been phenotypically validated [15]. For example, 27 genes identified

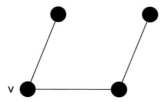

GDV(v)=(2,1,1,0,0,1,0,...0)

Figure 15.11 A small 4-node network. The graphlet degree vector of node *v* is (2,1,1,0,0,1,0 ...), because *v* is touched by two edges, the end of one 3-node path, the middle of another 3-node path, and the middle of a 4-node path.

as negative regulators of melanogenesis by an RNAi functional genomics approach were also identified as cancer gene candidates based on their GD-signature similarities [15]. Of these 27 genes, 85%, i.e. 23 of them, were validated in the literature as cancer-associated genes. Interestingly, 20 of these 27 genes are kinases, enzymes that are known to dynamically regulate the process of cellular transformation. Several of these kinases are known regulators of melanogenesis. Also, from the topology around nodes in PPI networks described by GD-signatures, by finding nodes that have GDVs similar to GDVs of nodes that are known regulators of melanogenesis in the human PPI network, novel regulators of melanogenesis in human cells were successfully identified and validated by systems-level functional genomics RNAi screens [15].

Similarly, GD-signatures were used to establish a link between network topology around a node in a PPI network and homology [32]. The GDV similarity of homologous proteins in a PPI network has been shown to be statistically significantly higher than that of non-homologous proteins. When this topological similarity is compared with their sequence identity, it has been shown that network similarity uncovers almost as much homology as sequence identity. Hence, it has been argued that genomic sequence and network topology are complementary sources of biological information for homology detection, as well as for analyzing evolutionary distance and functional divergence of homologous proteins.

A related topic is that of network-based approaches to *systems pharmacology*. Network analyses of drug action are starting to be used as part of this emerging field that aims to develop an understanding of drug action across multiple scales of organismal complexity, from cell to tissue to organism [33]. Biochemical interaction networks, such as PPI networks, have been linked into a "super-network" with networks of drug similarities, interactions, or therapeutic indications. For example, a network connecting drugs and drug targets (proteins affected by a drug) was constructed and used to generate two "network projections:" (1) a network in which nodes are drugs

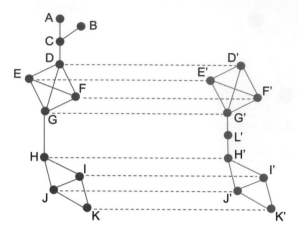

Figure 15.12 An example of an alignment of two networks.

and they are connected if they share a common target; and (2) a network in which nodes are targets and they are connected if they are affected by the same drugs [34]. By analyzing these two network projections, conclusions have been made about existing drugs affecting few novel targets, as well as about drug targets having higher degrees than non-targets in the PPI network. Again, the latter might be an artifact of disease-related parts of the PPI network being more studied. A survey of network-based analyses in systems pharmacology can be found in [33].

5 Network alignment

Analogous to genetic sequence alignment, network alignment is expected to have a deep impact on biological understanding. Network alignment is the general problem of finding the best way to "fit" graph G into graph H. Note that in biological networks, it is unlikely that G would exist as an exact subgraph of H due to noise in the data (e.g. missing edges, false edges, or both) and also due to biological variation. For these reasons, it is not obvious how to measure the "goodness" of this fit. A simple example illustrating network alignment is presented in Figure 15.12. Analogous to genomic sequence alignments, biological network alignments can be useful for knowledge transfer, since we may know a lot about some nodes in one network and almost nothing about aligned, topologically similar nodes in the other network. Also, network alignments can be used to measure the global similarity between biological networks of different species, and the resulting matrix of pairwise global network similarities can be used to infer phylogenetic relationships [35]. However, unlike with the sequence

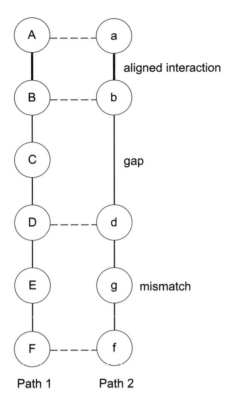

Figure 15.13 An illustration of an *aligned interaction*, a *gap*, and a *mismatch* in a pathway alignment. Vertical lines represent PPIs, horizontal dashed lines represent alignment between proteins with significant sequence similarity (BLAST E-value $\leq E_{cutoff}$). Adapted from [40].

alignment, the problem of network alignment is computationally infeasible to solve exactly. Hence, approximate solutions are being sought.

Analogous to sequence alignments, there exist *local* and *global* network alignments. Local alignments map independently each local region of similarity. For example, in Figure 15.12, nodes D, E, F, G from the black network could simultaneously be aligned to nodes D′, E′, F′, G′ as well as to nodes H′, I′, J′, K′ in the orange network. Thus, such alignments can be ambiguous, since one node can have different pairings. On the contrary, a global network alignment uniquely maps each node in the smaller network to only one node in the larger network, as illustrated in Figure 15.12. However, this may lead to suboptimal matchings in some local regions. For biological networks, the majority of currently available methods used for alignment have focused on local alignments (e.g. [36, 37]. Generally, local network alignments are not able to identify large subgraphs that have been conserved during evolution (e.g. [35]). Global network alignments have also been proposed (e.g. [35, 38, 39], but most of the existing methods incorporate some *a priori* information about nodes, such as sequence similarities of

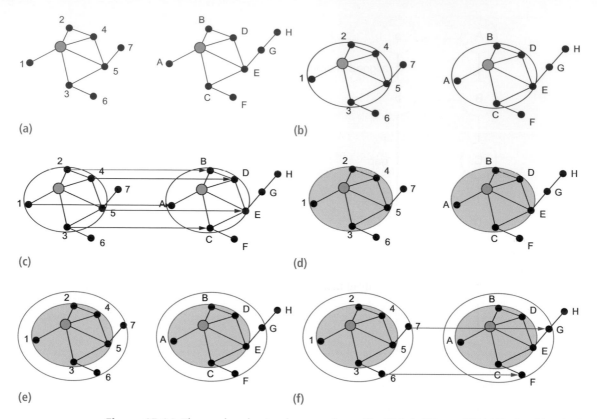

Figure 15.14 The seed-and-extend approach used in GRAph ALigner (GRAAL) algorithm [35]. (a) The green nodes are chosen as seed nodes and aligned based on their GDV similarity score. (b) The neighbors of seed nodes in the two networks are considered. (c) The neighbors of seed nodes in the two networks are greedily aligned. (d) The shaded area represents the aligned parts of the two networks. (e) The neighbors of aligned nodes in the two networks are considered. (f) The neighbors of aligned nodes in the two networks are greedily aligned.

proteins in PPI networks (see below), or they use some form of learning on a set of "true" alignments [38].

There are two main issues in each of the network alignment algorithms. First, how to define similarity scores between nodes from different networks. Second, how to quickly identify high-scoring alignments among the exponentially many possible alignments. For PPI networks, the first issue is usually addressed by designing a node similarity measure as a function of protein sequence similarity and some sort of their topological similarity in the network (see below). The second issue is often solved by greedy algorithms to reduce the computational time; a *greedy algorithm* makes locally optimal choices at each step of its execution hoping to find the global optimum (but usually with no proven guarantee of achieving it, so actual performance must be tested empirically). There exist many network alignment algorithms, so giving the details of each is out of the scope of this chapter. Hence, we illustrate them on a couple of examples.

In the simplest case, we can define similarity between a protein pair solely by their sequence similarity. This is typically done by applying BLAST to perform all-to-all alignment between sequences of proteins from two different networks. Then the simplest network alignment would correspond to interactions across PPI networks involving pairs of proteins in one species and their best sequence-matched proteins in the other. However, network alignment algorithms go beyond this simple identification of conserved protein interactions to identify large and complex network subgraphs that have been conserved across species. Usually, this is done by having the highest-scoring node pair between two networks aligned and used as an "anchor" or "seed" for the search algorithm that extends around these seed nodes in a greedy way in each of the networks looking for larger optimal network alignments (Figure 15.15). In the remainder of this section, we describe algorithms illustrating these concepts.

The earliest network alignment algorithm, called PathBLAST, searches for high-scoring pathway alignments between two networks [36, 40]. The alignments are scored via the product of the probability that each aligned protein pair is truly homologous (based on BLAST E-value of aligning the protein sequences) and that each aligned PPI is a true interaction (based on false-positive rates associated with interactions). This method has identified orthologous pathways between baker's yeast and bacterium *Helicobacter pylori* and 150 high-scoring pathway alignments of length four (four proteins per path) were identified. Although the number of interactions that were conserved between the two species was low, the use of "gaps" and "mismatches" in a pathway (see Figure 15.16) allowed for detection of larger network regions that were generally conserved. A *gap* occurs when a PPI in one path "skips over" a protein in the other path; a *mismatch* is defined to occur when aligned proteins do not share sequence similarity (Figure 15.13). As a validation that the identified aligned pathways corresponded to conserved cellular functions, it was shown that the aligned network regions were significantly enriched in certain biological processes.

A global network alignment algorithm that uses only network topology to score node alignments is called GRAph ALigner (GRAAL) [35]. Since it uses only network topology, it can be applied to any networks, not just biological ones. The alignments of nodes are scored based on their GDV similarity described in Section 4, and do not use the protein sequence information. The seed-and-extend approach used in GRAAL works as follows (illustrated in Figure 15.14). The highest-scoring node pair (i.e. the one with the highest GDV similarity) is used as a seed pair around which the greedy algorithm "extends" trying to find the largest possible (in terms of the number of nodes and edges) high-scoring aligned subgraphs. After the seed nodes are aligned (green

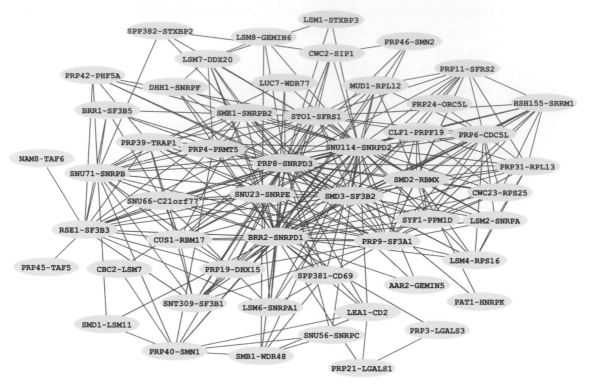

Figure 15.15 GRAAL's alignment of yeast and human PPI networks. Each node corresponds to a pair of yeast and human proteins that are aligned. Alignment is determined based on GDV similarity of the two proteins, without using sequence similarity. An edge between two nodes means that an interaction exists in both species between the corresponding protein pairs. Thus, the displayed networks appear, in their entirety, in the PPI networks of both species. The second largest CCS consists of 286 interactions amongst 52 proteins; this subgraph shows very strong enrichment for the same biological function (splicing) in both yeast and human PPI networks. The figure is taken from [35].

nodes in Figure 15.14a), the neighbors of aligned nodes are considered (Figure 15.14b) and aligned so that the score of the newly aligned nodes is maximized, i.e. pairs of nodes with the highest GDV similarity are greedily aligned. In the illustration in Figure 15.14c, this corresponds to node 1 being aligned with node A, node 2 to node B, node 3 to node C, node 4 to node D, and node 5 to node E. Next, the neighbors of aligned nodes that are not aligned yet are found (Figure 15.14e) and aligned using the same principle. This is repeated until all nodes that can be reached are aligned. However, this may result in some unaligned nodes in both networks. Also, to allow for gaps and mismatches, GRAAL repeats this seed-and-extend approach on modified networks: in each of the networks, edges are added to link nodes at distance $\leq p$, first for $p = 2$ and after aligning such modified networks, then the same is repeated for $p = 3$. This

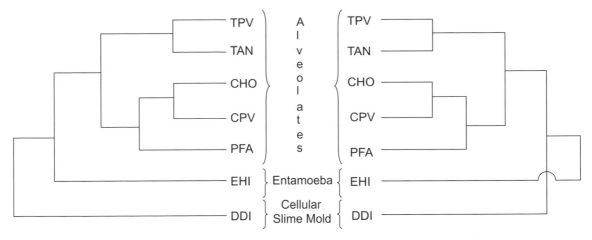

Figure 15.16 Comparison of the phylogenetic trees for protists obtained by genetic sequence alignments (left) and GRAAL's metabolic network alignments (right). The following abbreviations are used for species: CHO, *Cryptosporidium hominis*; DDI, *Dictyostelium discoideum*; CPV, *Cryptosporidium parvum*; PFA, *Plasmodium falciparum*; EHI, *Entamoeba histolytica*; TAN, *Theileria annulata*; TPV, *Theileria parva*; the species are grouped into "Alveolates," "Entamoeba," and "Cellular Slime mold" classes [35].

allows for a path of length p in one network to be aligned to a single edge in the other, which is analogous to allowing insertions and deletions in sequence alignment.

When applied to human and baker's yeast PPI networks, GRAAL exposes regions of network similarity about an order of magnitude larger than other algorithms. The algorithm aligns network regions of yeast and human in which a large percentage of proteins perform the same biological function in both species. For example, GRAAL [35] aligns a 52-node subnetwork between yeast and human in which 98% of yeast and 67% of human proteins are involved in splicing (Figure 15.15). This result is encouraging, since splicing is known to be conserved even between distant eukaryotes. Because the algorithm aligns functionally similar regions, it is further used to transfer biological knowledge from annotated to unannotated parts of aligned networks.

Furthermore, analogous to sequence alignment, GRAAL is also used to infer phylogeny, with the intuition that species with more similar network topologies should be closer in the phylogenetic tree. The algorithm has been used to infer phylogenetic trees for protists and fungi from the alignments of their metabolic networks, and the resulting trees show a striking resemblance to the trees obtained by sequence comparisons (Figure 15.16) [35]. Hence, network alignments in general could potentially provide a new, independent source of biological and phylogenetic information.

The reason for developing methods that rely on topology only for aligning large biological networks is twofold. While genetic sequences describe a part of biological information, so too do biological networks. Sequence and network topology have been shown to provide complementary insights into biological knowledge [32]. Sequence alignment algorithms do not use biological information external to sequences to perform alignments. Analogously, using only topology for network alignment might be appropriate, since using biological information external to network topology might hinder the discovery of biological information that is encoded solely in network topology. We need to design reliable algorithms for purely topological network alignments first and then integrate them with other sources of biological information.

DISCUSSION

In this chapter, we reviewed currently available methods for graph-theoretic analysis and modeling of biological network data. Even though network biology is still in its infancy, it has already provided insights into biological function, evolution, and disease. The impact of the field is likely to increase as more biological network data of high quality becomes available and as better methods for their analysis are developed. Synergy between biological and computational scientists is necessary for advancing this nascent research field.

QUESTIONS

(1) Why do we use network properties?
(2) Name network properties and describe how they can be computed.
(3) Name three high-throughput methods for protein–protein interaction detection.
(4) Describe the sources of bias introduced in the protein–protein interaction network data that were obtained by "pull-down" experiments.

REFERENCES

[1] N. Simonis, J.-F. Rual, A.-R. Carvunis, et al. Empirically controlled mapping of the Caenorhabditis elegans protein–protein interactome network. Nature Meth., 6(1):47–54, 2009.

[2] N. J. Krogan, G. Cagney, H. Yu, et al. Global landscape of protein complexes in the yeast Saccharomyces cerevisiae. Nature, 440:637–643, 2006.

[3] A. H. Y. Tong, G. Lesage, G. D. Bader, *et al.* Global mapping of the yeast genetic interaction network. *Science*, 303:808–813, 2004.

[4] J.-F. Rual, K. Venkatesan, T. Hao, *et al.* Towards a proteome-scale map of the human protein–protein interaction network. *Nature*, 437:1173–1178, 2005.

[5] B. Titz, S. V. Rajagopala, J. Goll, *et al.* The binary protein interactome of *Treponema pallidum* – the syphilis spirochete. *PLoS One*, 3:e2292, 2008.

[6] M. D. Dyer, T. M. Murali, and B. W. Sobral. The landscape of human proteins interacting with viruses and other pathogens. *PLoS Pathogens*, 4:e32, 2008.

[7] J. D. H. Han, D. Dupuy, N. Bertin, M. E. Cusick, and M. Vidal. Effect of sampling on topology predictions of protein–protein interaction networks. *Nature Biotechnol.*, 23:839–844, 2005.

[8] O. Kuchaiev, A. Stevanovic, W. Hayes, and N. Pržulj. GraphCrunch 2: Software tool for network modeling, alignment and clustering. *BMC Bioinform.*, 12:24, 2011.

[9] D. B. West. *Introduction to Graph Theory*, 2nd edn. Prentice Hall, Upper Saddle River, NJ, 2001.

[10] M. E. J. Newman. The structure and function of complex networks. *SIAM Rev.*, 45(2):167–256, 2003.

[11] N. Pržulj, D. G. Corneil, and I. Jurisica. Modeling interactome: Scale-free or geometric? *Bioinformatics*, 20(18):3508–3515, 2004.

[12] R. Milo, S. S. Shen-Orr, S. Itzkovitz, N. Kashtan, D. Chklovskii, and U. Alon. Network motifs: Simple building blocks of complex networks. *Science*, 298:824–827, 2002.

[13] N. Pržulj. Biological network comparison using graphlet degree distribution. *Bioinformatics*, 23:e177–e183, 2007.

[14] Y. Artzy-Randrup, S. J. Fleishman, N. Ben-Tal, and L. Stone. Comment on "Network motifs: Simple building blocks of complex networks" and "Superfamilies of evolved and designed networks". *Science*, 305:1107c, 2004.

[15] T. Milenković, V. Memisević, A. K. Ganesan, and N. Pržulj. Systems-level cancer gene identification from protein interaction network topology applied to melanogenesis-related interaction networks. *J. R. Soc. Interf.*, doi:10.1098/rsif.2009.0192, 2009.

[16] V. Memisević, T. Milenković, and N. Pržulj. An integrative approach to modeling biological networks. *J. Integr. Bioinform.*, 7(3):120, 2010.

[17] R. Pastor-Satorras, E. Smith, and R. V. Sole. Evolving protein interaction networks through gene duplication. *J. Theor. Biol.*, 222:199–210, 2003.

[18] N. Pržulj, O. Kuchaiev, A. Stevanovic, and W. Hayes. Geometric evolutionary dynamics of protein interaction networks. In: *2010 Pacific Symposium on Biocomputing (PSB)*, 2010.

[19] A. S. Schwartz, J. Yu, K. R. Gardenour, R. L. Finley Jr., and T. Ideker. Cost-effective strategies for completing the interactome. *Nature Meth.*, 6(1):55–61, 2009.

[20] M. Lappe and L. Holm. Unraveling protein interaction networks with near-optimal efficiency. *Nature Biotechnol.*, 22(1):98–103, 2004.

[21] N. Pržulj, D. G. Corneil, and I. Jurisica. Efficient estimation of graphlet frequency distributions in protein–protein interaction networks. *Bioinformatics*, 22(8):974–980, 2006. doi:10.1093/bioinformatics/btl030.

[22] O. Kuchaiev, M. Rasajski, D. Higham, and N. Pržulj. Geometric de-noising of protein–protein interaction networks. *PLoS Comput. Biol.*, 5:e1000454, 2009.

[23] R. Sharan, I. Ulitsky, and R. Shamir. Network-based prediction of protein function. *Mol. Syst. Biol.*, 3(88):1–13, 2007.

[24] R. Sharan and T. Ideker. Protein networks in disease. *Genome Res.*, 18:644–652, 2008.

[25] H. Jeong, S. P. Mason, A.-L. Barabási, and Z. N. Oltvai. Lethality and centrality in protein networks. *Nature*, 411(6833):41–42, 2001.

[26] H. Yu, P. Brawn, M. A. Yildirim, *et al.* High-quality binary protein interaction map of the yeast interactome network. *Science*, 322:104–110, 2008.

[27] M. Stumpf, W. P. Kelly, T. Thorne, and C. Winf. Evolution at the systems level: The natural history of protein interaction networks. *Trends Ecol. Evol.*, 22:366–373, 2007.

[28] N. Pržulj, D. Wigle, and I. Jurisica. Functional topology in a network of protein interactions. *Bioinformatics*, 20(3):340–348, 2004.

[29] H. N. Chua, W. K. Sung, and L. Wong. Exploiting indirect neighbours and topological weight to predict protein function from protein–protein interactions. *Bioinformatics*, 22:1623–1630, 2006.

[30] E. Nabieva, K. Jim, A. Agarwal, B. Chazelle, and M. Singh. Whole-proteome prediction of protein function via graph-theoretic analysis of interaction maps. *Bioinformatics*, 21:i302–i310, 2005.

[31] A. D. King, N. Pržulj, and I. Jurisica. Protein complex prediction via cost-based clustering. *Bioinformatics*, 20(17):3013–3020, 2004.

[32] V. Memisević, T. Milenković, and N. Pržulj. Complementarity of network and sequence information in homologous proteins. *J. Integr. Bioinform.*, 7(3):135, 2010.

[33] S. I. Berger and R. Iyengar. Network analyses in systems pharmacology. *Bioinformatics*, 25:2466–2472, 2009.

[34] M. A. Yildirim, K. I. Goh, M. E. Cusick, A. L. Barabási, and M. Vidal. Drug–target network. *Nature Biotechnol.*, 25:1119–1126, 2007.

[35] O. Kuchaiev, T. Milenković, V. Memisević, W. Hayes, and N. Pržulj. Topological network alignment uncovers biological function and phylogeny. *J. R. Soc. Interf.*, 2010. doi:10.1098/rsif.2010.0063.

[36] B. P. Kelley, Y. Bingbing, F. Lewitter, R. Sharan, B. R. Stockwell, and T. Ideker. Path-BLAST: A tool for alignment of protein interaction networks. *Nucl. Acids Res.*, 32:83–88, 2004.

[37] J. Flannick, A. Novak, S. S. Balaji, H. M. Harley, and S. Batzglou. Graemlin general and robust alignment of multiple large interaction networks. *Genome Res.*, 16(9):1169–1181, 2006.

[38] J. Flannick, A. F. Novak, C. B. Do, B. S. Srinivasan, and S. Batzoglou. Automatic parameter learning for multiple network alignment. In: *RECOMB '08*, Proceedings of the 12th Annual International Conference on Research in Computational Molecular Biology. Springer-Verlag, Heidelberg, 214–231, 2008.

[39] C.-S. Liao, K. Lu, M. Baym, R. Singh, and B. Berger. Isorankn: Spectral methods for global alignment of multiple protein networks. *Bioinformatics*, 25(12):i253–i258, 2009.

[40] B. P. Kelley, R. Sharan, R. M. Karp, *et al.* Conserved pathways within bacteria and yeast as revealed by global protein network alignment. *Proc. Natl. Acad. Sci. U S A*, 100:11,394–11,399, 2003.

CHAPTER SIXTEEN

Regulatory network inference

Russell Schwartz

Identifying the complicated patterns of regulatory interactions that control when different genes are active in a cell is a challenging problem, but one essential to understanding how organisms function at a systems level. In this chapter, we will examine the role of computational methods in making such inferences by studying one particularly important version of this problem: the inference of genetic regulatory networks from gene expression data. We will first briefly cover some necessary background on the biology of genetic regulation and technology for measuring the activities of distinct genes in a sample. We will then work through the process of how one can abstract the biological problem of finding interactions among genes into a precise mathematical formulation suitable for computational analysis, starting from very simple variants and gradually working up to models suitable for analysis of large-scale networks. We will also briefly cover key algorithmic issues in working with such models. Finally, we will see how one can transition from simplified pedagogical models to the more detailed, realistic models used in actual research practice. In the process, we will learn about some key concepts in computer science and machine learning, consider how computational scientists think about solving a problem, and see why such thinking has come to play an essential role in the emerging field of systems biology.

Introduction

Each cell in a biological organism depends on the coordinated activity of thousands of different kinds of proteins occurring in potentially millions of variations. To function properly, the cell must ensure that each of these proteins is present in the specific places

Bioinformatics for Biologists, ed. P. Pevzner and R. Shamir. Published by Cambridge University Press.
© Cambridge University Press 2011.

it is needed, at the proper times, and in the necessary quantities. An exquisitely complicated network of regulatory interactions ensures these conditions are met throughout the cell's lifetime. Such regulatory interactions include mechanisms for controlling when DNA molecules are properly primed to produce RNA, how often RNA molecules are produced from DNA, how long RNA molecules persist in cells, how often the RNA molecules give rise to proteins, how the proteins are shuttled about the cell, how they are chemically modified at any given time, with which other proteins they are associated, and when they are degraded. These various kinds of regulation are carried out and interconnected through an array of specialized regulatory proteins.

In a regulatory network inference problem, one seeks to infer these complex sets of interactions using indirect measurements of the activities of the individual components of the system. Identifying how genes regulate one another is a fundamental problem in basic biological research into how organisms function, develop, and evolve. Regulatory networks also have important practical applications in helping us to interpret large-scale genomic data and to use them to understand how organisms respond to disease, potential treatments, and other environmental influences. While we cannot hope to do justice to such a complicated problem in one chapter, we can look at one special case of the problem that will illustrate the general principles behind a broad array of work in the field. We will specifically examine the problem of how one can infer transcriptional regulatory networks – networks describing regulatory behavior that act by controlling when RNA is transcribed from DNA – using measurements of RNA expression levels.

The problem of regulatory network inference is interesting not only for its intrinsic scientific merit but also as a model for several important themes in how modern computational biology is practiced and how one reasons about computational inferences from complex biological data sets in general. First, regulatory network inference provides an example of how computational biology intersects with another major trend in modern biological research: systems biology. Systems biology arose out of the realization that one cannot hope to understand the complicated networks of interactions typical of real biological systems by looking at just one or a few components at a time, as was long the standard in biological research. Rather, to infer the overall behavior of a system, researchers must build unified models of the interactions of many components, often using large, noisy data sets. This sort of inference critically depends on computer science methods to enumerate over large numbers of possible models of a given system and weigh the plausibility of each model given the available data. Such systems-level thinking increasingly drives research in biology and has vastly increased the need for computer science expertise in the biological world.

More fundamentally, regulatory network inference is a great example of a problem in machine learning, a subdiscipline of computer science concerned with inferring probabilistic models of complex systems from exactly the kinds of large, error-prone data sets one increasingly encounters in biological contexts. Machine learning has thus emerged as one of the key technologies behind modern high-throughput biology. If we want to understand current directions in computational biology, we need to understand how a researcher thinks about a machine learning problem and some of the basic ways he or she poses and solves such a problem.

Furthermore, regulatory network inference is a problem whose solution critically depends on a careful matching of the class of models one wishes to solve with the data one has available to solve them. It therefore provides a great case study for thinking about the general topic of designing mathematical models for problems in the real world, which is really the beginning of any work in computational biology. The network inference problem is perhaps unusual among those covered in this text in that the hardest, and perhaps most interesting, part of solving it is simply formalizing the problem we wish to solve. This chapter will therefore focus primarily on the issue of formulating the problem mathematically and less so on the details of how one actually solves it.

1.1   The biology of transcriptional regulation

Before we can consider computational approaches to regulatory network inference, we need to know something about the biology of transcriptional regulation. At a high level, a transcriptional regulatory network can be understood in terms of the interactions of two elements: transcription factors and transcription factor binding sites. A transcription factor is a specialized protein that controls when a gene is transcribed to produce RNA. A transcription factor binding site is a small segment of DNA recognized by a particular transcription factor. Transcription factor binding sites are usually, but not exclusively, found near a region called a promoter that occurs near the start of each gene. A promoter serves to recruit the polymerase complex that will read the DNA to produce an RNA transcript. When the transcription factor is present, and perhaps appropriately activated, it will physically bind to its transcription factor binding sites wherever they are exposed in the DNA. The presence of the transcription factor then influences how the transcriptional machinery of the cell acts on the corresponding gene. A given transcription factor can facilitate the recruitment of the polymerase, causing the target gene to be transcribed at a higher level when the transcription factor is present, or it can interfere with the recruitment of the polymerase, reducing expression of the target gene. Furthermore, transcription factors may act in groups, with a specific gene's activity level dependent on the levels of several different transcription factors to different degrees. Figure 16.1 illustrates the concept of transcription factor binding.

Transcription factors are themselves proteins transcribed from genes, and a transcription factor may therefore help to control the expression of another transcription

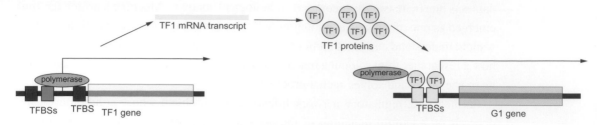

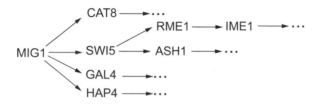

Figure 16.1 Illustration of how transcription factors regulate gene expression. A transcription factor gene (left) produces an mRNA transcript, which in turn produces a protein, TF1, that will bind to transcription factor binding sites (TFBSs) in the promoter regions of other genes, such as the target gene G1 (right). The presence of TF1 is here depicted as blocking recruitment of the RNA polymerase to G1, inhibiting its production of mRNA transcripts.

Figure 16.2 Example of a small section of a transcriptional regulatory network from *Saccharomyces cerevisiae* taken from Guelzim *et al.* [1], involved in regulating the response of cell metabolism to stresses such as lack of nutrients. A central "hub" gene, MIG1, responds to the availability of glucose in the cell. It in turn regulates several other transcription factors, including SWI5, which helps to control cell division, and CAT8, GAL4, and HAP4, which regulate various aspects of cell metabolism. SWI5 itself regulates the transcription factors RME1, which helps control meiosis, and ASH1, which regulates genes involved in more specific steps of cell division. RME1 regulates the transcription factor IME1, which regulates its own subset of meiosis-specific genes. Each of these transcription factors regulates various other downstream targets with more specific functions.

factor or even itself. Transcription factors are typically organized into complicated networks of transcription factors regulating other transcription factors, which regulate others, which regulate others, and so forth, before finally activating modules of non-regulatory genes to perform various biological functions. Figure 16.2 shows an example of a small subset of a real regulatory network from the yeast *Saccharomyces cerevisiae* [1].

There are many sources of experimental data by which one might infer a regulatory network and we will primarily confine ourselves to one particular such source of data: gene expression measurements. To date, most such expression data come from *microarrays*. A microarray is a small glass plate covered with thousands or millions of tiny spots, each made up of many copies of a single short DNA strand called a "probe." When one exposes a purified sample of nucleic acid (DNA or RNA) to a microarray, pieces of the nucleic acid from the sample will anneal to those spots whose DNA sequences are complementary to the sample sequences. To use this principle to quantify RNA in a sample, one will typically convert the RNA into complementary DNA strands (called cDNAs) through the process of reverse transcription, break the cDNAs into small pieces, and then fluorescently label the pieces by attaching a small molecule to each piece of cDNA whose presence one can measure by light emissions. When the labeled sample is run over the microarray and then washed away, we expect to find fluorescence only on those spots to which some sample has annealed and roughly in direct proportion to how much sample has annealed there. We can thus use these fluorescence intensities to give us a quantitative measure of how much RNA complementary to each probe was present in the sample. Figure 16.3 shows an example of a microarray. A typical expression microarray may have a few probes each for every known gene in a given organism's genome, as well as potentially others to detect non-coding genes and other non-genic sources of transcribed RNA. For our purposes, we will simplify a bit and assume that a microarray gives us a measure of how much RNA from each gene is present, or *expressed*, in a given sample.

In a typical microarray experiment, one will use several copies of a given microarray and apply them to a collection of samples gathered under different conditions. These conditions may correspond to different time points, different individuals from whom a tissue sample has been taken, different nutrients or drugs that have been applied to samples, or any other sort of variation that might be expected to change the activities of genes. The data from each gene across all samples are commonly normalized relative to some control sample (typically a pooled mixture of all conditions), giving a measure of the expression level of that gene in each condition relative to the control. Thus, we can think of an array as providing us with a matrix of relative expression data, in which we have one column of data for each condition and one row for each gene. We will assume that this matrix of gene expression measurements represents the data from which we wish to infer the regulatory network.

The preceding description of the problem and the data available to solve it omits many details, as one always must in posing a computational problem, but it provides a reasonable beginning for formulating a mathematical model of the network inference problem. In the remainder of this chapter, we will survey the basic ideas behind how one can go from gene expression measurements to inferred regulatory networks. We will seek to build an intuition for the problem by starting with a simple variant and gradually moving toward a realistic model of the problem in practice. We will conclude with some discussion of the further complications that come up in real-world systems and how the interested reader can learn more about these topics.

Figure 16.3 A microarray slide showing relative levels of nucleic acid in two samples that are complementary to a set of probes [2]. The two samples are labeled in red and green, producing yellow spots when the samples show similar expression levels and red or green spots when one sample shows substantially different expression than the other.

 2   Developing a formal model for regulatory network inference

2.1 Abstracting the problem statement

If we want to develop a computational method for the regulatory network inference problem, we need to begin by developing an abstraction of the problem, i.e. a formal mathematical description of what we will consider the inputs and outputs of the problem to be. Abstracting a problem requires precisely defining what data we assume we have available to us and how we will represent those data, as well as what an answer to

	C1	C2	C3	C4	C5	C6	C7	C8
G1	1	1	0	0	1	1	1	0
G2	0	1	0	1	1	1	1	0
G3	0	0	1	0	0	0	0	1
G4	0	0	0	0	0	1	0	1

Figure 16.4 A toy example of a discretized gene expression data set describing the activities of four genes (G1–G4) in eight conditions (C1–C8). Each row of the matrix (running left to right) describes the activity of one gene under all conditions and each column (running top to bottom) describes the activity of all genes under one condition.

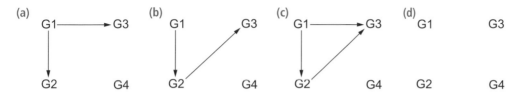

Figure 16.5 A set of possible networks for the expression data of Figure 16.4.

the problem will look like and how we will choose among possible answers. To help us develop an intuition for posing such a problem, we will start with a very simple abstraction of transcriptional regulatory network inference.

We will first develop an abstraction of the input data. We can begin by assuming that the only data we have available to us are a set of microarray measurements comprising a matrix in which each element describes the expression level of one gene in one condition. To keep things simple for the moment, we will further assume that each data point takes on one of two possible values: "1" if the gene is expressed at a higher than average level (informally, that the gene is "on" or "active") and "0" if the gene is expressed at a lower than average level (informally, that the gene is "off" or "inactive"). We are thus making the decision for this level of abstraction to discard the true continuous (real-valued) data that would be produced by the microarray in order to derive a more conceptually tractable model. Figure 16.4 shows a hypothetical example of such an input data set for four genes in eight conditions.

We must also define some formalized statement of the output of a network inference algorithm. In a generic sense, our output should be a model of a network identifying pairs of genes that appear to regulate one another. In this simple version of the problem, we will pick a binary output as well: for any ordered pair of genes, G1 and G2, we will say that either G1 regulates G2 or G1 does not regulate G2. We can represent the output of the inference problem by the set of ordered pairs of genes corresponding to regulatory relationships. This representation of the output can be visualized as a network, also called a *graph*, consisting of a set of vertices with pairs of vertices (or nodes) joined by edges. Here, we create one node for each gene and place a directed edge between any pair of genes Gi and Gj for which Gi regulates Gj. Figure 16.5 shows a few examples of possible networks for the data of Figure 16.4 according to this particular representation of the model.

In choosing this particular representation, we are again making some assumptions about what we will and will not consider important in a model. We are choosing to use a model that represents directionality of regulation; "G1 regulates G2" means something different in our model than "G2 regulates G1." A regulatory network inference algorithm need not distinguish between those possibilities. On the other hand, we are choosing to ignore the fact that regulation can be positive (activation) or negative (repression). We could alternatively have chosen to maintain a sign on each regulatory relationship to distinguish these possibilities, as is typically done in network models. We are similarly ignoring the fact that regulatory relationships could have different strengths (G1 might regulate G2 strongly or weakly), something that is certainly true and which one might denote by placing a numerical weight on each edge. Regulatory relationships could in fact be described by essentially arbitrary functions of expression. We will also assume that genes cannot self-regulate and that we do not have directed cycles, which are paths in the network that lead from a gene back to itself. These assumptions are not, in fact, accurate, but help us establish a more conceptually simple model. Making such trade-offs, in a way that is appropriate to the data available to us and the uses to which we want to put them, is one of the hardest but most important issues in developing a formal model. Our goal in developing the present model is to help us understand the inference problem and so we favor a relatively simple model, but we might favor a very different model if we had some other goal in mind.

The two formalizations defined in this section – a formal representation of the input to the problem and a formal representation of the output to the problem – are two of the main ingredients in a formal problem statement. There is a third component we will need, though: a formal specification of how we will judge any given output for a given input. This measure of the quality of a possible output, known as an *objective function*, is not so easy to define for a complicated problem like this. We will spend the next few subsections showing how to define a precise objective function for the regulatory network inference problem, starting with some intuition behind the problem and building up to a general formulation.

2.2 An intuition for network inference

A good starting point for an objective function is to consider informally how we can reason about the evidence available to us to develop a plausible model.[1] We can see

[1] The terminology here may be confusing to readers previously familiar with mathematical modeling, as the term "model" has a different meaning in the mathematical modeling community than it does in the machine

at an intuitive level how one might evaluate possible regulatory networks for a given data set by closely examining the data of Figure 16.4. We can observe that genes G1 and G2 are generally, although not always, active and inactive in the same conditions. We might therefore guess that G1 regulates G2, and specifically that G1 activates G2. G1 and G3 are generally active in opposite conditions. This, too, might be seen as evidence of regulation, in this case perhaps that G1 represses G3. G4's activity appears unrelated to that of G1, G2, or G3 and we might therefore conclude that it is probably not in a regulatory relationship with any of them. We therefore might conjecture that Figure 16.5a provides a good model of the regulatory network we want to infer.

Intuition can only take us so far, though. The same reasoning that led us to the network of Figure 16.5a could just as easily lead us to Figure 16.5b or Figure 16.5c. For that matter, we do not know if the correlations we think we see in the data are sufficiently well supported by the data that we should believe them. Perhaps Figure 16.5d (no regulation) is the true network and the apparent correlations arose from random chance. If we want to be able to choose among these possibilities, we will need to be a bit more precise about how we we will decide what makes for a "plausible" model.

2.3 Formalizing the intuition for an inference objective function

To go from intuition to a formal computational problem, we will need to come up with a way of specifying precisely how good one model is relative to another. A common way of accomplishing this for noisy data inference problems is to define the problem in terms of probabilities. We will use a particular variant of a probabilistic model, known as a likelihood model, in which we judge a model by how probable we think it is that the observed data could have been generated from that model. This probability is known as the *likelihood* of the model. We then seek the model that gives us the greatest likelihood, known as the *maximum likelihood model*.

To put the intuitive problem into a formal framework, we first need to develop some notation. As in Figure 16.4, we will assume our input is a matrix, which we will call D. We will refer to each row of the matrix, corresponding to a single gene, as a vector $\mathbf{d}_i$. So for example, the row for gene G1 is represented by the vector $\mathbf{d}_1 = [11001110]$. Each element of each row is represented by a single scalar (non-vector) value d_{ij}. For example, the expression of gene G1 in condition C2 is given by $d_{12} = 1$.

We will also need a notation to refer to our output, i.e. the regulatory network we would like to infer. As discussed in the preceding section, our output can be represented

learning community. We will follow machine learning practice in using "model" to refer to a particular output of the network inference problem, i.e. a network modeling the regulatory interactions among the input genes. In mathematical modeling terminology, a "model" of the problem would refer instead to what we have here called the "formal problem statement."

by a graph, which we can call G. Any given G is itself defined by a set of vertices V, with one vertex per gene, and a set of edges E, with potentially one edge for each pair of genes. Thus, for example, we can refer to the model of Figure 16.5a by the graph

$$G = (V, E) = (\{v_1, v_2, v_3, v_4\}, \{(v_1, v_2), (v_1, v_3)\}). \tag{16.1}$$

The vertex set contains four vertices, one for each of the four genes, and the edge set contains two edges, one for each of the two posited regulatory relationships.

We will be working specifically with probability models, which will require that our models include some additional information to let us determine how likely the model is to produce a given set of expression data. We will defer the details of these probabilities for the moment and just declare that we have some additional set P of probability parameters contained in the model. For a maximum likelihood model, we define those additional values contained in P to be whatever will make the likelihood function as large as possible. The exact contents of P will depend on the graph elements V and E, as we will see shortly. For our formal purposes, then, an output model M consists of the elements (V, E, P) defining the proposed regulatory relationships and the probability of outputting any given expression matrix D from that model M. This probability, called the likelihood of the model, is denoted by the probability function $Pr\{D|M\}$, read as "the probability of D given M." Our goal will be to find

$$\max_{M} Pr\{D|M\},$$

i.e. the maximum likelihood model over all possible models M for a given data set D. We still have more work to do, though, to define precisely what it means mathematically to find the M maximizing $Pr\{D|M\}$.

2.3.1 Maximum likelihood for one gene

We next need to specify how one actually evaluates the function $Pr\{D|M\}$ for a known D and M. We can start by considering just one gene, G1, whose expression is described by the vector $\mathbf{d}_1 = [11001110]$. Since we are now assuming that there is only one gene, we cannot have any regulatory relationships. Therefore, we have only one possible graph G for our model: $G = (V, E) = (\{v_1\}, \{\})$, a vertex set of one node and an empty edge set. To determine the likelihood of the model, we will need to evaluate $Pr\{\mathbf{d}_1|(V, E, P)\}$, the probability that the model $M = (V, E, P)$ would lead to the output vector $\mathbf{d}_1$. It is a universal law of probability that the probability of a pair of independent outcomes is the product of the probabilities of the individual outcomes. Therefore, if we assume that each condition represents an independent experiment then the probability of outputting the complete vector $\mathbf{d}_1$ will be given by the product of probabilities of outputting each element of that vector. Thus, if we knew the probability that G1 was active in a given condition given our model M ($Pr\{d_{1i} = 1|M\}$, which we will call $p_{1,1}$) and the probability that G1 was inactive in a given condition given model M ($Pr\{d_{1i} = 0|M\}$, which we will call $p_{1,0}$) then we could determine the probability of the whole vector as follows:

$$Pr\{\mathbf{d}_1 = [11001110]|M\} = Pr\{d_{11} = 1|M\} \times Pr\{d_{12} = 1|M\} \times Pr\{d_{13} = 0|M\}$$

$$\times Pr\{d_{14} = 0|M\} \times Pr\{d_{15} = 1|M\} \times Pr\{d_{16} = 1|M\}$$

$$\times Pr\{d_{17} = 1|M\} \times Pr\{d_{18} = 0|M\}$$

$$= p_{1,1} \times p_{1,1} \times p_{1,0} \times p_{1,0} \times p_{1,1} \times p_{1,1} \times p_{1,1} \times p_{1,0}. \qquad (16.2)$$

For this particular model, $p_{1,1}$ and $p_{1,0}$ are precisely the additional model parameters P that we need to know to finish formally specifying the model.

As noted above, those additional values contained in P must be whatever will make the likelihood function as large as possible. Fortunately, those maximum likelihood values are easy to determine, at least for this model. The values that will give the maximum likelihood are given by the fractions of observations corresponding to each given probability in the observed data. In other words, we observe that G1 is active in five conditions out of eight, giving a maximum likelihood estimate of $p_{1,1} = 5/8$. G1 is inactive in three conditions out of eight, giving a maximum likelihood estimate of $p_{1,0} = 3/8$. This procedure for learning optimal parameters of P then lets us complete the formal specification of our model M as follows:

$$M = (V, E, P) = \left(\{v_1\}, \{\}, \left\{Pr\{d_{1i} = 1|M\} = \frac{5}{8}, Pr\{d_{1i} = 0|M\} = \frac{3}{8}\right\}\right). \quad (16.3)$$

We also now have all the tools we need to come up with a precise quantitative statement of the likelihood of the data given the model for this simple one-gene case:

$$Pr\{\mathbf{d}_1 = [11001110]|M\} = p_{1,1} \times p_{1,1} \times p_{1,0} \times p_{1,0} \times p_{1,1} \times p_{1,1} \times p_{1,1} \times p_{1,0}$$

$$= \frac{5}{8} \times \frac{5}{8} \times \frac{3}{8} \times \frac{3}{8} \times \frac{5}{8} \times \frac{5}{8} \times \frac{5}{8} \times \frac{3}{8} \approx 0.00503. \qquad (16.4)$$

This number is not too useful to us when we only have one model to consider, but will become our measure for evaluating possible models with more complicated examples.

2.3.2 Maximum likelihood for two genes

Now that we know how to evaluate a likelihood function for one gene, we will move on to considering two genes, G1 and G2, simultaneously. There are now three possible hypotheses we can consider: neither G1 nor G2 regulates the other, G1 regulates G2, or G2 regulates G1. Each of these hypotheses can be converted into a formal model using the concepts laid out above. We will want to determine which of these three models maximizes the likelihood of both genes given the model:

$$\max_{M} Pr\{\mathbf{d}_1 = [11001110], \mathbf{d}_2 = [01011110]|M\}. \tag{16.5}$$

To keep the notation from getting too cumbersome, we will henceforth abbreviate the above likelihood as $Pr\{\mathbf{d}_1, \mathbf{d}_2|M\}$.

Our first model, which we will call M_1, assumes that neither G1 nor G2 regulates the other. Formally, $M_1 = (V_1, E_1, P_1) = (\{v_1, v_2\}, \{\}, P_1)$, where we will again defer defining P_1 precisely until we see how we will use it. For this model, we can treat the outputs $\mathbf{d}_1$ and $\mathbf{d}_2$ as independent sets of data since we assume neither gene regulates the other. As we noted above, the assumption that two variables are independent means that we can derive their joint probability by multiplying their individual probabilities:

$$Pr\{\mathbf{d}_1, \mathbf{d}_2|M_1\} = Pr\{\mathbf{d}_1|M_1\} \times Pr\{\mathbf{d}_2|M_1\}. \tag{16.6}$$

We can then evaluate each of these two probabilities exactly as we did in the one-gene case. The additional probability parameters P_1 that we will need to know are the probability G1 is active or inactive independently of G2 and the probability G2 is active or inactive independently of G1. Extending our notation from the one-gene case, $P_1 = \{p_{1,1}, p_{1,0}, p_{2,1}, p_{2,0}\}$. We can derive maximum likelihood estimates for these probabilities as above by observing the fraction of outputs that are 1 or 0 for each gene. As before, we can estimate $p_{1,1} = 5/8$ and $p_{1,0} = 3/8$. We similarly observe five 1s and three 0s for G2, so we estimate $p_{2,1} = 5/8$ and $p_{2,0} = 3/8$. We then get the following estimate for the likelihood of G1's outputs:

$$Pr\{\mathbf{d}_1|M_1\} = p_{1,1} \times p_{1,1} \times p_{1,0} \times p_{1,0} \times p_{1,1} \times p_{1,1} \times p_{1,1} \times p_{1,0}$$

$$= \frac{5}{8} \times \frac{5}{8} \times \frac{3}{8} \times \frac{3}{8} \times \frac{5}{8} \times \frac{5}{8} \times \frac{5}{8} \times \frac{3}{8} \approx 0.00503, \tag{16.7}$$

and the following for G2's outputs:

$$Pr\{\mathbf{d}_2|M_1\} = p_{2,0} \times p_{2,1} \times p_{2,0} \times p_{2,1} \times p_{2,1} \times p_{2,1} \times p_{2,1} \times p_{2,0}$$

$$= \frac{3}{8} \times \frac{5}{8} \times \frac{3}{8} \times \frac{5}{8} \times \frac{5}{8} \times \frac{5}{8} \times \frac{5}{8} \times \frac{3}{8} \approx 0.00503. \tag{16.8}$$

Thus,

$$Pr\{\mathbf{d}_1, \mathbf{d}_2|M_1\} = \left(\frac{5}{8}\right)^5 \times \left(\frac{3}{8}\right)^3 \times \left(\frac{5}{8}\right)^5 \times \left(\frac{3}{8}\right)^3 \approx 2.53 \times 10^{-5}. \tag{16.9}$$

Things get trickier when we move to a model assuming some regulation. We will now consider the possibility that G1 regulates G2. For this model, $M_2 = (V_2, E_2, P_2) = (\{v_1, v_2\}, \{(v_1, v_2)\}, P_2)$. That is, the model assumes a single regulatory edge running from v_1 to v_2 representing the assumption that G2's expression is a function of G1's expression. As before, we can assume G1's expression is an independent random variable:

$$Pr\{\mathbf{d}_1|M_2\} = p_{1,1} \times p_{1,1} \times p_{1,0} \times p_{1,0} \times p_{1,1} \times p_{1,1} \times p_{1,1} \times p_{1,0}$$

$$= \frac{5}{8} \times \frac{5}{8} \times \frac{3}{8} \times \frac{3}{8} \times \frac{5}{8} \times \frac{5}{8} \times \frac{5}{8} \times \frac{3}{8} \approx 0.00503. \qquad (16.10)$$

We must, however, assume that G2's expression depends on G1's. More formally, our likelihood function will need a term of the form $Pr\{\mathbf{d}_2|M_2, \mathbf{d}_1\}$, which we read as "the probability of $\mathbf{d}_2$ given M_2 and $\mathbf{d}_1$." This function will depend on a model of how likely it is that d_{2i} is 1 when d_{1i} is 1 as well as how likely it is that d_{2i} is 1 when d_{1i} is 0. We will therefore need to specify four probability parameters:

- $p_{2,0,0}$: the probability $d_{2i} = 0$ when $d_{1i} = 0$
- $p_{2,0,1}$: the probability $d_{2i} = 0$ when $d_{1i} = 1$
- $p_{2,1,0}$: the probability $d_{2i} = 1$ when $d_{1i} = 0$
- $p_{2,1,1}$: the probability $d_{2i} = 1$ when $d_{1i} = 1$

P_2 is defined by the probabilities we need to evaluate $Pr\{\mathbf{d}_1|M_2\}$ and those we need to evaluate $Pr\{\mathbf{d}_2|M_2, \mathbf{d}_1\}$, so $P_2 = \{p_{1,1}, p_{1,0}, p_{2,0,0}, p_{2,0,1}, p_{2,1,0}, p_{2,1,1}\}$. As before, we can derive maximum likelihood estimates of these parameters by counting the fraction of times we observe each value of G2 for each value of G1. We have five instances in which G1 is 1 and four of these five also have G2 = 1. Thus, $p_{2,1,1} = 4/5$ and $p_{2,0,1} = 1/5$. Similarly, we have three instances in which G1 = 0 and two of these three have G2 = 0. Thus, $p_{2,0,0} = 2/3$ and $p_{2,1,0} = 1/3$. Therefore,

$$Pr\{\mathbf{d}_2|M_2, \mathbf{d}_1\} = p_{2,0,1} \times p_{2,1,1} \times p_{2,0,0} \times p_{2,1,0} \times p_{2,1,1} \times p_{2,1,1} \times p_{2,1,1} \times p_{2,0,0}$$

$$= \frac{1}{5} \times \frac{2}{3} \times \frac{1}{3} \times \frac{4}{5} \times \frac{4}{5} \times \frac{4}{5} \times \frac{4}{5} \times \frac{2}{3} \approx 0.0121. \qquad (16.11)$$

The complete likelihood for this model is then given by

$$Pr\{\mathbf{d}_1, \mathbf{d}_2|M_2\} = Pr\{\mathbf{d}_1|M_2\}Pr\{\mathbf{d}_2|\mathbf{d}_1, M_2\} \approx 0.00503 \times 0.0121 \approx 6.10 \times 10^{-5}.$$

We can therefore conclude that M_2 is a more likely explanation for the data than M_1.

Evaluating the final model for two genes, $M_3 = (V_3, E_3, P_3) = (\{v_1, v_2\}, \{(v_2, v_1)\}, P_3)$, proceeds analogously to the evaluation of M_2:

$$Pr\{\mathbf{d}_1, \mathbf{d}_2|M_2\} = Pr\{\mathbf{d}_2|M_3\}Pr\{\mathbf{d}_1|\mathbf{d}_2, M_2\}, \qquad (16.12)$$

i.e. the model is the product of a term accounting for the independent likelihood of G2 and the likelihood of G1 given that it is a function of G2. We can evaluate $Pr\{\mathbf{d}_2|M_3\}$

as we did for M_1:

$$Pr\{\mathbf{d}_2|M_3\} = p_{2,0} \times p_{2,1} \times p_{2,0} \times p_{2,1} \times p_{2,1} \times p_{2,1} \times p_{2,1} \times p_{2,0}$$

$$= \frac{3}{8} \times \frac{5}{8} \times \frac{3}{8} \times \frac{5}{8} \times \frac{5}{8} \times \frac{5}{8} \times \frac{5}{8} \times \frac{3}{8} \approx 0.00503. \tag{16.13}$$

We can also evaluate $Pr\{\mathbf{d}_1|\mathbf{d}_2, M_3\}$ as we did for $Pr\{\mathbf{d}_2|\mathbf{d}_1, M_2\}$. We define a new set of parameters:

- $p_{1,0,0}$: the probability $d_{1i} = 0$ when $d_{2i} = 0$
- $p_{1,0,1}$: the probability $d_{1i} = 0$ when $d_{2i} = 1$
- $p_{1,1,0}$: the probability $d_{1i} = 1$ when $d_{2i} = 0$
- $p_{1,1,1}$: the probability $d_{1i} = 1$ when $d_{2i} = 1$

We estimate the parameters by identifying all occurrences of $G2 = 0$ and $G2 = 1$ and, for each, counting how often $G1 = 0$ and $G1 = 1$: $p_{1,0,0} = 1/3$, $p_{1,1,0} = 2/3$, $p_{1,0,1} = 4/5$, $p_{1,1,1} = 1/5$. These probabilities collectively define $P_3 = \{p_{2,1}, p_{2,0}, p_{1,0,0}, p_{1,0,1}, p_{1,1,0}, p_{1,1,1}\}$. Then,

$$Pr\{\mathbf{d}_2|M_3, \mathbf{d}_1\} = p_{1,1,0} \times p_{1,1,1} \times p_{1,0,0} \times p_{1,0,1} \times p_{1,1,1} \times p_{1,1,1} \times p_{1,1,1} \times p_{1,0,0}$$

$$= \frac{1}{5} \times \frac{2}{3} \times \frac{1}{3} \times \frac{4}{5} \times \frac{4}{5} \times \frac{4}{5} \times \frac{4}{5} \times \frac{2}{3} \approx 0.0121. \tag{16.14}$$

Putting it all together gives us the full model likelihood

$$Pr\{\mathbf{d}_1, \mathbf{d}_2|M_2\} \approx 0.00503 \times 0.0121 \approx 6.10 \times 10^{-5}. \tag{16.15}$$

Thus, M_3 has the same likelihood as M_2.

If we had just the two genes to consider then we could run through these possibilities and come to the final conclusion that M_1 is a poorer model of the data, while M_2 and M_3 are better models than M_1 and equally good to one another.

It is worth noting that it is not a coincidence that M_2 and M_3 yield identical likelihoods. In fact, the problem as we posed it guarantees that the likelihood of any model will be identical to that of a mirror image model, in which the directionality of all edges is reversed. We might therefore conclude that our formalization of the problem was, in this respect, poorly matched to our data and that we should have posed the problem in terms of finding undirected networks. Alternatively, we might consider ways of adding additional information by which we might disambiguate the directions of regulatory edges, a topic we will consider later in the chapter. For now, however, we will ignore this issue and continue working through the problem as we have formalized it.

2.3.3 From two genes to several genes

The mathematics became fairly complicated when we moved from one to two genes, so one might expect that moving to three or four will be much harder. In fact, though, it is not much more difficult to reason about four genes, or forty thousand, than it is to reason about two. The number of models one can potentially consider goes up rapidly with increasing numbers of genes, but evaluating the likelihood of any given model is not that much harder conceptually. To see why, let us consider just three of the possible models of all four genes from Figure 16.4.

One model we might wish to consider is that no gene regulates any other. We can call this model M_1', which would correspond to the assumption that

$$Pr\{\mathbf{d}_1, \mathbf{d}_2, \mathbf{d}_3, \mathbf{d}_4 | M_1'\} = Pr\{\mathbf{d}_1 | M_1'\} \times Pr\{\mathbf{d}_2 | M_1'\} \times Pr\{\mathbf{d}_3 | M_1'\} \times Pr\{\mathbf{d}_4 | M_1'\}. \tag{16.16}$$

We can evaluate each of these terms just as we did when we considered two genes. For example, to evaluate $Pr\{\mathbf{d}_1 | M_1'\}$, we define variables $p_{1,0}$ and $p_{1,1}$ representing the probabilities G1 is 0 or 1, estimate these probabilities by counting the fraction of occurrences of G1 = 0 and G1 = 1, and multiply probabilities across conditions:

$$Pr\{\mathbf{d}_1 | M_1'\} = p_{1,1} \times p_{1,1} \times p_{1,0} \times p_{1,0} \times p_{1,1} \times p_{1,1} \times p_{1,1} \times p_{1,0}$$

$$= \frac{5}{8} \times \frac{5}{8} \times \frac{3}{8} \times \frac{3}{8} \times \frac{5}{8} \times \frac{5}{8} \times \frac{5}{8} \times \frac{3}{8} \approx 0.00503. \tag{16.17}$$

Similarly,

$$Pr\{\mathbf{d}_2 | M_1'\} = p_{2,0} \times p_{2,1} \times p_{2,0} \times p_{2,1} \times p_{2,1} \times p_{2,1} \times p_{2,1} \times p_{2,0}$$

$$= \frac{3}{8} \times \frac{5}{8} \times \frac{3}{8} \times \frac{5}{8} \times \frac{5}{8} \times \frac{5}{8} \times \frac{5}{8} \times \frac{3}{8} \approx 0.00503,$$

$$Pr\{\mathbf{d}_3 | M_1'\} = p_{3,0} \times p_{3,0} \times p_{3,1} \times p_{3,0} \times p_{3,0} \times p_{3,0} \times p_{3,0} \times p_{3,1} \tag{16.18}$$

$$= \frac{6}{8} \times \frac{6}{8} \times \frac{2}{8} \times \frac{6}{8} \times \frac{6}{8} \times \frac{6}{8} \times \frac{6}{8} \times \frac{2}{8} \approx 0.0111,$$

$$Pr\{\mathbf{d}_4 | M_1'\} = p_{4,0} \times p_{4,0} \times p_{4,0} \times p_{4,0} \times p_{4,0} \times p_{4,1} \times p_{4,0} \times p_{4,1}$$

$$= \frac{6}{8} \times \frac{6}{8} \times \frac{6}{8} \times \frac{6}{8} \times \frac{6}{8} \times \frac{2}{8} \times \frac{6}{8} \times \frac{2}{8} \approx 0.0111.$$

The formal statement of the model is, then,

$$M_1' = (V_1', E_1', P_1') = (\{v_1, v_2, v_3, v_4\}, \{\}, \{p_{1,0}, p_{1,1}, p_{2,0}, p_{2,1}, p_{3,0}, p_{3,1}, p_{4,0}, p_{4,1}\}) \tag{16.19}$$

and the likelihood of the whole model is

$$Pr\{\mathbf{d}_1, \mathbf{d}_2, \mathbf{d}_3, \mathbf{d}_4 | M_1'\} \approx 0.00503 \times 0.00503 \times 0.0111 \times 0.0111 \approx 3.00 \times 10^{-9}.$$

$$(16.20)$$

We might alternatively consider a model M_2' in which G1 regulates G2, G2 regulates G3, and nothing regulates G1 or G4. M_2' corresponds to the assumption that

$$Pr\{\mathbf{d}_1, \mathbf{d}_2, \mathbf{d}_3, \mathbf{d}_4 | M_2'\} = Pr\{\mathbf{d}_1 | M_2'\} \times Pr\{\mathbf{d}_2 | \mathbf{d}_1, M_2'\} \times Pr\{\mathbf{d}_3 | \mathbf{d}_2, M_2'\} \times Pr\{\mathbf{d}_4 | M_2'\}.$$

$$(16.21)$$

The G1 and G4 terms can be evaluated just as with model M_1':

$$Pr\{\mathbf{d}_1 | M_2'\} = p_{1,1} \times p_{1,1} \times p_{1,0} \times p_{1,0} \times p_{1,1} \times p_{1,1} \times p_{1,1} \times p_{1,0}$$

$$= \frac{5}{8} \times \frac{5}{8} \times \frac{3}{8} \times \frac{3}{8} \times \frac{5}{8} \times \frac{5}{8} \times \frac{5}{8} \times \frac{3}{8} \approx 0.00503, \qquad (16.22)$$

$$Pr\{\mathbf{d}_4 | M_2'\} = p_{4,0} \times p_{4,0} \times p_{4,0} \times p_{4,0} \times p_{4,0} \times p_{4,1} \times p_{4,0} \times p_{4,1}$$

$$= \frac{6}{8} \times \frac{6}{8} \times \frac{6}{8} \times \frac{6}{8} \times \frac{6}{8} \times \frac{2}{8} \times \frac{6}{8} \times \frac{2}{8} \approx 0.0111.$$

The G2 term can be handled just as when we considered G1 and G2 alone:

$$Pr\{\mathbf{d}_2 | M_2', \mathbf{d}_1\} = p_{2,0,1} \times p_{2,1,1} \times p_{2,0,0} \times p_{2,1,0} \times p_{2,1,1} \times p_{2,1,1} \times p_{2,1,1} \times p_{2,0,0}$$

$$= \frac{1}{5} \times \frac{2}{3} \times \frac{1}{3} \times \frac{4}{5} \times \frac{4}{5} \times \frac{4}{5} \times \frac{4}{5} \times \frac{2}{3} \approx 0.0121. \qquad (16.23)$$

Finally, the G3 term can be handled analogously to the G2 term:

$$Pr\{\mathbf{d}_3 | \mathbf{d}_2, M_2'\} = p_{3,0,0} \times p_{3,0,1} \times p_{3,1,0} \times p_{3,0,1} \times p_{3,0,1} \times p_{3,0,1} \times p_{3,0,1} \times p_{3,1,0}$$

$$= \frac{1}{3} \times \frac{5}{5} \times \frac{2}{3} \times \frac{5}{5} \times \frac{5}{5} \times \frac{5}{5} \times \frac{5}{5} \times \frac{2}{3} \approx 0.148. \qquad (16.24)$$

We thus get the complete likelihood:

$$Pr\{\mathbf{d}_1, \mathbf{d}_2, \mathbf{d}_3, \mathbf{d}_4 | M_2'\} = 0.00503 \times 0.0121 \times 0.148 \times 0.0111 \approx 1.00 \times 10^{-7}.$$

$$(16.25)$$

We can therefore conclude that M_2' has a substantially higher likelihood than M_1'.

We can also consider models in which a given gene is a function of more than one regulator. For example, suppose we consider a model M_3' in which G1, G2, and G4 are unregulated but G3 is regulated by both G1 and G2. For this model, we assume that

$$Pr\{\mathbf{d}_1, \mathbf{d}_2, \mathbf{d}_3, \mathbf{d}_4 | M_3'\} = Pr\{\mathbf{d}_1 | M_3'\} \times Pr\{\mathbf{d}_2 | M_3'\} \times Pr\{\mathbf{d}_3 | \mathbf{d}_1, \mathbf{d}_2, M_3'\} \times Pr\{\mathbf{d}_4 | M_3'\}.$$

$$(16.26)$$

We can evaluate the G1, G2, and G4 terms exactly as with model M_1' above:

$$Pr\{\mathbf{d}_1|M_3'\} = p_{1,1} \times p_{1,1} \times p_{1,0} \times p_{1,0} \times p_{1,1} \times p_{1,1} \times p_{1,1} \times p_{1,0} \approx 0.00503.$$

(16.27)

Similarly,

$$Pr\{\mathbf{d}_2|M_3'\} = p_{2,0} \times p_{2,1} \times p_{2,0} \times p_{2,1} \times p_{2,1} \times p_{2,1} \times p_{2,1} \times p_{2,0} \approx 0.00503,$$

$$Pr\{\mathbf{d}_4|M_3'\} = p_{4,0} \times p_{4,0} \times p_{4,0} \times p_{4,0} \times p_{4,0} \times p_{4,1} \times p_{4,0} \times p_{4,1} \approx 0.0111.$$

(16.28)

To evaluate the G3 term, however, we will need to consider its dependence on states of both G1 and G2. We can capture this dependence with the following set of probability parameters:

- $p_{3,0,0,0}$: the probability $d_{3i} = 0$ when $d_{1i} = 0$ and $d_{2i} = 0$
- $p_{3,1,0,0}$: the probability $d_{3i} = 1$ when $d_{1i} = 0$ and $d_{2i} = 0$
- $p_{3,0,0,1}$: the probability $d_{3i} = 0$ when $d_{1i} = 0$ and $d_{2i} = 1$
- $p_{3,1,0,1}$: the probability $d_{3i} = 1$ when $d_{1i} = 0$ and $d_{2i} = 1$
- $p_{3,0,1,0}$: the probability $d_{3i} = 0$ when $d_{1i} = 1$ and $d_{2i} = 0$
- $\ldots$

We can then say

$$Pr\{\mathbf{d}_3|\mathbf{d}_1, \mathbf{d}_2, M_3'\} = p_{3,0,1,0} \times p_{3,0,1,1} \times p_{3,1,0,0} \times p_{3,0,0,1} \times p_{3,0,1,1} \times p_{3,0,1,1}$$

$$\times p_{3,0,1,1} \times p_{3,1,0,0}.$$

(16.29)

To estimate the probability parameters, we need to count values of G3 for each combination of values of G1 and G2. For example, to evaluate $p_{3,0,1,1}$ (the probability G3 $=0$ given that G1 $=1$ and G2 $=1$), we note that there are four conditions in which G1 $=1$ and G2 $=1$ and all four have G3 $= 0$. Thus, $p_{3,0,1,1} = 4/4$. Similarly, we estimate $p_{3,0,1,0} = 1/1$ and $p_{3,1,0,0} = 2/2$. We would then conclude that

$$Pr\{\mathbf{d}_3|\mathbf{d}_1, \mathbf{d}_2, M_3'\} = \frac{1}{1} \times \frac{4}{4} \times \frac{2}{2} \times \frac{1}{1} \times \frac{4}{4} \times \frac{4}{4} \times \frac{4}{4} \times \frac{2}{2} = 1.$$

(16.30)

Putting together all of the terms, we get

$$Pr\{\mathbf{d}_1, \mathbf{d}_2, \mathbf{d}_3, \mathbf{d}_4|M_3'\} \approx 0.00503 \times 0.00503 \times 1 \times 0.0111 \approx 2.81 \times 10^{-7}.$$ (16.31)

Thus, this new model M_3' has the highest likelihood of the three we have considered. We could repeat the analysis above for every possible model of the four genes G1–G4 and thereby find the maximum likelihood model.

2.4 Generalizing to arbitrary numbers of genes

The above examples cover essentially all of the complications we would encounter in evaluating the likelihood of any network model for these genes or any set of genes for the present level of abstraction. In particular, if we understand the three examples in the preceding section, we understand all of the concepts we need to evaluate networks of arbitrary complexity, at least at a simple level of abstraction. We will now see how to complete the generalization to arbitrary numbers of genes.

Suppose now that instead of four genes assayed in eight conditions, we have n genes assayed in m conditions. We can then represent our input matrix D as the set of vectors $\mathbf{d}_1, \ldots, \mathbf{d}_n$, each of length m. Any given model M will still have the form (V, E, P), where $V = \{v_1, \ldots, v_n\}$ now contains one element for each of the n genes and $E \subset V \times V$, i.e. the set of edges is a subset of the set of pairs of genes. (In reality, E will generally be much smaller than $V \times V$ due to the restriction that the graph does not contain directed cycles.) Defining P is a bit more complicated, as we require one probability parameter for each gene, each possible expression level of that gene, and each possible expression level of each of its regulators. More formally, for any given gene i regulated by a set of genes $R_i = \{j | (v_j, v_i) \in E\}$ (read as "the set of values j such that (v_j, v_i) is in set E") of size $m_i = |R_i|$ (the number of elements in set R_i), we require a model variable $p_{i,b_i,b_{i1},\ldots,b_{im_i}}$ for each $b_i, b_{i1}, \ldots, b_{im_i} \in \{0, 1\}$. This results in a set of 2^{m_i+1} parameters in P for gene i defining the probability of each possible state of gene i given each possible state of the genes that regulate it. Collectively, these sets $p_{i,b_i,b_{i1},\ldots,b_{im_i}}$ over all genes i define the probability parameter set P. We can find the maximum likelihood estimate for each such parameter $p_{i,b_i,b_{i1},\ldots,b_{im_i}}$, just as we did in the previous cases, by finding the observations in which genes $i_1, \ldots, i_{m_i}$ have values $b_{i_1}, \ldots, b_{i_{m_i}}$ and determining the fraction of those observations for which gene i has value b_i.

Evaluating the probability of an input matrix D given any particular model $M = (V, E, P)$ then follows analogously to the derivations for fixed n in the preceding sections. We can evaluate the likelihood of any particular expression vector $\mathbf{d}_i$ given the model M and the remaining expression matrix $D/\mathbf{d}_i = [\mathbf{d}_1, \mathbf{d}_2, \ldots, \mathbf{d}_{i-1}, \mathbf{d}_{i+1}, \ldots, \mathbf{d}_n]$ (i.e. the portion of D remaining when we remove $\mathbf{d}_i$) by taking the product over the probabilities of the observed output values:

$$Pr\{\mathbf{d}_i | D/\mathbf{d}_i, M\} = \prod_{j=1}^{m} p_{i,d_{ij},d_{r_{i1},j},\ldots,d_{r_{im_i},j}} \qquad (16.32)$$

where the indices $r_{i1}, \ldots, r_{im_i}$ come from the set R_i of inputs to gene i. While the notation gets complicated, intuitively this product simply expresses the idea that we can evaluate the probability of the gene's observed output vector by multiplying independent contributions from each condition.

Similarly, we can accumulate the likelihood function across all output genes i to get the full likelihood of input data D given model M:

$$Pr\{D|M\} = \prod_{i=1}^{n} Pr\{\mathbf{d}_i|D/\mathbf{d}_i, M\} = \prod_{i=1}^{n}\prod_{j=1}^{m} p_{i,d_{ij},d_{r_{i1},j},\ldots,d_{r_{im_i},j}} \tag{16.33}$$

where the r_{ik} values are again drawn from the set R_i. While the notation is again complex, the concept is simple. We can evaluate the probability of the entire data set by accumulating a product across all data points, evaluating each data point by the conditional probability of its observed value given the observed values of all of its input genes. Manually evaluating the likelihood of such a model for more than a few variables would be tedious but it is easily handled by a computer program.

 3 **Finding the best model**

The astute reader might notice that we have not yet mentioned any algorithms in this chapter. We know how to compare different models, but we may have a very large number of possible models to consider. Finding the best of all possible models will therefore require a more sophisticated approach than simply evaluating the likelihood for every possibility and picking the best one. Finding the best of all possible models is an example of a machine learning problem. Machine learning problems like this are very different from standard discrete algorithm problems in that we do not generally have a library of problem-specific algorithms with definite run times from which to draw. Rather, there are a host of generic learning methods that work broadly for problems posed with this sort of probabilistic model. Solving a machine learning problem often involves selecting some such generic algorithm and then tuning it to work especially well given the details of the particular inference being conducted. Actually solving real-world versions of the regulatory network inference problem is not trivial and requires expertise in statistics and machine learning beyond what we assume for readers of this text. In this section, though, we will very briefly consider some general strategies we can use to find a reasonable solution in practice.

For relatively small data sets, a variety of simple solutions are available. For the simplest instances of such a problem, one can try a brute-force search of all possible solutions. The four-gene example we considered, for instance, has a few thousand possible models and we could run through all of them in a reasonable time, evaluating the likelihood of each and finding the global maximum likelihood model. We could extend that brute-force approach to perhaps five or six genes, but not much farther. One alternative for larger networks is to use a *heuristic*, which is a method that provides no guarantees of good performance but tends to give at least a pretty good answer in a reasonable amount of time in practice. One such heuristic strategy is *hill-climbing*. With a hill-climbing heuristic, we start with an initial guess as to the network (perhaps assuming no regulation or using a best guess derived from the literature) and then pick a random potential edge to examine. If that edge is present in the network, we remove it, and if it is not present, we add it. We then evaluate the likelihoods of both the original and the modified networks; whichever network has a higher score is retained. (Note that if we wish to keep the restriction that the network has no cycles then we must test for cycles after each proposed change and assign likelihood zero to any network that has a cycle.) This process continues until we find a network whose likelihood cannot be improved by adding or removing any single edge. Many other generic optimization heuristics like hill-climbing can also be adapted to this problem.

There are also various heuristics specific to the network inference problem. For example, the guilt-by-association (GBA) method [3] suggests that we shrink the universe of possible models by only allowing edges between genes when there is a strong correlation between those genes' expression vectors. This improvement greatly reduces the search space of possible models and allows us to extend other optimization heuristics to much larger gene sets.

For more challenging data sets, a standard approach is to use a *Markov chain Monte Carlo method*, which is essentially a randomized version of the hill-climbing approach. The most widely used such method is the Metropolis–Hastings algorithm [4]. With a Metropolis–Hastings approach to the network inference problem, we can begin just as with hill–climbing, choosing a random edge and creating a version of the model in which that one edge is added if it was not present or removed if it was present. We then again evaluate the likelihood of the model in the original form, which we will call L_1, and in the modified form, which we will call L_2. If $L_2 > L_1$ then we make the change, just as with hill-climbing. If, however, $L_2 < L_1$, we still allow some chance of making the change, with probability L_2/L_1. While this may seem like a minor difference, it actually makes for a far more useful algorithm. We can use this Metropolis–Hastings approach to explore possible models and pick the best, but it also gives us quite a bit of useful information about distributions of models that we can use to assess confidence in the model chosen or specific features of that model. A similar alternative to Metropolis–Hastings is Gibbs sampling [5], which uses essentially the same algorithm for this problem except that on each step one either keeps the modified model with probability $L_2/(L_1 + L_2)$ or the original model with probability $L_1/(L_1 + L_2)$. There is an enormous literature on more sophisticated variants on Markov chain Monte Carlo methods and such methods are often effective for quite difficult problem instances.

For the most difficult data sets, we are likely to need more advanced methods than we can reasonably cover in this text. There is now a large literature on optimization methods for machine learning to which one can refer for solving the hardest problems. Some references to this literature are provided in the concluding section below.

4   Extending the model with prior knowledge

We have now seen a very basic version of how to evaluate possible models of a regulation of a genetic regulatory network, but what we have seen so far is still not likely to lead to accurate inferences from real data. There are simply too many possible models and too little data from which to learn them to hope that such a naïve approach will work well. If we want a genuinely useful method, the most important missing piece to our initial approach is some way of using what is already known or suspected about the system to constrain our inferences. This sort of external knowledge about a problem is generally encoded in a *prior probability*, also known simply as a *prior*. A prior probability is an estimate of how plausible we believe a variable or parameter of the model is independent of the data from which we are formally learning the model. It gives us a way to incorporate into our analysis whatever we know, or think we know, about the system being modeled.

To see how one can use a prior probability, let us suppose we already have a general idea of what the network we are inferring looks like. Perhaps we have referred to prior literature on the genes of interest to us and seen several papers reporting that G1 regulates G2 and a single paper reporting that G2 regulates G3. We might, on that basis, have some prior expectation that our model should include those regulatory relationships. Perhaps we decide that we are 90% confident that G1 regulates G2 and 50% confident that G2 regulates G3. We might also have some prior expectation that our network should be sparse, i.e. that most edges for which there is no literature support should not be present. We might then decide on a generic confidence of 10% that any other given regulatory relationship not mentioned in the literature is present. A prior probability gives us a rigorous way of building these estimates into our inferences. For example, let us consider model M_1' from Section 2.3 with the following likelihood function:

$$Pr\{\mathbf{d}_1, \mathbf{d}_2, \mathbf{d}_3, \mathbf{d}_4 | M_1'\} = Pr\{\mathbf{d}_1 | M_1'\} \times Pr\{\mathbf{d}_2 | M_1'\} \times Pr\{\mathbf{d}_3 | M_1'\} \times Pr\{\mathbf{d}_4 | M_1'\}.$$

(16.34)

We can incorporate our prior expectations into the network inference problem by changing our objective function from the above likelihood to the probability

$$Pr\{\mathbf{d}_1, \mathbf{d}_2, \mathbf{d}_3, \mathbf{d}_4 | M_1'\} \times Pr\{M_1'\},$$

where $Pr\{M_1'\}$ is a probability function over possible models that provides an estimate of how intrinsically plausible we believe each model to be independent of the data. To evaluate that prior probability, we need to consider each edge that might be present in M_1'. If we define $\bar{e}$ to mean the event that edge e is not present in the model, then

$$Pr\{M_1'\} = Pr\{\overline{(v_1, v_2)}\} \times Pr\{\overline{(v_1, v_3)}\} \times Pr\{\overline{(v_1, v_4)}\} \times Pr\{\overline{(v_2, v_1)}\} \times Pr\{\overline{(v_2, v_3)}\} \times \cdots$$

(16.35)

Since we believe that (v_1, v_2) is present with confidence 90%, we would say $Pr\{\overline{(v_1, v_2)}\} = 1 - 0.9 = 0.1$. Similarly, since we have 50% confidence that (v_2, v_3) is present, $Pr\{\overline{(v_2, v_3)}\} = 1 - 0.5 = 0.5$. For all other edges (v_i, v_j), $Pr\{\overline{(v_i, v_j)}\} = 1 - 0.1 = 0.9$. There are a total of 12 possible edges for models of 4 genes, so

$$Pr\{M_1'\} = 0.1 \times 0.5 \times (0.9)^{10} \approx 0.0174.$$

(16.36)

Adding in this prior knowledge, we can revise our estimate of the plausibility of model M_1' to:

$$Pr\{\mathbf{d}_1, \mathbf{d}_2, \mathbf{d}_3, \mathbf{d}_4 | M_1'\} Pr\{M_1'\} \approx 3.00 \times 10^{-9} \times 0.0174 \approx 5.23 \times 10^{-11}.$$

(16.37)

We can similarly incorporate this prior knowledge into our consideration of the alternative models. For M_2', we proposed that G1 regulates G2, which we believe with confidence 90%; G2 regulates G3, which we believe with confidence 50%; and that there are no other edges, which we believe each with confidence 90%. Thus, the prior probability for M_2' is

$$Pr\{M_2'\} = 0.9 \times 0.5 \times (0.9)^{10} \approx 0.141$$

(16.38)

and therefore

$$Pr\{\mathbf{d}_1, \mathbf{d}_2, \mathbf{d}_3, \mathbf{d}_4 | M_2'\} Pr\{M_2'\} \approx 0.141 \times 1.00 \times 10^{-7} \approx 1.41 \times 10^{-8}.$$

(16.39)

For M_3', we proposed that G1 does not regulate G2, an event we believe has probability 10%; that G1 does regulate G3, which we also believe has probability 10%; that G2 regulates G3, which we believe has probability 50%; and that no other genes regulate one another, which we believe with probability 90% for each such possible edge. Thus, we derive the prior probability

$$Pr\{M_3'\} = 0.1 \times 0.1 \times 0.5 \times 0.9^9 \approx 1.94 \times 10^{-3}.$$

(16.40)

Therefore, our complete objective value for that model is

$$Pr\{\mathbf{d}_1, \mathbf{d}_2, \mathbf{d}_3, \mathbf{d}_4 | M_3'\} Pr\{M_3'\} \approx 2.81 \times 10^{-7} \times 1.94 \times 10^{-3} \approx 5.44 \times 10^{-10}.$$

(16.41)

By comparing the three models, we can see that adding prior knowledge can substantially change our assessments about the relative merits of the models. We previously concluded that M_3' was the best of the three models we considered. M_3' shows poor agreement with our prior expectations, though, while M'_2 shows very good agreement. With this prior knowledge, M'_2 now stands out as the best of the models. This kind of use of prior knowledge is one of the most important factors in effectively handling complex model-inference problems in practice. There is an enormous amount of information available in the biological literature and making good use of that information is one of the key features likely to distinguish an accurate from an inaccurate inference.

Even when we lack real knowledge about a problem, some generic prior probabilities can be very helpful in achieving good results. One important special case of this is the use of prior probabilities to penalize model complexity. One might note that before we started considering prior knowledge, the more complicated models we considered generally outperformed the simpler ones. That phenomenon will occur even when the added complexity has no real biological basis because a maximum likelihood model will exploit every chance correlation occurring in the data to achieve a slightly better fit. In model inference, this phenomenon is known as *overfitting* and needs to be controlled. Prior probabilities provide a way to control for overfitting, by allowing us to specifically penalize more complicated models. Our decision above to assign a 10% prior probability to regulatory edges for which there was no prior evidence is a crude example of an anti-complexity prior. That assumption will tend to favor models having fewer regulatory relationships unless those additional relationships lead to significant improvements in the likelihood of the data being generated from the model. Some more mathematically principled ways to set an anti-complexity prior have also been developed. One such method is the Bayesian information criterion (BIC) [6], in which we set the prior probability of each inferred edge to be the inverse of the number of observed data points. Thus, we would penalize each edge by a factor of $1/8$ in our example.

5   Regulatory network inference in practice

We have now covered the major concepts one needs in order to pose and solve a basic version of the regulatory network inference problem, but there are still quite a few details that separate the methods above from the methods likely to be encountered in the current scientific literature. In this section, we will briefly consider a few extensions of the problem that will bring it much closer to those in use for challenging problem instances in practice. We will first consider how we can drop the assumption of discretization we made at the beginning of the chapter, making full use of real-valued expression data. We will then examine how the model can be extended to allow for additional sources of data beyond gene expression levels, as is commonly done in practice. While we cannot cover these extensions in detail, we can see how these

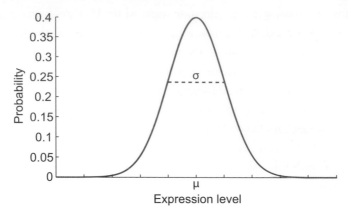

Figure 16.6 Example of a Gaussian curve commonly used as a model of real-valued expression data.

seemingly large changes to the problem actually follow straightforwardly from the principles we have already covered.

5.1 Real-valued data

One of the most dramatic simplifications we made in our toy model was the decision to discretize the data, taking data that are generally real-valued and converting them to binary active/inactive data. It is a minor change to use a more complex discretization – for example, having three labels to represent normal, overexpressed, and underexpressed genes – and we should be able to work out how to extend the concepts we have already covered to any discretized data set. It is possible, however, to work directly with continuous data by adding an assumption about the probability distributions from which data are generated.

It is common to assume that data are *normally distributed*, i.e. described by a Gaussian bell curve as in Figure 16.6. This curve is one example of a probability density function, which describes how likely it is for a given random variable to take on any given possible value. The density curve is highest around the value μ, indicating that the random variable will often be near μ, and is low for values far from μ, indicating that the random variable will rarely be much higher or lower than μ. For a Gaussian random variable, the peak value μ is the average value of the random variable, also known as its *mean*. The width of the bell is controlled by a parameter called its standard deviation (denoted σ). The Gaussian probability density is described by the function

$$Pr\{G = g\} = \frac{1}{\sqrt{2\pi}\sigma}e^{-(g-\mu)^2/(2\sigma^2)} \tag{16.42}$$

where G is the random variable (e.g. expression of gene G1) and g is a particular instance of that random variable (e.g. expression of gene G1 in condition C2).

We can convert our discretized approach above into an approach for real-valued data by using that Gaussian function in place of our previous discrete probability parameters. That is, if we know that the actual real expression value measured by the microarray for some gene i has mean μ_i and standard deviation σ_i, then we can say a given observed value d_{ij} of that gene has likelihood

$$Pr\{d_{ij}|M\} = \frac{1}{\sqrt{2\pi}\sigma_i}e^{-(d_{ij}-\mu_i)^2/(2\sigma_i^2)}. \tag{16.43}$$

The likelihood of a full expression vector $\mathbf{d}_i$ over m different conditions would then be given by

$$Pr\{\mathbf{d}_i|M\} = \prod_{j=1}^{m}\frac{1}{\sqrt{2\pi}\sigma_i}e^{-(d_{ij}-\mu_i)^2/(2\sigma_i^2)}. \tag{16.44}$$

To evaluate this likelihood for a specific data set, though, we need to know μ_i and σ_i. For a gene with no regulators, we will commonly pre-normalize the expression vector $\mathbf{d}_i$ by the formula

$$\hat{d}_{ij} = (d_{ij} - \mu_i)/\sigma_i, \tag{16.45}$$

which will produce a vector of $\hat{d}_{ij}$ values with mean 0 and standard deviation 1. We can then use this normalized vector in place of the raw d_{ij} values. For regulated genes, we will generally assume that μ is a function of the expression levels of its regulators. The most common assumption is that the mean μ_{ij} of a regulated gene i in condition j is a linear function of the expression levels of its regulators in that condition. That is, if we have a gene i regulated by genes $1, \ldots, k$, then we would assume that

$$\mu_{ij} = a_{i1}d_{1j} + a_{i2}d_{2j} + \ldots + a_{ik}d_{kj} \tag{16.46}$$

where each a_{ij} value is a constant that is part of our model.

Finding the maximum likelihood set of a_{ij} values is known as a *regression* problem, and specifically a *linear regression* problem for a linear model like that above. In the interest of space, we will not attempt to explain regression here, only note that finding the maximum likelihood a_{ij} values is a problem we can solve with some basic linear algebra.

5.2 Combining data sources

Another big difference between our toy model above and a real-world method is that an effective method in practice is likely to make use of far more data than just gene expression levels.

Some data sets will inherently have additional information we might use to improve the model. For example, if the data come from experiments at different points in time, we may be able to make a more effective model by assuming expression is a function of time. If the data come from samples subjected to drug treatments, then we may get a more accurate inference by assuming expression is a function of the concentration of drug applied to a given sample. More complicated models are often needed, specialized to the specific kind of data available, but the basics of evaluating and learning those models are not substantially different from what we covered above.

Making accurate predictions will often involve reference to an entirely different data set than the expression data we considered above. For example, we may have DNA sequence data available for the promoters of our genes, which we can examine for likely transcription factor binding sites. We may have direct experimental measurements of which transcription factors bind to which genes. We could treat such data as prior knowledge, building it into our model priors in an *ad-hoc* fashion. A more general approach, however, is to extend the likelihood model to account for multiple experimental measures.

To illustrate this approach, suppose that in addition to the expression data D, we also have a matrix of binding data B, in which an element b_{ij} is 1 if the product of gene i is reported to bind to the promoter of gene j. We can augment our prior likelihood formula for the expression data D to create one evaluating the model as a source for both D and B. If we assume the expression and binding data are independent outputs of a common model, then we can say

$$Pr\{D, B|M\}Pr\{M\} = Pr\{D|M\}Pr\{B|M\}Pr\{M\}.$$

We can evaluate $Pr\{D|M\}$ and the model prior $Pr\{M\}$ just as before.

The same concepts we used to derive a probabilistic model of D can then be used to derive a probabilistic model of B. To account for the possibility of errors in B, we can propose that data in B is a probabilistic function of the regulatory relationships in M. We can use four probability parameters to capture the possible relationships between B and M: $p_{b,0,0}$, the probability B reports no binding given that there is no binding; $p_{b,0,1}$, the probability B reports no binding given that there is binding; $p_{b,1,0}$, the probability B reports binding given that there is no binding; and $p_{b,1,1}$, the probability B reports binding given that there is binding. These four parameters would then augment the probability parameters P for our model $M = (V, E, P)$. Given some such model M we can then say:

$$Pr\{B|M\} = (p_{b,0,0})^{n_{0,0}} (p_{b,0,1})^{n_{0,1}} (p_{b,1,0})^{n_{1,0}} (p_{b,1,1})^{n_{1,1}} \tag{16.47}$$

where $n_{0,0}$ is the number of pairs of genes i and j for which $b_{ij} = 0$ and $(v_i, v_j) \notin E$, $n_{0,1}$ is the number of pairs of genes i and j for which $b_{ij} = 0$ and $(v_i, v_j) \in E$, $n_{1,0}$ is the number of pairs of genes i and j for which $b_{ij} = 1$ and $(v_i, v_j) \notin E$, and $n_{1,1}$ is the number of pairs of genes i and j for which $b_{ij} = 1$ and $(v_i, v_j) \in E$.

The same general ideas can be extended to much more complicated data sets. We can similarly add in any other independent data sources we want by adding an additional multiplicative term to the likelihood for each such data source. Matters get somewhat more complicated if we assume that some data sources are related to one another; for example, if we want to combine two different measures of gene expression. In such cases, we cannot assume distinct measures are independent of one another and therefore cannot simplify our likelihood functions as easily. Nonetheless, similar concepts and methods to those covered above will still apply even if the likelihood formulae are somewhat more complicated.

DISCUSSION AND FURTHER DIRECTIONS

We conclude this chapter with a brief summary and a discussion of where interested readers can go to learn more about the topics covered here. We have seen in this chapter how one can reason about the problem of regulatory network inference. Starting with a simple variant of the problem, we have seen how one can take the real biological problem and abstract it into a precise mathematical framework. In particular, we explored how maximum likelihood inference can be used to frame the regulatory network inference problem. We have further seen some basic methods one can use to find optimal models for that framework. We have, finally, seen how we can take this initial simplified view of the problem and extend it to yield sophisticated models that are not far from those used in practice for difficult real-world network inference problems.

In the process of learning a bit about how regulatory network inference is solved, we have also encountered some of the major paradigms by which computational biologists today think about hard inference problems in general. For example, we saw how to reason about model design, and in particular how one can think about the issue of abstraction in modeling and the kinds of trade-offs different abstractions involve. We saw how probabilistic models, and likelihood models in particular, can provide a general framework for inferring complex models from large, noisy data sets. In the process, we saw an example of how one conceptualizes a problem through the lens of machine learning, for example through reasoning about prior probabilities. These basic concepts in posing and solving for models of large data sources are central to much current work in high-throughput and systems biology. It does not take much imagination to see how the same basic ideas can apply to many other inference problems in biology.

In the space of one chapter, we can only receive a brief exposure to the many techniques upon which the regulatory network inference problem draws; we will therefore conclude with a short discussion of where the interested reader can learn more about the issues discussed here. The specific problem of analyzing gene expression microarrays has been intensively studied and several good texts are available. The beginning reader might refer to Causton *et al.* [7] while those looking for a more advanced treatment might refer to Zhang [8]. More generally, though, the methods described here are fundamental to the fields of statistical inference and machine learning; anyone looking to do advanced work in computational biology would be well advised to seek a strong grounding in those areas. There are numerous texts to which one can refer for statistics training. Wasserman [9, 10] provides a very readable introduction for the beginner. Mitchell [11] provides an excellent introduction to the fundamentals of machine learning and Hastie *et al.* [12] to more advanced topics in statistical machine learning. The specific kind of model we covered here is known as a Bayesian model (or Bayesian network model or Bayesian graphical model). There are many treatments one can reference on that class of statistical model specifically, such as Congdon [13], Gelman *et al.* [14], and Neapolitan [15]. We largely glossed over here the details of algorithms for solving for difficult Bayesian models. The above texts will provide more in-depth coverage of the general algorithmic techniques outlined above. For a deeper coverage of Markov chain Monte Carlo methods, one may refer to Gilks *et al.* [16]. We did not provide any coverage here of more advanced methods in optimization, an important area of expertise for those working on state-of-the-art methods. Optimization is a big field and no one text will do the whole area justice, but those looking for training on advanced optimization might consider Ruszczyński [17] and Boyd and Vandenberghe [18]. Curious readers may also refer to the primary scientific literature for seminal papers that introduced some of the major concepts sketched out there [19, 20].

QUESTIONS

(1) Construct a graph describing the regulatory relationships among four genes, one of which is the sole regulator of the other three.
(2) Provide a likelihood function for regulation of the genes described in Question 1.
(3) How might we change a likelihood function to model a more error-prone expression data source versus a less error-prone expression data source?
(4) How would we need to modify the likelihood function for expression of a single unregulated gene if we assume three different expression levels (high, medium, and low) instead of two (on and off)?

REFERENCES

[1] N. Guelzim, S. Bottani, P. Bourgine, and F. Képès. Topological and causal structure of the yeast transcriptional regulatory network. *Nature Genet.*, 31:60–63, 2002.

[2] National Human Genome Research Institute. Image provided for free public use through the US National Institutes of Health Image Bank as NHGRI press gallery photo 20018.

[3] M. B. Eisen, P. T. Spellman, P. O. Brown, and D. Botstein. Cluster analysis and display of genome-wide expression patterns. *Proc. Natl. Acad. Sci. U S A*, 95:14,863–14,868, 1998.

[4] N. Metropolis, A. W. Rosenbluth, M. N. Rosenbluth, A. H. Teller, and E. Teller. Equation of state calculation by fast computing machines. *J. Chem. Phys.*, 21:1087–1092, 1953.

[5] S. Geman and D. Geman. Stochastic relaxation, Gibbs distributions, and the Bayesian restoration of images. *IEEE Trans. Pattern Anal. and Machine Intell.*, 6:721–741, 1984.

[6] G. E. Schwarz. Estimating the dimension of a model. *Ann. Stat.*, 6:461–464, 1978.

[7] H. Causton, J. Quackenbush, and A. Brazma. *Microarray Gene Expression Data Analysis: A Beginner's Guide*. Blackwell Science, Malden, MA, 2003.

[8] A. Zhang. *Advanced Analysis of Gene Expression Microarray Data*. World Scientific Publishing, Toh Tuck Link, Singapore, 2006.

[9] L. Wasserman. *All of Statistics*. Springer, New York, 2004.

[10] L. Wasserman. *All of Non-Parametric Statistics*. Springer, New York, 2006.

[11] T. M. Mitchell. *Machine Learning*. WCB/McGraw-Hill, Boston, MA, 1997.

[12] T. Hastie, R. Tibshirani, and J. Friedman. *The Elements of Statistical Learning: Data Mining, Inference, and Prediction*. Springer-Verlag, New York, 2001.

[13] P. Congdon. *Applied Bayesian Modelling*. John Wiley and Sons, Chichester, 2003.

[14] A. Gelman, J. B. Carlin, H. S. Stern, and D. B. Rubin. *Bayesian Data Analysis*. CRC Press, Boca Raton, FL, 2004.

[15] R. E. Neapolitan. *Learning Bayesian Networks*. Pearson Prentice Hall, Upper Saddle River, NJ, 2004.

[16] W. R. Gilks, S. Richardson, and D. J. Spiegelhalter. *Markov Chain Monte Carlo in Practice*. Chapman and Hall/CRC, Boca Raton, FL, 1996.

[17] A. Ruszczyński. *Nonlinear Optimization*. Princeton University Press, Princeton, NJ, 2006.

[18] S. Boyd and L. Vandenberghe. *Convex Optimization*. Cambridge University Press, New York, 2004.

[19] P. Dhaseleer, S. Liang, and R. Somogyi. Genetic network inference: From co-expression clustering to reverse engineering. *Bioinformatics*, 16:707–726, 2000.

[20] N. Friedman, M. Linial, I. Nachman, and D. Pe'er. Using Bayesian networks to analyze expression data. *J. Comp. Biol.*, 7:601–620, 2000.

GLOSSARY

Adjacency: Defined by two synteny blocks that are adjacent to each other in two species.

Alignment: A correspondence between symbols in two sequences. Symbols without corresponding symbols are said to correspond to a **gap**. Each pair of corresponding symbols is given a **weight** dependent on whether it is a match (positive weight) or a mismatch (negative weight or a penalty), and each gap is assigned a penalty dependent on its length. The **alignment score** is the total of all weights. The **optimal alignment** has the highest score.

Alignment Score: See "Alignment."

Allele: One of the alternative forms of a gene at a specific location. It can also refer to the specific nucleotide (A,C,G,T) if that position varies among individuals in a population.

Anagram: A word or phrase formed by rearranging the characters of another word or phrase. For example, "eleven plus two" can be rearranged into the new phrase "twelve plus one."

Ancestral Genome Reconstruction: The attempt to restore the genomic events (substitutions, insertions, deletions, genome rearrangements, and duplications) that happened during evolution.

Bipartition: A division of the vertices of a tree into two subtrees.

Bitstring: A string consisting of 0s and 1s which is used to represent binary numbers or the presence/absence of a feature of interest.

Bootstrap Support: A measure of the reliability of internal nodes in a tree.

Breakpoint: Defined by two synteny blocks that are adjacent in one species and separate in another.

Child Node: See "Tree."

Cis Regulatory Module: A genomic cluster of binding sites for multiple transcription factors. The presence of such clusters may indicate interactive binding of multiple transcription factors that synergystically regulate gene transcription.

Coevolution: The genetic change of one species in response to the change in another.

Complete Subtree: A subtree consisting of a node and all its descendents (children, children of children, etc.).

Conditional Probability: The probability of a state of interest (s) computed only on the subset of cases where a specified condition (c) is true. Denoted by $\Pr(s|c)$.

Consensus Binding Site: Given a set of k-nucleotide long binding sites for a transcription factor, the consensus binding site is a sequence of k nucleotides comprised of the most frequent nucleotide at each position among the known binding sites.

Contingency Table: In statistics, a contingency table is used to display the frequency of two or more variables in a matrix format.

Cospeciation: In the study of cophylogeny, a cospeciation event corresponds to contemporaneous speciation events in the host and parasite trees.

Cumulative Skew: The sum of skew values across thinly sliced adjacent sequence windows.

Cumulative Skew Diagram: A plot of cumulative skew along the length of a genome.

Degree: The degree of a node is the number of edges touching the node.

Degree Distribution: A distribution of the degrees of all nodes in a given network.

Depth: See "Tree."

Duplication: In the study of cophylogeny, a duplication event corresponds to a speciation event in the parasite tree that is not contemporaneous with a speciation event in the host tree. In genomics, a duplication of a genomic region creates an additional copy of that region.

Dynamic Programming: An efficient algorithmic technique for solving a wide range of problems without direct enumeration of all possible solutions.

Edge: See "Network."

Eulerian Cycle: A cycle in a graph which traverses each edge exactly once.

Eulerian Cycle Problem (ECP): The computational problem of finding an Eulerian cycle in an arbitrary graph or proving that such a cycle does not exist in the graph.

Evolutionary Tree: See "Phylogeny."

Fisher's Exact Test: A statistical test used to analyze the significance of a contingency table.

Fragment Assembly: The computational stage of genome sequencing, which consists of using generated reads to assemble the genome.

Gap: See "Alignment."

GC-content: The proportion of all nucleotides in a DNA molecule that are either guanine or cytosine.

GC-skew: A measure of guanine excess (equivalently, cytosine depletion) on one strand of a DNA sequence as compared to its complementary strand.

Gene Expression: The amount of RNA corresponding to a given gene; commonly used as a measure of the gene's level of activity.

Gene Recognition: Identification of the protein-coding regions in a DNA sequence.

Genome Rearrangement: A mutation that affects a large portion of a given genome. A genome rearrangement occurs when one or two chromosomes break and the fragments are reassembled in a different order. In general, these rearrangements are comprised of inversions, translocations, fusions, and fissions.

Genome Sequencing: The process of determining an organism's complete genome.

Genotype: The combination of alleles that describe the genetic makeup of an individual.

Glycan: In biochemistry, the carbohydrates (sugars) linked to other molecules (such as proteins or lipids) are called glycans. Glycans are components of glycoconjugates, such as glycoproteins and glycolipids. There exist many different glycans on the cell surface, some of which share similar structures.

Glycan Array: A glycan array comprises a library of synthetic (thus structurally known) glycans that are automatically printed on a glass slide, which is a platform to simultaneously assay the interaction between a glycan-binding protein and hundreds of its potential glycan ligands. A glycan array experiment can detect the subset of glycans that interact with the glycan-binding protein being assayed.

Graph: See "Network."

Graphlet: A small induced subgraph of a large network, in which an induced subgraph refers to a subgraph which contains every edge from the original graph that connects two vertices of the subgraph.

Hamiltonian Cycle: A cycle in a graph which visits every vertex exactly once.

Hamiltonian Cycle Problem (HCP): The computational problem of finding a Hamiltonian cycle in an arbitrary graph or proving that such a cycle does not exist in the graph. The HCP is NP-Complete.

Haplotype Block: A high LD-region in a genome.

Hash Table: A data structure that uses a hashing function to store information based on (key, value) pairs.

Hemagglutin (HA): A kind of membrane protein attached on the surface of the influenza virion. Hemagglutinin can recognize the glycans and glycoproteins on the surface of the host cells and therefore induce the infection of influenza virus.

Horizontal Gene Transfer: The transfer of genes between organisms of different species or strains.

Host Switch (also known as horizontal transfer): In the study of cophylogeny, a host switch event corresponds to a parasite species switching from one host lineage to another.

Infinite Sites Assumption: The hypothesis that a given genome is large enough relative to mutation rates such that any site mutates at most once in the genealogical history of the population.

Influenza Virus: Influenza virus is the cause of influenza. It belongs to the family Orthomyxoviridae of RNA viruses and has three subtypes (A, B, and C, respectively). The influenza virion is a globular particle protected by a lipid bilayer, which infects epithelial cells of the host respiratory systems.

Inversion: See "Reversal."

***l*-mer:** A sequence of l nucleotides, which is represented by the orderings of the letters A, G, C, and T.

***l*-mer Multiplicity:** The number of times that an l-mer occurs in a given genome or in a set of reads.

Leaf Node: See "Tree."

Likelihood: The conditional probability of a set of observations given a specified model.

Likelihood Function: A mathematical function describing the probability of any possible set of observations of a system, commonly representing the visible experimental outputs of a system in terms of a set of parameters describing a model of the system.

Linear Programming: A general formulation of problems involving maximizing or minimizing a linear objective function subject to certain linear constraints.

Link: See "Network."

Linkage Disequlibrium (LD): See "Linkage Equilibrium."

Linkage Equilibrium: The random assortment of alleles at different loci due to historical recombination events. If the loci are spatially close with a small number of recombination events between them, the alleles may be correlated, resulting in **linkage disequilibrium**.

Locus: A location on the genome. It can refer to a specific genomic coordinate, or a genetic marker such as a gene in the region.

Loss: In the study of cophylogeny, a loss event occurs when a parasite species moves from a host lineage to its child without speciating. (Technically, this may be due to a failure to speciate or one of several other processes, such as extinction or sampling failure.)

Maximum Parsimony Problem: A computational problem for computing phylogenies from a set of sequences, where the objective is a tree with the sequences at the leaves, with additional sequences at the internal nodes in the tree, so that a minimum number of substitutions occurs in the tree.

Mutation: A change in the order or composition of the nucleotides in a DNA sequence.

Mutualism: A relationship between two species that benefits both species.

Network (also known as graph): a set of objects, called **nodes**, along with pairwise relationships that link the nodes, called **links** or **edges**.

Network Motif: A subgraph recurring in a network at frequencies much higher than those found in randomized networks.

Network Property: An easily computable approximate measure of network topology that is commonly used for comparing large networks.

Node: See "Network."

NP-complete: A classification of problems in computer science that are all equivalent to each other. No efficient algorithm to any NP-complete problem has ever been found, although neither have NP-complete problems been proven to be intractable.

NP-hard: The NP-hard problems are the hardest problems within the set NP of computational problems. The set NP consists of all decision problems (Yes/No questions, such as "can we split this group of people into two sets so that no two people in the same set know each other?") for which we can verify a "Yes" answer in polynomial time. To say that a computational problem is NP-hard means that if we could solve this problem in polynomial time, then all problems that are known to be NP-hard could also be solved exactly in polynomial time. To date, no one knows whether it is possible to solve any NP-hard problem in polynomial time.

Observable Variable: A variable that can be measured without uncertainty.

Optimal Alignment: See "Alignment."

Parent Node: See "Tree."

Phenotype: The observable biochemical and physical traits of an individual. For example, height, weight, and eye color are all phenotypes, as are more complex quantities such as blood pressure.

Phylogenetic Footprint: A non-protein-coding region in a genome that has been conserved throughout the course of evolution. Evolutionary conservation is indicative of a regulatory role for the region.

Phylogenetic Tree: See "Phylogeny."

Phylogeny (also called an evolutionary tree, or a phylogenetic tree): This is typically a rooted, binary tree, so that each internal node has exactly two children.

Point Mutation: A DNA mutation in which only a single nucleotide is changed.

Polytene Chromosome: A giant chromosome that originates from multiple rounds of replication (without cell division) in which the individual replicated DNA molecules remain fused together.

Positional Weight Matrix (PWM): A construction commonly used to represent the DNA binding specificity of a transcription factor. For a k-nucleotide long binding site, the PWM has four rows for each of the four nucleotides and k columns for the k binding site positions. Each column of the PWM includes the frequencies with which each of the four bases are observed at the specific binding site position among the known binding sites of the transcription factor.

Posterior: The resulting probability of a model or hidden parameter value based on computing Bayes' Law for the available observations; specifically, the conditional probability of the model given the observations.

Prior: The unconditional probability of a model or hidden parameter value prior to taking any observations into consideration.

Prior Probability: A probability assigned to possible values of a variable in a system independent of the specific data available for a given analysis problem; often used in statistical modeling to encode a bias towards model features we expect to find based on prior knowledge of a system.

Protein–Protein Interaction (PPI) Network: A network in which proteins are modeled as nodes and edges exist between pairs of nodes corresponding to proteins that can physically bind to each other.

Read: See "Read Generation."

Read Generation: The experimental stage of genome sequencing, which amounts to identifying small pieces of the genome, called **reads**.

Recombination Hotspot: A low-LD region of a genome.

Replication/Transcription Bubble: The separation of two complementary strands of a double-stranded DNA molecule to allow for synthesis of nascent DNA/RNA.

Replication Origin/Terminus: The position in a genome where replication starts/ends.

Reversal: An important type of genome rearrangement. A reversal (also called an **inversion**) occurs when a segment of a chromosome is excised and then reinserted with the opposite orientation and with the forward and reverse strands exchanged.

Root Node: See "Tree."

Single Nucleotide Polymorphism (SNP): A single nucleotide variation in a genome that recurs in a significant proportion of the population of the associated species. Pronounced "snip."

Subgraph: A subgraph of a graph G is a graph whose nodes and edges belong to G.

Subtree: A subtree of a tree is a tree consisting of a subset of connected nodes in the original tree.

Synteny Block: A set of clustered genomic markers with an evolutionarily conserved order.

Systematic Evolution of Ligands by Exponential Enrichment (SELEX): An *in-vitro* technique to determine the DNA binding specificity of a protein.

Tag SNP: A member of a set of SNPs which when taken together are sufficient to distinguish the patterns within a haplotype block.

Transcription Bubble: See "Replication/Transcription Bubble."

Transcription Factor (TF): A protein that interacts with the gene transcription machinery of a cell to regulate the expression levels of genes.

Transcriptional Regulatory Network: A mathematical model of the influence of genes in a common cell upon one another's expression levels. Consists of nodes representing individual genes or gene isoforms and edges representing the influence exerted by a source gene on the expression level of a target gene.

Tree: A tree is a *directed* (*rooted*) graph with no cycles, in which each node has zero or more **children nodes** and at most one **parent node**. The nodes having no child are called the **leaf nodes**. The only node in a tree with zero parent is called the **root node**. The **depth** of a node is defined as the length (i.e. the number of edges) of the path from that node to the root. Both the nodes and edges in a tree can be labeled. For example, the nodes in a glycan tree are labeled by the monosaccharide residues, and the edges in a glycan tree are labeled by the linkage type.

Treelet: Given a labeled tree, an *l*-treelet is a subtree with *l* nodes. Notably, a treelet is a subgraph of a tree if and only if both their topology and node/edge labels match.

Tree of Life: A tree that depicts the evolutionary relationships between all cellular life forms.

Tree Topology: The branching order in a phylogeny.

INDEX

Entries in bold text refer to a section of the book.